D0892053

ADRENERGIC RECEPTORS IN MAN

RECEPTORS AND LIGANDS
IN INTERCELLULAR COMMUNICATION

edited by
Bernhard Cinader

Institute of Immunology
University of Toronto
Toronto, Ontario, Canada

1. Multiple Dopamine Receptors: Receptor Binding Studies in Dopamine Pharmacology
   Milton Titeler

2. Structure and Function of Fc Receptors
   edited by Arnold Froese and Frixos Paraskevas

3. Cell Surface Dynamics: Concepts and Models
   edited by Alan S. Perelson, Charles DeLisi, and Frederik W. Wiegel

4. Polypeptide Hormone Receptors
   edited by Barry I. Posner

5. Recognition and Regulation in Cell-Mediated Immunity
   edited by James D. Watson and John Marbrook

6. Insulin: Its Receptor and Diabetes
   edited by Morley D. Hollenberg

7. Parasite Antigens: Toward New Strategies for Vaccines
   edited by Terry W. Pearson

8. Adrenergic Receptors in Man
   edited by Paul A. Insel

Other Volumes in Preparation

ADRENERGIC RECEPTORS IN MAN

edited by

Paul A. Insel

School of Medicine
University of California, San Diego
La Jolla, California

MARCEL DEKKER, INC.   New York and Basel

Library of Congress Cataloging-in-Publication Data

Adrenergic receptors in man.

(Receptors and ligands in intercellular communication ;
v. 8)
    Includes index.
    1. Adrenergic receptors. I. Insel, Paul A. [date]
II. Series. [DNLM: 1. Receptors, Adrenergic--physiology.
W1 RE107LM v. 8 / WL 102.8 A2418]
QP364.7.A37 1987   612'.89   86-24008
ISBN 0-8247-7629-1

MARCEL DEKKER, INC.
270 Madison Avenue, New York, New York 10016

Current printing (last digit):
10 9 8 7 6 5 4 3 2 1

PRINTED IN THE UNITED STATES OF AMERICA

Series Introduction

Cells communicate with one another to bring about orderly differentiation and regeneration. Receptor-ligand interaction can occur via
secreted ligands; it can also occur between membranes of different
cell types, i.e., via adhesion molecules which play a role in the
structural development of organs, exemplified by neural cell adhesion
and embryological development under the influence of "master" cells.
Communication during adult life maintains coordination and balance
in the multicellular organism. The words of this communication are
molecules, i.e., factors, that convey signals by combination with
receptors. These signals initiate further metabolic events and these,
in turn, can give rise to the production of other factors and thus
to the sentences of the intercellular language; the resulting intercommunication is intense and continuous. Some diseases are attributable
to a localized defect in this intercommunication. The intensity of the
communication declines in the last half of life; however, its redundancy
permits a certain latitude for continued function. The polymorphism
of the rate of change and in its onset is the basis for the gerontology
and geriatrics of the future. Many disease entities can be defined in
terms of defects in one component and, ultimately, could be treated
by replacing the missing component.

This series is devoted to the study of biological language, and
will cover receptor-ligand interaction in health and disease. Special
volumes are dedicated to the development of drugs modeled to fit
receptors, to the understanding of the role of receptors in host-
parasite interactions, and to receptor polymorphism, blockade, and
activation.

Books will be authored individually or will consist of contributions
from specialized investigators. This volume in our series is devoted

to adrenergic receptors in man, a topic in which biology, pathophysiology, and therapeutic strategies have undergone an intimate process of interlinked growth. The book will, therefore, interest both the biologist and the practicing physician.

Bernhard Cinader

Foreword

Although the concept of multiple adrenergic receptors was proposed almost forty years ago, advances in our knowledge about the biochemistry of these receptors were relatively slow. About ten years ago there was a rapid growth in our understanding of adrenergic receptors, especially at the molecular level. This was mainly due to the introduction of radiolabeled adrenergic ligands of high specific activity that recognized and bound to tissue receptors, as well as the availability of numerous selective agonists and antagonists. Subtypes of adrenergic receptors that interact with epinephrine, norepinephrine, and many drugs have been described. These subtypes of receptors, named α_1, α_2, β_1, and β_2, are present in almost every tissue in the body in various combinations. When norepinephrine, epinephrine, or certain drugs bind to these subtypes of adrenergic receptors, they produce diverse stimulatory or inhibitory physiological responses.

Receptors on the cell surface are just the tip of the iceberg. Upon stimulation they trigger a cascade of biochemical events, first in the plasma membrane, and then in the interior of the cell. In many but not all systems yet studied, adrenergic receptors are able to transduce their messages through proteins in the plasma membrane that bind to guanine nucleotides. These transducing proteins then stimulate or inhibit effector systems such as adenylate cyclase, phospholipase A_2, phospholipase C, or ion channels. Depending on the subtype of adrenergic receptor and the properties of the transducing and effector protein of a cell, second messengers such as cyclic AMP, inositol triphosphate, diacylglycerol, and arachidonic acid and its many metabolites are either generated or prevented from forming. In addition, stimulation of adrenergic receptors can produce changes in intracellular ions, in particular, calcium. These second messengers then initiate numerous complex intracellular biochemical reactions to produce a biological response characteristic of the target cell.

The multitude of actions of adrenergic receptors has important implications for biochemistry, pharmacology, physiology, endocrinology, immunology, neurobiology and almost all branches of medicine. Clinicians are aware of the impact that drugs affecting α-and β-adrenergic receptors have had on the treatment of cardiovascular diseases. With the sheer quantity of papers about adrenergic receptors published in diverse journals, it is difficult for clinicians and basic scientists to keep up with the new developments. This volume gathers together many new advances in the study of adrenergic receptors and should be of interest to both clinicians and basic scientists.

Julius Axelrod

Preface

In the current era of "publication metastasis" in the biomedical sciences, I feel it is important to justify the publication of a new book. I decided to undertake the task of assembling this compendium because I sensed a growing gap between those investigators actively studying catecholamines and the response of target cells to these compounds and other scientists and physicians who utilize the results of such research.

Several developments have contributed to the increasing difficulty of staying well-informed about these topics. Foremost among these is the increased attention of researchers to the biochemical mechanisms mediating catecholamine response. Some of the advanced concepts and methodological approaches necessary for these biochemical studies are not as easily understood as are earlier observations, such as the results of assays of plasma catecholamines. A further problem is the increasing breadth of the data base in the numerous disciplines involved. Adrenergic agonists are able to regulate physiological events in virtually every organ in the body. For this reason and because pharmacological agents that block adrenergic receptors are now widely used both in research and in clinical medicine, it is very difficult to keep abreast of the field. If one desires to obtain a thorough analysis of catecholamine action, examination of literature that is rather far removed from one's usual area of interest is required. In addition, this research area has attracted the attention of investigators who are involved in either "basic research," "clinical investigation," or both. These two groups do not always communicate effectively.

Several different scientific approaches have been used to study catecholamines and catecholamine action: physiology, pharmacology, biochemistry, biophysics, anatomy, and (in recent years) cell and molecular biology. It has become increasingly difficult for investigators in any one of these areas to develop a general overview of all aspects of the action of catecholamines. Scientists or physicians whose interest lies in adrenergic action in human beings have been at a loss to find a single reference that incorporates detailed information from the several disciplines involved in those studies. The goal of this book is to provide such a reference, focusing on studies in man and the application of information obtained in experimental animals and other model systems to humans.

I set out to involve a group of investigators who had each contributed important information in the field of adrenergic receptors in man and asked them to assemble material that would be useful to colleagues with whom they would be likely to interact as well as to those in more "distant" areas. This volume thus attempts to be didactic as well as to present reviews of particular aspects of the field. My bias is that an integrated approach is required to avoid the parochial view with which specialists sometimes approach this topic.

The text has been organized so that the reader will initially review the methods that can be used to study adrenergic receptors in vivo, the physiological responses that these approaches have defined, and the clinical pharmacology of alpha- and beta-adrenergic blocking drugs. We next consider methods that can be used to examine adrenergic receptors in vitro, in particular, radioligand binding techniques, and the new biochemical insights that this research has provided. Physiologic and pharmacologic regulation of adrenergic receptors are then discussed. The balance of the book examines whether alterations of adrenergic receptors contribute to pathophysiology in several different organ systems.

I should point out two caveats. The authors prepared their materials between 1983 and 1984, during a period of major growth in terms of new information and new ideas. It is possible that in spite of efforts to update the material prior to publication, knowledgable readers will detect minor errors of omission and commission. A second caveat relates to topics that I decided would not be presented. This book discusses alpha- and beta-adrenergic receptors and does not attempt to include information on dopamine receptors. Also, certain areas that may be important clinically, but about which limited information has been available, have not been included. Examples include adrenergic receptors in neoplastic cells and cancer, in the eye, in various ocular diseases, and in gastrointestinal and renal disorders.

The goals that I set in preparing this work will be fulfilled if readers find that it is of use in developing new hypotheses for investigative studies or for therapeutic management of patients. I look forward to learning if we have succeeded in closing the gap that I described at the beginning of this preface.

Paul A. Insel, M.D.

Introduction

The catecholamines epinephrine (adrenaline, in Europe) and norepinephrine (noradrenaline) are widely recognized as hormones and neurotransmitters, respectively, that modulate function in many organ
systems. This idea has been a dominant theme in both clinical and
experimental medicine during the past two decades. The classification
by Ahlquist in 1948 of catecholamine (adrenergic) responses and tissue
receptors into alpha and beta types has promoted research on the
molecular mechanisms of hormone and neurotransmitter action as well
as providing a framework for the discovery of important pharmacological
agents.

Research during the 1950s and 1960s emphasized primarily physiological assays of tissue responses to catecholamines and provided the
framework for subtyping these responses and, in turn, adrenergic
receptors, into alpha$_1$, alpha$_2$, beta$_1$, and beta$_2$. Initially it was
suggested that this subtyping reflected precise anatomic and physiologic differences, i.e., the heart was thought to possess only beta$_1$-
adrenergic receptors that mediated effects of catecholamines on rate
and force of cardiac contraction; alpha$_1$-adrenergic and alpha$_2$-
adrenergic receptors were thought to be exclusively postsynaptic
and presynaptic, respectively. Within the last several years, this
notion has been disproved by many "exceptions to the rules." Thus,
myocardial cells have been shown to possess both beta$_1$- and beta$_2$-
adrenergic receptors and both of these subtypes may contribute to
chronotropic responses. Moreover, numerous examples have been
described of tissues that possess postsynaptic alpha$_2$-adrenergic
receptors (a partial list includes adipocytes, platelets, and renal
tubular epithelial cells).

Thus, it is now generally acknowledged that pharmacological
criteria provide the best means to characterize a particular physiological or biochemical response as alpha$_1$-, alpha$_2$-, beta$_1$-, or beta$_2$-
adrenergic. The availability of a wide number of agonists and antagonists with appropriate specificity for particular receptor subtypes
now forms the basis of this classification scheme (see Chapters 1 and

2). As new drugs are developed, it is possible, of course, that additional subclasses of adrenergic receptors will be identified.

Further refinement of this classification scheme will most likely require new antagonists that block responses produced by catecholamines or other "classical" agonists. Up until now, the discovery and development of adrenergic antagonists has played a major role not only in fundamental studies of physiology and pharmacology but also in clinical pharmacology and therapeutics (see Chapters 3 and 4). The beta-adrenergic antagonist propranolol was originally introduced for limited cardiovascular indications. However, astute clinicians subsequently realized the much broader utility of this class of drugs (see Chapter 3). This is perhaps the best example known of a drug that has acquired many more (rather than fewer) approved uses after entry into the pharmacopeia.

A detailed understanding of adrenergic receptors themselves is unlikely to come from assessment of physiological responses, because these responses usually require many intermediate steps between receptor occupancy and observed response. The development of methods to assess the "recognition function" of receptors (the binding sites of agonists and antagonists) and the initial biochemical signals for the various classes of adrenergic receptors have provided major breakthroughs in defining receptors in precise molecular terms (see Chapters 5 and 6). Radioligand binding has been a powerful technique for characterizing properties of receptors in intact cells and membrane preparations and for examining whether various physiological states, pharmacological treatments, or clinical disorders change adrenergic response by alteration of receptors rather than by changes in postreceptor events (see Chapters 5-13). Studies of this type with human material are rapidly accumulating. The use of photoaffinity probes for adrenergic receptors (to permit identification of receptors under denaturing conditions) and of techniques for receptor solubilization will facilitate the definition of even further molecular details regarding receptor structure and function. When these approaches are used together with immunological and, ultimately, molecular biological techniques, definitive understanding of the identity of adrenergic receptor subtypes, i.e., the relationship between "family members" of the adrenergic receptor proteins and molecular changes in receptors should be forthcoming. These latter techniques are unfortunately not yet available for studies of human adrenergic receptors.

The assessment of second messengers of adrenergic receptors has been best elucidated in human cells for beta-adrenergic receptors and alpha$_2$-adrenergic receptors (see Chapters 6 and 11). The human platelet has provided a particularly useful model system for examining the linkage of alpha$_2$-adrenergic receptors to inhibition of adenylate cyclase via a guanine nucleotide binding protein (G_i) that is distinct from the guanine nucleotide binding protein (G_S) that links beta-adrenergic (and other "stimulatory") receptors to the stimulation of adenylate cyclase and production of cyclic AMP. The evidence that at least one human disease—pseudohypoparathyroidism—can involve alterations in the G_S protein and in cyclic AMP formation indicates the potential importance of efforts directed at the identification of these proteins in human cells (see Chapters 6 and 10). The possibility that G proteins may link adrenergic receptors to second-messenger systems other than adenylate cyclase is another area of active investigation.

At the time this volume was prepared, the clinical importance of changes in adrenergic receptors in various physiological and pathophysiological settings was not yet adequately defined. Considerable evidence, largely indirect in nature, has accrued regarding possible alterations in these receptors in cardiovascular disorders (Chapter 8), allergic disorders (Chapter 9), endocrine-metabolic disorders (Chapter 10), autonomic dysfunction (Chapter 12), and psychiatric diseases (Chapter 13). The thorough investigation of possible abnormalities in adrenergic receptors in these settings by the use of state-of-the-art methods will probably be required in order to define the precise role of alterations in these receptors.

It seems appropriate to note that this volume has been written at about the end of the first decade during which sensitive biochemical techniques were available for analyses of second messengers and for direct identification of adrenergic receptors as discrete molecular entities. When viewed from that vantage point, this area of investigation is still rather "young" and almost certainly has not yet entered its adolescence. As such, material presented in this book should be viewed as a beginning phase of an area of clinical and basic research whose fully mature appearance is not yet clearly discernible. The improvement in techniques and in the precision with which questions may now be posed should facilitate a rapid maturation of knowledge regarding adrenergic receptors in man.

Paul A. Insel, M.D.

Contents

Series Introduction (Bernhard Cinader)   iii
Foreword (Julius Axelrod)   v
Preface   vii
Introduction   ix
Contributors   xv

1   In Vivo Methods for Studying Adrenergic Receptors   1
    Christopher P. Nielson and Robert E. Vestal

2   Defining the Role of Adrenergic Receptors in
    Human Physiology   37
    John P. Bilezikian

3   Adrenergic Receptors as Pharmacological Targets:
    The Beta-Adrenergic Blocking Drugs   69
    William H. Frishman

4   Adrenergic Receptors as Pharmacological Targets:
    The Alpha-Adrenergic Blocking Drugs   119
    William H. Frishman and Shlomo Charlap

5   In Vitro Methods for Studying Human Adrenergic
    Receptors: Methods and Applications   139
    Harvey J. Motulsky and Paul A. Insel

6   Biochemical Characterization of Human
    Adrenergic Receptors   161
    Ross Feldman and Lee E. Limbird

7   Physiologic and Pharmacologic Regulation of
    Adrenergic Receptors   201
    Paul A. Insel and Harvey J. Motulsky

8   Adrenergic Receptors in Cardiovascular Disease   237
     R. Wayne Alexander

9   Adrenergic Receptors in Allergic Disorders   259
     Diana L. Marquardt and Stephen I. Wasserman

10   Adrenergic Receptors in Endocrine and
     Metabolic Diseases   285
     Philip E. Cryer

11   Activation of Human Platelets by Epinephrine   303
     Sanford J. Shattil

12   Adrenergic Receptors in Neurological Disorders   339
     David Robertson and Alan S. Hollister

13   Adrenergic Receptors in Psychiatric Disease   353
     Thomas R. Insel and Robert M. Cohen

Index   377

Contributors

R. Wayne Alexander, M.D., Ph.D. Cardiovascular Division, Brigham and Women's Hospital, Harvard Medical School, Boston, Massachusetts

John P. Bilezikian, M.D. Departments of Medicine and Pharmacology, College of Physicians and Surgeons of Columbia University, New York, New York

Shlomo Charlap, M.D. Coronary Care Unit, Long Island College Hospital, and Department of Medicine, Downstate Medical Center, Brooklyn, New York

Robert M. Cohen, M.D. Laboratory of Cerebral Metabolism, National Institute of Mental Health, National Institutes of Health, Bethesda, Maryland

Philip E. Cryer, M.D. Department of Medicine, Washington University in St. Louis, St. Louis, Missouri

Ross Feldman, M.D. Department of Medicine, University of Iowa, Iowa City, Iowa

William H. Frishman, M.D. Department of Medicine, The Albert Einstein College of Medicine, Bronx, New York

Alan S. Hollister, M.D., Ph.D. Department of Medicine, Vanderbilt University Medical Center, Nashville, Tennessee

Paul A. Insel, M.D. Department of Medicine, School of Medicine, University of California, San Diego, La Jolla, California

Thomas R. Insel, M.D. Laboratory Clinical Science, National Institute of Mental Health, National Institutes of Health, Bethesda, Maryland

Lee E. Limbird, Ph.D. Department of Pharmacology, Vanderbilt University School of Medicine, Nashville, Tennessee

Diana L. Marquardt, M.D. Department of Medicine, University of California, San Diego Medical Center, San Diego, California

Harvey J. Motulsky, M.D. Department of Medicine, School of Medicine, University of California, San Diego, La Jolla, California

Christopher P. Nielson, M.D.* Veterans Administration Medical Center, Boise, Idaho

David Robertson, M.D. Departments of Medicine and Pharmacology, Vanderbilt University Medical Center, Nashville, Tennessee

Sanford J. Shattil, M.D. Hematology-Oncology Section, Department of Medicine, Hospital of the University of Pennsylvania, Philadelphia, Pennsylvania

Robert E. Vestal, M.D.* Veterans Administration Medical Center, Boise, Idaho

Stephen I. Wasserman, M.D. Department of Medicine, University of California, San Diego Medical Center, San Diego, California

---

**Present affiliation*: University of Washington School of Medicine, Seattle, Washington

1

In Vivo Methods for Studying Adrenergic Receptors

CHRISTOPHER P. NIELSON* and ROBERT E. VESTAL*
Veterans Administration Medical Center, Boise, Idaho

I. INTRODUCTION

In vivo investigation of adrenergic response is important in the classification of receptor subtypes and the characterization of pharmacological
agents. Since description of the beta-receptor antagonist dichloro-
isoproterenol in 1958 (1) an increasing number of selective alpha- and
beta-receptor agonists and antagonists have become available. These
agents are not only valuable in medical practice but also provide
powerful probes for investigation of autonomic function. Catecholamines
have prominent actions on smooth muscle, nerve terminals, and myocardium as well as conspicuous metabolic manifestations, including
hyperglycemia, hyperlipidemia, hypokalemia, and increased oxygen
consumption. Comparisons of the potency of norepinephrine, epinephrine, and isoproterenol in smooth muscle contraction or relaxation
by Ahlquist in 1948 led to the description of alpha and beta receptors
as sites of catecholamine action (2). On the basis of selective pharmacological response, both alpha and beta receptors have subsequently
been further subdivided. Organ responses are categorized according
to activity of receptor-specific agonists and antagonists.

Present affiliation: University of Washington School of Medicine,
Seattle, Washington

II. DOSE-RESPONSE EVALUATION

The dose-response relationship is fundamental for the determination of drug activity and organ effect. The existence of receptors and the description of drug effect according to the laws of mass action was suggested by Langley as early as 1878. Classical receptor theory, as developed by Clark in the 1920s, proposes that the magnitude of response is proportional to the number of drug-receptor complexes and that the maximal effect occurs only when all receptors are occupied (3). Under these assumptions, the dissociation constant (concentration required for 50% receptor occupancy) is equal to the concentration of drug producing a half maximal response. Although such a model may be true in some cases, most dose-response relationships are more complex. Because the tissue stimulus produced by receptor binding of different agonists is often not equal, responses vary despite equal numbers of drug-receptor complexes. Furthermore, the pathway leading from receptor to measured effect often includes multiple biochemical steps and mechanical factors. For example, smooth muscle relaxation induced by isoproterenol requires receptor binding, activation of adenylate cyclase by a guanine regulatory protein, cyclic adenosine monophosphate (AMP) activation of protein kinase, phosphorylation of enzymes that probably lead to decreased cytosolic calcium, and finally relaxation of myofibrils. Because the response may be limited at steps other than receptor binding, certain agonists are capable of producing a maximal effect with only a proportion of receptors occupied. Intrinsic activity describes the potential of a drug to produce a response after formation of a drug-receptor complex (4). Partial agonists have lower intrinsic activity (or less capacity to produce a response) than full agonists, and antagonists have no (or very low) intrinsic activity. Full agonists are usually capable of producing a maximal response at less than 100% receptor

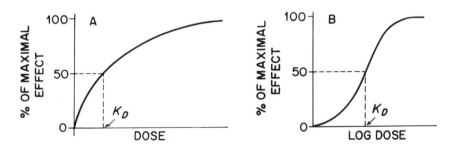

FIGURE 1   Response as a function of drug dose. (From Ref. 95, used with permission.)

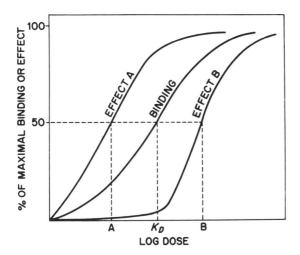

FIGURE 2 Relationship of drug effect to receptor binding. Effects A and B demonstrate high and low intrinsic activity, respectively. (From Ref. 95, used with permission).

occupancy (5). Receptors whose occupancy is not required for maximal response have been termed "reserve" or "spare" receptors. Characterization of the drug-receptor interaction requires determination of the dissociation constant (K_d, the ability of a drug to bind to the receptor) and the intrinsic activity (the ability of a drug to evoke a response). Accurate determination of these constants is most commonly performed using in vitro studies. Response as a function of dose is typically hyperbolic (Fig. 1A) and maximal effect is approached asymptomatically. When response is plotted as a function of log (dose), a sigmoidal curve results (Fig. 1B). Correlation of the dose-response curve with receptor binding depends at least in part upon the drug's intrinsic activity. If the intrinsic activity is high (reserve receptors may exist), then the response is shifted to the left of receptor binding (effect A in Fig. 2). If intrinsic activity is low, higher concentrations of drugs are required to induce equal response (effect B in Fig. 2).

Because results obtained using in vivo techniques are influenced by tissue delivery, clearance, and homeostatic mechanisms, dose-response curves may not demonstrate a typical sigmoidal or hyperbolic configuration. The receptor dissociation constant and drug intrinsic activity often cannot be determined from such data. Response is characterized by drug potency, slope of the dose-response curve, and maximum effect. Potency describes the response evoked by a specific quantity or concentration of drug. It is dependent upon

drug distribution, metabolism, and tissue response. Drugs are often
compared by relative potency, the ratio of equally effective doses.
Although tissue distribution and homeostatic responses may be important, the slope of the dose-response curve is primarily dependent
upon the drug's intrinsic activity. Lower intrinsic activity is generally
associated with a steeper slope (Fig. 2). Maximal response cannot
always be measured if adverse effects limit the administered dose.
Also, response is influenced by both drug-receptor interaction (full
or partial agonist) and effector site characteristics. Effects are often
elicited as a function of time. Since three-dimensional representations
that include dose and time are complicated, responses are often represented by the peak or equilibrium effect. For purposes of defining
receptor characteristics, dose is optimally represented as the concentration of drug at the effector site, a measurement which is difficult
to obtain. Blood levels may be useful, but factors influencing local
drug delivery and metabolism in the immediate environment of receptors
are often difficult to quantify. Since catecholamines affect vascular
smooth muscle tone and cardiac output, adrenergic agonists or antagonists may alter tissue distribution during the period of administration.
In addition, anatomical factors may influence response. Since postsynaptic alpha$_2$ receptors in the vasculature appear to be extrajunctional (outside the neuromuscular junction), and may therefore be
preferentially exposed to circulating drugs compared to intrajunctional
alpha$_1$ receptors, responses to exogenously administered drugs may
disproportionately represent alpha$_2$ effects (6). In vivo studies must
be interpreted with the knowledge of the relative specificity of pharmacological agents utilized, an understanding of potential interference
with drug delivery by anatomical factors (synaptic reuptake mechanisms), and the realization that observed responses may result from
stimulation of mixed populations of receptors that are in different
anatomical locations. Changes in adrenergic response during the
period of study must also be considered in the design of in vivo
experiments. The sensitivity of adrenergic receptors is regulated
by agonist stimulation. In vivo administration of beta-adrenoceptor
agonists induces a decrease in receptor numbers (7,8) and decreased
adenylate cyclase activity (9). Increased endogenous catecholamines
associated with salt restriction may also induce receptor down-
regulation (10). Isoproterenol infusion causes a rapid decrease in
adenylate cyclase activity without an immediate change in receptor
number (11). These studies suggest that desensitization is not a
uniform phenomenon. In vitro studies indicate that desensitization
may be apparent within minutes (12) of agonist exposure and is
associated with decreased agonist affinity, whereas chronic stimulation
by an agonist may induce a decrease in receptor number, perhaps
due to internalization of receptors (13). Alpha-adrenergic responses

may show similar regulation, although investigations have not been as extensive. Desensitization of receptor responses may therefore be a problem in the design of in vivo studies.

III. ADRENERGIC RECEPTOR SUBTYPES

Characterization of receptor subtypes which mediate physiological responses is dependent upon the use of selective agonists and antagonists whose properties have been previously determined. Once a specific response is associated with a receptor subtype, then it may be utilized to define drug activity. Alpha receptors were initially described as the site of phenylephrine-induced vasoconstrictive activity, and beta receptors as the site of isoproterenol-induced vasodilation (2). As adrenoceptor antagonists, increasingly specific agonists, and radioligands have been developed, receptor definitions have been progressively refined. Beta receptors are characterized by an isoproterenol response greater than the response to epinephrine, which in turn is greater than the response to norepinephrine. Conversely, alpha receptors show the greatest response to norepinephrine, then to epinephrine, and only weak or no response to isoproterenol. Both alpha and beta receptors have been subdivided on the basis of pharmacological response and anatomical location.

The two major subtypes of beta receptors were recognized by Lands in 1967 according to pharmacological responses (14). Epinephrine and norepinephrine have about equal potency at beta$_1$ receptors, whereas epinephrine is 10-100 times more potent at beta$_2$ sites. Multiple agonists and antagonists with some degree of selectivity at beta or alpha receptors have been identified (see Chap. 3). The degree of binding selectivity between drugs acting at receptor subtypes is variable. Although metoprolol and atenolol are both beta$_1$-selective antagonists, metoprolol has a 10- to 20-fold degree of selectivity, while atenolol is only threefold more potent at beta$_1$ receptors (15). It is important to recognize that the selectivity of these agents is not absolute, and with increasing drug concentrations, both receptor subtypes may be affected. Anatomical studies have demonstrated a predominance of beta$_1$ receptors in cardiac and adipose tissue, whereas beta$_2$ receptors are generally found in the lungs and vascular smooth muscle. Despite the predominance of one receptor subtype, both receptors coexist in most tissues.

Alpha receptors have more recently been subdivided. Alpha$_1$ receptors are primarily postsynaptic and located intrajunctionally. Alpha$_2$ receptors have been identified presynaptically and in extrajunctional sites postsynaptically (6) as well as in adipose tissue, pancreatic islets, platelets, and renal tubules. It appears that alpha$_1$

receptors are therefore primarily exposed to norepinephrine released from an adjacent nerve terminal and are relatively protected from circulating catecholamines by synaptic reuptake mechanisms.

IV. IN VIVO STUDY OF ADRENERGIC RESPONSE

Examination of adrenergic response in vivo can be undertaken using either endogenous catecholamines or by administration of pharmaceuticals. The former method is, in general, less invasive but, as will be described subsequently, may not be as easy to quantify, especially when comparing results among individuals. Both approaches will be discussed in the subsequent sections.

A. Endogenous Catecholamine Release

The sympathetic nervous system is activated under physiological conditions by exercise, standing, the Valsalva maneuver, and performance of mental arithmetic. Secondary release of neuronal norepinephrine may be induced with tyramine. Monoamine oxidase inhibitors will transiently increase levels of endogenously released catecholamines. Because the degree of stimulation may show significant interindividual variability, blood levels of catecholamines are useful. Although older fluorometric methods were unreliable, low concentrations of catecholamines may now be measured using catechol-O-methyl-transferase radioenzymatic assay or high-pressure liquid chromatography with electrochemical detection (16). Norepinephrine is primarily released from neurons and usually functions as an intrasynaptic neurotransmitter. Plasma concentrations therefore represent an "overflow" phenomenon and are indicative of release minus degradation and reuptake. Under conditions of increased neuronal activity, norepinephrine plasma concentrations can reach levels that have end organ effects, and in this circumstance, norepinephrine functions as a hormone. Levels must reach 1500-2000 pg/ml (10 nM, about 10 times basal level) to produce significant hemodynamic effects (17). Because epinephrine is secreted from the adrenal medulla and transported to receptor sites by the circulation, it is a hormone in the traditional sense. Infusion of epinephrine increases the heart rate and blood pressure at plasma concentrations of 50-100 pg/ml, levels commonly observed under physiological conditions (18). Although plasma catecholamine levels may be the best means of assessing receptor stimulation, the correlation with concentrations at receptors may be difficult to predict.

Clearance of plasma epinephrine and norepinephrine occurs rapidly by several mechanisms. Average plasma clearance for epinephrine is

about 50 ml/kg min at steady-state concentrations of 25-75 pg/ml (18).
Norepinephrine clearance is 25 ml/kg min at steady-state levels of
230-345 pg/ml (normal basal levels are 65-390 pg/ml) (17). The
clearance of both norepinephrine and epinephrine is increased at
higher plasma levels. Sympathetic postganglionic neurons have a
high-affinity, low-capacity storage system (uptake 1). Extraneuronal
cells primarily metabolize the catecholamines after uptake by a low-
affinity, high-capacity system (uptake 2). Uptake 1 has higher
affinity for norepinephrine, uptake 2 has greater affinity for epinephrine, and isoproterenol is only removed by uptake 2. Beta-
adrenergic stimulation appears to increase catecholamine clearance
(19). Beta-adrenoceptor blockade with propranolol has been shown
to decrease the clearance rate for epinephrine by more than 75% (20).
Alpha-adrenergic blockade does not have a similar effect on catecholamine clearance. At least one component of this effect may be due to
increased enzymatic degradation, since membrane-bound catechol-O-
methyl-transferase activity is increased by beta agonists in canine
myocardium (21). The acceleration of catecholamine degradation by
beta-adrenergic stimulation may have important ramifications in studies
of beta-adrenoceptor antagonists. Since increased catecholamine levels
result from beta-receptor blockade, nonantagonized receptor responses
may be increased. Administration of a $beta_1$-specific antagonist might
therefore increase alpha- and $beta_2$-adrenergic activity.

Since catecholamine levels change quickly in response to stress
and posture, blood samples must be obtained under reproducible
conditions. This may require a standard period in one position
(supine, sitting) and prior placement of an indwelling intravenous
cannula to eliminate stimulation from venipuncture. Because brachial
venous samples will include norepinephrine released in the forearm
and will be depleted of any epinephrine extracted (22), specimens
may reflect higher norepinephrine and lower epinephrine levels than
simultaneous arterial levels. As previously noted, circulating norepinephrine represents the overflow from synaptic junctions and may not
be representative of concentrations at $alpha_1$ receptors within the
junction. Measurement of dopamine beta-hydroxylase (DBH) has
been utilized as an indication of neuronal norepinephrine release.
Dopamine beta-hydroxylase is released at the synaptic terminal simultaneously with norepinephrine. Thus, changes in circulating levels
should parallel neuronal activity. Unfortunately, there is substantial
genetic variation in DBH levels and comparisons between individuals
are inaccurate (23). Chromagranin A or other measures of catecholamine storage vesicle release may prove to be better indices of neuronal
activity (24).

Endogenous release of catecholamines can be evoked by a number
of different maneuvers. Standardized methods include treadmill,

TABLE 1   Endogenous Release of Catecholamines

	Epinephrine (pg/ml) (mean ± SEM)	Norepinephrine (pg/ml) (mean ± SEM)
Heparin-lock sample	23 ± 4	201 ± 50
Valsalva	36 ± 4	224 ± 43
Hand grip	47 ± 8	275 ± 40
Cold pressor	51 ± 8	343 ± 64
Ambulatory	42 ± 12	443 ± 47
Treadmill	90 ± 20	896 ± 105

Source: From Ref. 25.

bicycle ergometry, and step test. Although the Valsalva maneuver, cold pressor testing, hand grip, and assumption of upright posture from supine position all elicit significant catecholamine increase, exercise is the most potent stimulus of sympathetic nervous system activity (Table 1). Some experimental designs involve combinations of stimuli. A common example is bicycle ergometry in which the primary stimulus is from exercise and a significant secondary response may occur related to hand grip. Robertson et al. (25) studied 15 subjects (age 18-54 years) who abstained from methylxanthines and were in sodium balance on a 150-mEq sodium diet. Plasma catecholamine measurements after static and dynamic forms of exercise (Table 1) showed treadmill testing to be the most potent stimulus. Catecholamine levels decrease on high sodium diets (10) and increase with methylxanthines (26,27). Basal and stimulated norepinephrine levels have usually been shown to increase with age (28,29). Investigations of basal levels, however, are not entirely in agreement (30).

1.   Responses to Endogenously Released Catecholamines

In vivo studies of adrenergic responses have been extensively utilized for determination of the pharmacological properties of adrenoceptor agonists and antagonists and a number of adrenoceptor responses have now been characterized (Table 2). Beta$_1$ receptors predominate in the heart where stimulation induces an increase in contractility, conduction velocity, and automaticity. Intestinal relaxation, coronary vasodilation, and lipolysis are also beta$_1$-receptor responses, at least to a moderate degree. Beta$_2$-receptor responses include glycogenolysis,

TABLE 2 In Vivo Tests of Beta-Adrenergic Receptors

Endogenous catecholamines
1. Exercise tachycardia
2. Exercise-induced bronchodilation

Exogenously-administered agonists
1. Isoproterenol-stimulated increases in heart rate (I_{25})
2. Isoproterenol-stimulated changes in blood pressure and blood flow
3. Finger tremor
4. Plasma cyclic AMP
5. Metabolic endocrine responses (hyperglycemia, hypokalemia, hypophosphatemia, hyperactatemia, hypomagnesemia, hyper-insulinemia, decreased plasma contisol, increased plasma renin.

insulin release, and smooth muscle relaxation in bronchi, uterus, and peripheral vasculature. Studies of alpha responses are less complete and have been primarily devoted to vasoconstrictive actions and blood pressure changes. We will discuss several types of responses to endogenously released catecholamines.

  a. Exercise-induced sympathetic nervous system activation: Exercise causes both sympathetic simulation and parasympathetic withdrawal. Although repeated testing and exposure to pharmacological agents may present problems owing to conditioning, learned behavior, and tolerance, multiple studies in one individual can provide useful results (31). The dose response curves to sotalol were similar whether determined in five subjects given six repeat doses or in 30 individuals each given one dose (31). Resting heart rate is primarily dependent upon parasympathetic innervation. With exercise, sympathetic stimulation increasingly predominates and as heart rate approaches 200 beats/min the parasympathetic component is negligible. Vagal response is blocked with atropine, 0.04 mg/kg i.v. (32). The vagal response to exercise appears to be reproducible in one individual. After experiments have been performed using atropine, it is possible to simply subtract the vagal element in subsequent studies (33). Although the heart rate response to isoproterenol is blocked with 0.2 mg/kg propranolol intravenously (34), exercise stimulation may require up to 0.6 mg/kg for near complete sympathetic blockade (35). Despite complete sympathetic and parasympathetic blockade, heart rate can still increase with exercise., Although the mechanism is unclear, this phenomenon has been confirmed in dogs after heart transplant and adrenalectomy and appears to be unrelated to catecholamines (36). Even so, potency of beta-receptor antagonists can be characterized by either absolute or percentage reduction in exercise-stimulated

heart rate. Measurements of maximum heart rate with exercise before
and after administration of a beta-receptor antagonist are adequate
to determine the achieved degree of sympathetic blockade.

Although exercise-induced tachycardia results from both beta-
adrenoceptor stimulation and vagal withdrawal, the reduction in
maximal rate following administration of beta-receptor antagonist
correlates well with degree of $beta_1$-receptor blockade. The fact
that $beta_1$ antagonists are more effective in blocking the tachycardia
associated with exercise than that induced by isoproterenol (37)
suggests that a component of the isoproterenol response may be due
to $beta_2$ receptor stimulation. Because plasma levels of beta antagonists
are variable following a fixed oral dose, the measurement of beta
blockade is of clinical importance. A 30% reduction in exercise heart
rate correlates with a high degree of beta-receptor blockade and
can be achieved with 100 ng/ml propranolol (38), 350 ng/ml metoprolol
(38), or 35 ng/ml timolol (39). A 15% reduction is achievable with
12 ng/ml propranolol, 35 ng/ml metoprolol, and 3.5 ng/ml timolol.
The oral doses that are typically equipotent are propranolol, 80 mg;
metoprolol, 100 mg; and timolol, 10 mg (40).

   b. *Exercise-induced bronchodilation*: Because bronchospasm is
a serious side effect of beta-receptor antagonists, a means of evaluating
the relative effects of different agents on bronchial smooth muscle
would be desirable. Exercise induces mild bronchodilation with a
measurable increase in peak flow rate (PFR). The increment in PFR
is, however, only 5% (30 L/min with baseline of 550 L/min) of resting
levels with significant scatter between measurements under similar
conditions. The small response to exercise may be the reason that
differences between beta-receptor antagonists have been difficult to
demonstrate with this technique. Propranolol, metoprolol, practolol,
and oxprenolol in doses producing similar blockade of heart rate have
not been established to have a consistent and major effect on resting
or exercise PFR when given orally or intravenously to normal subjects
(41). All agents except oxprenolol show a trend toward decreased
exercise PFR. A study in asthmatics showed significant decreases
compared to baseline FEV_1 (forced expiratory volume over the first
second of expiration) of 89% for propranolol (100 mg), 94% for atenolol
(100 mg), 94% for pindolol (5 mg), and 94% for atenolol (300 mg) (42).
Propranolol and pindolol at these doses blocked the bronchodilatory
response to isoproterenol to a greater degree than acebutolol or atenolol.
The doses used were not tested for equal $beta_1$ receptor blockade.
Exercise-stimulated peak flow rate and FEV_1 studies are difficult to
interpret because of small changes and interindividual variability.
Patients suffering from bronchospasm may be more suitable subjects
than normals, but the risk of adverse reactions would be expected
to be increased.

  c. *Hypoglycemic stimulation of endogenous catecholamine release:*
Hypoglycemia induces increases in plasma epinephrine and norepinephrine (Fig. 3). The effects of catecholamines in homeostatic responses
to hypoglycemia can be studied with beta-blockers. Schluter et al.
studied the metabolic response to insulin (0.1 U/kg)-induced hypoglycemia in eight young male subjects after propranolol (160 mg p.o.),
metoprolol (200 mg p.o.), or pindolol (15 mg p.o.) (43). Serum
growth hormone after hypoglycemia (plasma glucose 30-40 mg/dl)
increases to approximately 25 ng/ml in controls or after pindolol and
increases to a significantly greater degree (approximately 35 ng/ml)
after propranolol or metoprolol. Plasma norepinephrine response to
hypoglycemia is also increased by propranolol or metoprolol. Adrenocorticotropic hormone (ACTH) increase was greater after propranolol

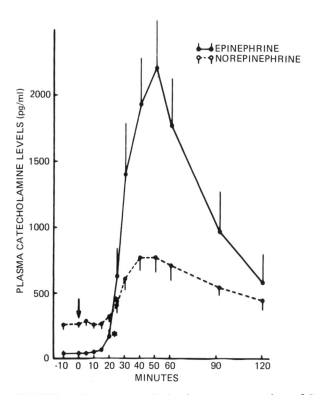

FIGURE 3 Plasma catecholamine concentrations following intravenous
injection of 0.15 U/kg of crystalline insulin at time O. Values are the
means ± SEM from six healthy adult males. (From Ref. 96, used with
permission.)

or metoprolol administration, but the differences were not significantly different than after placebo. Hypoglycemia persists for a longer period following propranolol. Heart rate response to hypoglycemia is blocked by all three beta-blockers, although subjects still recognized symptoms of hypoglycemia, including diaphoresis, hunger, and a "heavy" sensation. Asymptomatic hypoglycemia did not occur. Although adrenergic regulation of carbohydrate metabolism is complex and specific receptor responses are difficult to define, the clinical relevance of pharmacological investigation is obvious.

2. Responses to Exogenously Administered Agonists

An alternative to stimulation of endogenous catecholamine release is the administration of adrenergic receptor agonists. The forms of stimulation are clearly not analogous and, as previously noted, responses will frequently differ even if circulating blood levels are equal. Isoproterenol sensitivity testing has been standardized and is widely utilized for nonselective beta-receptor stimulation. Prenalterol (beta$_1$), terbutaline (beta$_2$), metaproterenol (beta$_2$), salbutamol (beta$_2$) and epinephrine (alpha and beta$_2$) are other available beta-receptor agonists. Norepinephrine (nonselective alpha), phenylephrine (alpha$_1$), and clonidine (alpha$_2$) are commonly used alpha-receptor agonists.

a. Isoproterenol sensitivity testing—increases in heart rate: Standardized methodology for evaluation of heart rate response using bolus intravenous isoproterenol was described in 1972 by Cleveland et al. (44) and George et al. (45). Some minor differences in technique involving volume and rate of isoproterenol administration have been shown to be inconsequential (46). Increasing doses of isoproterenol are rapidly infused and heart rate response is measured (Fig. 4). Isoproterenol-induced decreases in systolic and diastolic blood pressure may be evaluated with similar techniques. Subjects are tested in the supine position in a quiet room. They are instructed that isoproterenol may cause a desire to inhale deeply and that they should attempt to breathe regularly, since fluctuation in respiratory rate may affect the heart rate. Isoproterenol is prepared in appropriate dilutions with a preservative such as 0.1% sodium metabisulfite. An initial test dose of 0.1 µg is administered intravenously into a rapidly flowing saline infusion. Heart rate is monitored by electrocardiogram. Peak heart rate is calculated from the three shortest R-R intervals or with continuous ratemeter. Maximum response is typically noted at approximately 50 sec. Subsequent doses are doubled and administered after delays of 5-10 min to allow heart rate to return to baseline. The test is continued until peak heart rate is increased by at least 30 beats/min. This dose range will include a linear section

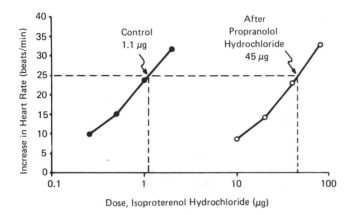

FIGURE 4 Heart rate response to intravenous isoproterenol before and after propranolol. (From Ref. 45, used with permission.)

of the dose-response curve. The dose required to increase heart rate 25 beats/min is a commonly used standardization of response and is termed the I_{25}. It is usually calculated from linear regression analysis of response as a function of the log isoproterenol dose. The test is fairly reproducible, although a coefficient of variation of 20% occurs with repeated studies in the same subject (44). Significant atrial or ventricular ectopy is an indication to discontinue testing. Since sinus arrhythmia can interfere with the acquisition of accurate measurements, screening electrocardiograms may be of value.

Marked variation in isoproterenol sensitivity occurs between individuals. A decrease in sensitivity occurs with aging (Fig. 5) (47,48). Changes in receptor density on circulating blood elements correlate with I_{25} (10). Both receptor density and I_{25} are altered by physiological variation in catecholamines (10). The interindividual variation in isoproterenol sensitivity may therefore, in part, be due to dietary factors (including sodium), stress, and age. Racemic isoproterenol has been used for most human studies. Although some in vitro investigations have suggested that d-isoproterenol may have beta$_2$-blocking activity (49), no evidence has been presented of different in vivo responses to l-isoproterenol compared to dl-isoproterenol.

Isoproterenol sensitivity may also be determined using continuous infusion. Responses appear to include a greater component owing to vagal withdrawal than with bolus injections (50). Tolerance may occur and facilitation of neuronal norepinephrine release may be induced by stimulation of presynaptic beta$_2$ receptors (51,52). Graded isopro-

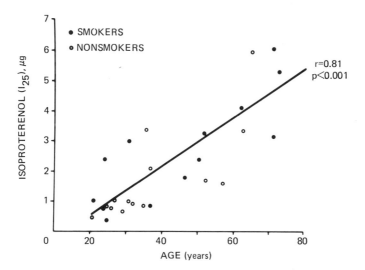

FIGURE 5 Isoproterenol dose required to increase heart rate by
25 beats/min as a function of age. (From Ref. 47, used with permission.)

terenol infusions up to 35 ng/kg min increase plasma norepinephrine
concentrations (Fig. 6) (53). Responses (Fig. 7) could therefore
reflect a component of alpha-adrenergic stimulation. These complicating
factors are less apparent with bolus administration of isoproterenol.
    Although the use of small, rapidly administered isoproterenol
doses decreases the activation of homeostatic mechanisms, heart rate
response appears to be mediated by both direct cardiac beta adrenoceptor and by a reflex decrease in vagal tone in response to peripheral
vasodilation (54). Peripheral vascular $beta_2$-adrenoceptor stimulation
produces vasodilation, decreased blood pressure, and consequent
decreased vagal activity. This mechanism was initially suggested by
studies using antagonists of isoproterenol-induced tachycardia.
Increased blockade of the isoproterenol effect by practolol ($beta_1$
antagonist) was demonstrable after atropine administration (55). It
was hypothesized that atropine eliminated vagal withdrawal induced
by $beta_2$-receptor activation. Since practolol only antagonized isoproterenol at $beta_1$ receptors, the blockade was more prominent when
$beta_2$ responses were minimized. Recently, $beta_2$ receptors have been
identified in the myocardium by radioligand studies, so direct activation
may also be important (56,57). Propranolol, which blocks both $beta_1$
and $beta_2$ receptors, antagonizes isoproterenol more effectively than
practolol even in studies using atropine (58). The chronotropic effect

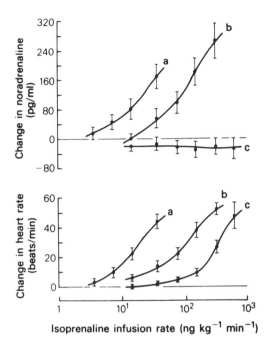

FIGURE 6 Changes in plasma norepinephrine and heart rate during isoproterenol infusion. Ten subjects with borderline hypertension studied before beta-adrenoceptor blocker treatment (a) and after treatment with atenolol (b) or propranolol (c). Atenolol was administered orally 100 mg/day and propranolol was given 80 mg four times daily for 1 week. (From Ref. 53, used with permission.)

of isoproterenol may be related to direct myocardial $beta_1$- and $beta_2$-receptor stimulation as well as to a component of vagal withdrawal from peripheral $beta_2$-receptor stimulation. Atropine is unnecessary in most studies unless vagal withdrawal is expected to interfere with interpretation of results.

 b. Isoproterenol sensitivity testing—changes in blood pressure and blood flow: Blood pressure responses to isoproterenol may be evaluated using similar experimental protocols, although intra-arterial monitoring is necessary. Both infusion (Fig. 7) and bolus techniques of isoproterenol demonstrations have been utilized. Results vary somewhat, probably because of factors such as norepinephrine release (53) and receptor regulation. Maximum hemodynamic response occurs at the same time as maximum heart rate (about 50 sec) following bolus administration. Diastolic blood pressure is primarily dependent upon

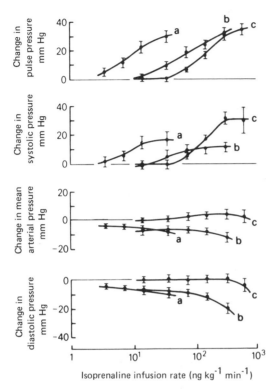

FIGURE 7 Changes in arterial pressures following isoproterenol
infusion in 10 subjects with borderline hypertension before beta-blocker
(a) and after treatment with atenolol (b) or propranolol (c). Treatment
protocols the same as in Figure 6. (From Ref. 53, used with permission.)

peripheral vascular resistance and is decreased by $beta_2$-adrenoceptor-
mediated vasodilation. Isoproterenol typically causes little change in
systolic blood pressure, which is dependent upon stroke volume and
heart rate as well as peripheral vascular resistance. Mean arterial
pressure is decreased. A recent series of experiments by Arnold
and McDevitt (58) investigated hemodynamic responses to bolus
administration of isoproterenol using propranolol and practolol.
Propranolol (19.09 μg/kg min for 15 min, followed by 1.07 μg/kg min
i.v.) and practolol (152.8 μg/kg min for 15 min, followed by 2.5
μg/kg min i.v.) were calculated to provide levels of 75 ng/ml and
1.25 μg/ml, respectively (59,60). Exercise testing was performed

to demonstrate similar degrees of $beta_1$-receptor blockade (26% and 21% decrease in exercise tachycardia, respectively). Propranolol and practolol increased the dose of isoproterenol required to decrease diastolic pressure 10 mm Hg from 0.93 to 9.0 µg and 1.8 to 2.43 µg, respectively. The respective dose ratios (ratio of agonist doses [A'/A] that induce equal response with antagonist [A'] and without antagonist [A]) were therefore 97.5 and 1.3. These results suggest greater antagonism of $beta_2$ receptors by propranolol than by practolol, with a consequently greater dose of isoproterenol being required to induce peripheral vasodilation and decreased diastolic pressure. The doses of isoproterenol required to increase heart rate by 25 beats/min (I_{25}) after placebo, propranolol, or practolol were 1.45 µg, 65.0 µg, 6.17 µg, respectively. Concomitant decreases in diastolic blood pressure for these doses of isoproterenol were 13.0, 5.6, and 20.7 mm Hg, respectively. The changes in systolic pressure were -3.2, +16.2, and -16.6 mm Hg, respectively. The dose ratio for diastolic blood pressure ($beta_2$ response) after propranolol was 97.5, whereas that for heart rate was 43.7 (primarily $beta_1$ response). The antagonism of isoproterenol by propranolol may therefore be proportionately greater at $beta_2$ than at $beta_1$ receptors. Propranolol antagonized the decrease in diastolic pressure and led to an increase in systolic pressure, whereas practolol at doses causing similar $beta_1$ blockade had minimal $beta_2$ effect and facilitated isoproterenol-mediated vasodilation.

Isoproterenol administration by the bolus technique can be used to more directly investigate $beta_2$-receptor response by measuring forearm blood flow. It must be recognized that peripheral blood flow is dependent on cardiac output, so a component of $beta_1$-receptor activity cannot be eliminated. Plethysmography is utilized to determine changes in forearm volume per unit time after venous occlusion (61). A venous cuff is inflated to 60 mm Hg for 10 sec and deflated for 5 sec. The blood flow into veins is measured during the period of occlusion as volume per 100 ml forearm tissue per minute. Increasing doses of isoproterenol are administered until a desired response or excessive heart rate occurs. Using regression analysis, the isoproterenol dose required to achieve a given response such as 5 ml/100 ml min (IF_5) may be calculated. Arnold et al. (62) have used this technique to investigate the blockade of isoproterenol response by propranolol and practolol. The heart rate responses (I_{25}) after placebo, 40 mg oral propranolol, or 200 mg oral practolol, were 1.06 µg, 27.67 µg (dose ratio 27.1), and 5.32 µg (dose ratio 5.2), respectively. The forearm blood flow response described by the IF_5 after placebo, propranolol, or practolol was 0.5 µg, 25.48 µg (dose ratio 55), and 1.51 µg (dose ratio 3), respectively. At the doses utilized, propranolol more effectively antagonized forearm blood flow response than did

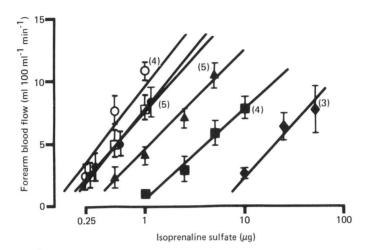

FIGURE 8 Isoproterenol-induced increase in forearm blood flow in
the presence of placebo (○,□); practolol, 50 mg (●); practolol,
200 mg (▲); propranolol, 10 mg (■); and propranolol, 40 (♦). Each
point shows the mean ± SEM; n = 6 unless indicated. (From Ref. 62,
used with permission.)

practolol (Fig. 8). This is compatible with the hypothesis that the
response is predominantly beta₂-receptor mediated.

Peripheral blood flow may also be evaluated using measurements
of skin temperature or radioisotope techniques. Skin temperature of
an extremity is determined using a thermistor after acclimatization
in a room at constant temperature. In a study of 10 normal subjects,
skin temperature was 31.6°C, with a standard deviation of 0.34°C
(63). Repeat measurement 90 min after 100 mg metoprolol showed a
temperature decrease of 0.15°C (not significant), whereas following
80 mg propranolol, the decrease was 1.3°C (p < 0.05). Initial decay
of radioactivity following intradermal or intramuscular injection of
radioisotope is proportional to blood flow (63). Skin and muscle blood
flows using this technique are not significantly changed by metoprolol
(100 mg) but are reduced 32 and 20%, respectively, by propranolol
(80 mg). Reports of cold extremities, Raynaud's syndrome (64),
and gangrene (65) in association with beta-blockers are indications
of the potential clinical relevance of peripheral blood flow studies.
Forearm plethysmography, skin temperature, and radioisotope studies
all appear to be reproducible means of evaluating beta₂-receptor
modulation of peripheral blood flow.

c. Isoproterenol sensitivity testing—further quantitative aspects:
In vitro studies have demonstrated decreased beta-adrenoceptor
responses with aging. Investigation of heart rate response to isoproterenol and propranolol has provided in vivo evidence of decreased
beta-receptor sensitivity in the elderly (47). Isoproterenol sensitivity
testing reveals an increasing I_{25} (Fig. 4) with aging. Isoproterenol
responses repeated after propranolol (10 mg at 1.1 mg/min followed
by 55 µg/min infusion) allow calculation of dose ratios. For a known
concentration of antagonist (P), the antagonist dissociation constant
(K_d) may then be determined from the equation:

$$\log (IP/I - 1) = \log P - \log K_d$$

where IP is agonist dose with antagonist, and I is agonist dose without
antagonist. Since multiple receptors and probably variable drug
concentrations at each receptor occur with such an in vivo study,
the K_d is at best simply proportional to propranolol activity (concentration required to produce half maximal response or K_m) rather than
a true dissociation constant. This constant (K_m) was demonstrated
to increase with age (Fig. 9) indicating that a larger concentration
of propranolol is required to antagonize isoproterenol-induced tachycardia. Although usually impractical in in vivo studies, this form of
analysis can be extended if multiple doses of antagonist are utilized.
A Schild plot (66) of log (IP/I - 1) as a function of log (P) is a linear

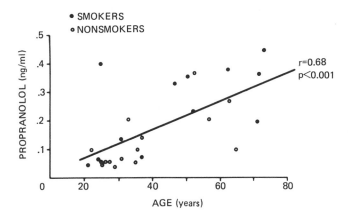

FIGURE 9 Concentration of propranolol required to produce half
maximal response (K_m) as a function of age. (From Ref. 47, used
with permission.)

expression with slope of unity if the antagonist is competitive. The intercept on the ordinate is the pKa (negative common logarithm of the antagonist dissociation constant), or as it is commonly referred to, the pA_2. The Schild plot has the advantage of multiple measurements which can be evaluated with regression analysis.

d. *Tremor—a beta$_2$ receptor-mediated response to adrenergic agonists*: Physiological tremor is proportional to beta-adrenoceptor activity. Finger tremor has a characteristic frequency of 8-12 Hz. The usual low amplitude is enhanced by infusion of catecholamines (Fig. 10). Isoproterenol increases the rate of contraction and relaxation in slow skeletal muscle fibers, although the opposite effect occurs in fast muscle fibers (49,67). Slow fibers are effected by lower concentrations of catecholamines than fast fibers, and are believed to be responsible for enhanced tremor (67). The tremor is measured with an accelerometer taped to the dorsum of the middle finger while the arm is comfortably supported on an arm rest and splint. The accelerometer output is amplified, usually with a high-frequency filter, and displayed on a chart recorder. Amplitude of the tremor is measured and expressed as a percentage change from control level. Using bolus administration of isoproterenol, the dose required to increase heart rate 25 beats/min correlates well with that required to increase tremor by 150% (68). Terbutaline-induced increase of heart rate and tremor is effectively blocked by propranolol, but only the heart rate change is blocked by metoprolol, suggesting that tremor is a beta$_2$-receptor-mediated response (69). Arnold et al. and McDevitt have shown that dose ratios (DR) of isoproterenol obtained using propranolol (40 mg) are similar when measuring heart rate (DR 18.2) and tremor (DR 17.1). Practolol (120 mg), a beta$_1$-specific receptor blocker, has a greater effect on heart rate (DR 4.8) than on tremor (DR 2.1) (68). Physiological tremor therefore appears to be a potentially useful parameter for the evaluation of beta$_2$-receptor responses in vivo.

e. *Cyclic AMP response to isoproterenol*: Plasma cAMP has been shown to increase after isoproterenol administration. Vascular endothelial cells are probably a major source of the increment in cAMP, although this is not well established. Cyclic AMP may be a more direct measure of beta-receptor activation than responses such as heart rate. Results are, however, complicated by factors influencing cAMP diffusion from cells and degradation. Kaliner et al. (70) have published responses to isoproterenol infusion at increasing concentration from 6 to 21 ng/kg min (3 ng/kg min increments). Infusion was maintained for 10 min, followed by a 20-min rest period before the next higher concentration was initiated. Plasma cAMP was measured using radioimmunoassay before and after each infusion. The concentration of isoproterenol required to increase cAMP level by 50% was

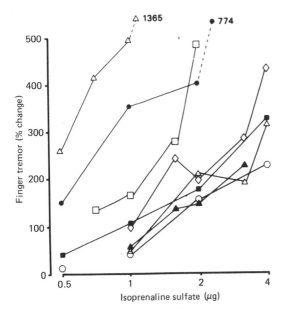

FIGURE 10 Finger tremor amplitude response to isoproterenol infusion. (From Ref. 68, used with permission.)

used as a measure of beta-receptor response. In 25 normal subjects this concentration was 8.08 ± 0.62 ng/kg min. Using this method of isoproterenol infusion, the concentration of isoproterenol causing a 50% increase in plasma cAMP induced an increase in pulse pressure of approximately 22 mm Hg (70).

f. Metabolic responses to beta-receptor agonists: Terbutaline (0.1-1 mg over 1-hr infusion), salbutamol (0.1-0.7 µg/kg min), and rimiterol (0.11-0.44 µg/kg min) have been utilized to determine cardiovascular and metabolic responses to beta$_2$ adrenoceptor stimulation (71,72). Subjects experience marked anxiety and tremor with terbutaline doses greater than 0.5 mg. Increases in heart rate and systolic blood pressure with decreased diastolic blood pressure occurs with all three agents. Serum glucose is increased and serum potassium decreased by doses of terbutaline greater than 0.1 mg. Maximal metabolic effects using salbutamol and rimiterol (72) include hyperglycemia (7.2 mM/L increase), hypokalemia (1.8 mM/L decrease), hypophosphatemia (0.45 mM/L decrease), hyperlactatemia (2 mM/L increase), ketonemia (2 mM/L increase), hypocalcemia (0.118 mM/L decrease), hypomagnesemia (0.053 mM/L decrease), increased plasma insulin concentration (50 µU/ml), decreased cortisol (2.8 nM/L), and increased plasma renin (3.72 ng/ml hr). Prominent variation occurs in individual susceptibility to the effects of beta-adrenoceptor agonists.

Metabolic responses are frequently complex and reflect characteristics of multiple receptors at more than one effector site. Plasma renin activity (PRA) is increased by stimulation of renal sympathetic nerves (73) or administration of beta agonists. Studies with specific beta-receptor agonists and antagonists suggest components of both $beta_1$- and $beta_2$-receptor response. Prenalterol, a $beta_1$-specific agent, at doses sufficient to increase systolic pressure and heart rate (75 µg/kg) decreases PRA (74). The effects of blood pressure change on renin release complicate the interpretation of these results. $Beta_2$-receptor agonists, including fenoterol (75,76), salbutamol (72), and rimiterol (72), increase PRA in vivo. $Beta_2$-adrenoceptor blockers are also more effective inhibitors of isoproterenol-induced increases in renin activity than are $beta_1$-adrenoceptor antagonists (76). Although decreases in PRA are noted with beta-blockers, total plasma renin has not been found to be significantly changed (77). The renin concentration in plasma treated at pH 3.3 is considered total renin whereas measurements using plasma treated at pH 4.5 are referred to as active renin (PRA). The total renin is believed to include prorenin (78). Metoprolol (100 mg) decreases PRA from 1.42 to 0.30 ng/ml hr with a concommitant increase in total renin of 22.96-27.46 ng/ml hr (77). Plasma angiotensin II decreases from 49.44 to 16.19 pg/ml (77). Exercise-evoked increases in both active and inactive renin measurements are reduced by metoprolol. Adrenergic modulation of renin therefore appears to involve both renin release and metabolism. Although a $beta_2$-receptor-mediated mechanism may be of primary importance, there is evidence of a $beta_1$-mediated-receptor component as well. Plasma renin responses are therefore difficult to interpret as a means of in vivo receptor investigation.

Beta-adrenergic responses are similarly complex in carbohydrate and lipid metabolism. As previously described, hypoglycemia induces release of endogenous catecholamines. Experiments using anesthetized dogs reveal that isoproterenol (0.015 µg/kg min) and salbutamol ($beta_2$ agonist, 0.02 µg/kg min) provoke insulin release without increasing the blood glucose (79). Propranolol (0.3 mg/kg i.v.) blocks the response but practolol ($beta_1$ antagonist, 0.8 mg/kg i.v.) has no effect. The response to salbutamol and lack of effect from practolol is suggestive that a $beta_2$ adrenoceptor mediates the insulin response. Studies in man (80) have demonstrated an increase in insulin during isoproterenol (threefold increase using 0.06 µM/kg min) or terbutaline (twofold increase using 2.3 nM/kg min) infusion that is completely blocked by propranolol (34 µM i.v.) but only partially decreased by metoprolol (44 µM i.v.). Insulin release induced by terbutaline ($beta_2$ agonist) and lack of complete blockade by metoprolol ($beta_1$ antagonist)

are compatible with a beta$_2$ response. Insulin responses can be difficult to evaluate in man, since blood glucose increases with beta-
adrenergic stimulation. A mild increase in blood glucose (1.44 mM/L)
after terbutaline infusion (500 μg) is blocked by propranolol (0.19
mM/L after 40 mg p.o) more effectively than by metoprolol (0.47 mM/L
after 50 mg p.o.) (81). These results suggest that hepatic gluconeogenesis is stimulated by beta$_2$-adrenoceptor activation. Glycogenolysis,
which may also contribute to an increase in glucose after sympathetic
stimulation, appears to be mediated by alpha receptors (82). Glucose-
induced increase in serum insulin and C-peptide is mildly decreased
by propranolol (160 mg) or metoprolol (200 mg) but not by pindolol
(15 mg) (43). If insulin is administered intravenously, however,
the clearance is prolonged after beta-blockers. Beta-receptor
antagonists therefore appear to inhibit both insulin release and
metabolism, although the clinical significance of the effects are unclear.

Hypokalemia occurs following beta-adrenergic stimulation. This
may result from beta$_2$-adrenoceptor-mediated stimulation of membrane
sodium-potassium ATPase (83), with resultant cellular influx of
potassium and efflux of sodium. Since blood glucose and insulin
increase with beta-adrenergic stimulation, potassium influx associated
with glucose transport may contribute to observed hypokalemia (72).
The effect is clinically important, since a decrease in serum potassium
from 4.06 to 3.22 mM/L occurs with plasma epinephrine levels (4.67
nM/L) commonly found following myocardial infarction (84). Electrocardiographic changes associated with the hypokalemia include QT
prolongation and T wave flattening. Atenolol (beta$_1$ antagonist,
50 mg b.i.d.) reduces the hypokalemia (decrease from 4.06 to 3.67
mM/L) and timolol (nonspecific, 10 mg b.i.d.) increases serum
potassium from 4.06 to 4.25 mM/L despite epinephrine administration.
Hypokalemia has been reported after salbutamol, rimiterol (Fig. 11),
or terbutaline administration (81). The response to terbutaline is
also antagonized more effectively by nonspecific beta-receptor antagonist (propranolol, 40 mg p.o.) than beta$_1$-specific antagonist (metoprolol, 50 mg p.o.). The decrease in serum potassium therefore
appears to be primarily mediated by beta$_2$ adrenoceptors. The partial
antagonism by atenolol and metoprolol indicates that the agents are
either not entirely selective at the doses utilized or that the hypokalemia is partly due to beta$_1$ response. Nevertheless, the marked
differences in beta$_1$-specific and nonspecific beta-receptor blockers
suggest that hypokalemic responses may be useful for defining the
degree of selectivity of beta-receptor antagonists (81).

3. Alpha-Receptor Responses

Alpha-adrenergic effects on blood pressure, cutaneous blood
flow, or pupillary dilatation may be utilized to study receptor characteristics (Table 3). Incremental infusion of norepinephrine (0.01-1.0

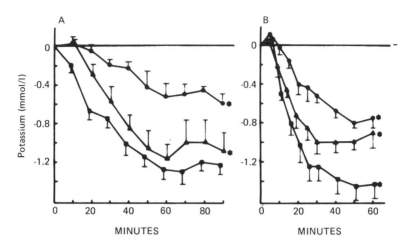

FIGURE 11 Plasma potassium concentrations during (A) salbutamol
(●) 0.1 μg/kg min, (▲) 0.4 μg/kg min, (■) 0.7 μg/kg min, or (B)
rimiterol (●) 00.1 μg/kg min, (▲) 0.2 μg/kg min, (■) 0.44 μg/kg min.
Mean ± SEM, n = 4. (From Ref. 72, used with permission.)

μg/kg min) or phenylephrine (0.5-10 μg/kg min) induce increases in
systolic and diastolic blood pressure (85). Heart rate and blood
pressure measurements are recorded each minute to determine steady-
state response at each dose level. Infusion rate is increased at 5-min
intervals until a predetermined maximum response is achieved.
Regression analysis of the linear section of the log dose-response
curve may then be used to determine the dose of agonist required
to raise the blood pressure a given amount (e.g., 20 mm Hg).
Norepinephrine often elicits a biphasic peripheral vascular response
with increased blood flow at low doses and vasoconstriction with
larger doses (85). Vasoconstrictor responses are antagonized by
alpha-blockers, while beta-adrenergic blockade abolishes vasodilator
components of the response. The ability of classic alpha-adrenergic
agonists to act as beta-adrenergic agonists is thus apparent and makes
careful interpretation of results important.

TABLE 3 In Vivo Tests of Alpha-Adrenergic Receptors

Blood pressure or blood flow response to infused agonist

Cutaneous blood flow response to intradermally injected agonist

Pupillary response to conjunctivally instilled agonist

The increasing availability of alpha-receptor subtype-specific agonists and antagonists may allow more extensive in vivo definition of receptor response. Experience is limited at this time and experimental design may be complicated by opposing cardiovascular effects of receptor subtypes. Specific agonists include phenylephrine (alpha$_1$) and clonidine (alpha$_2$). Prazosin (alpha$_1$) and yohimbine and Rx 78-1094 (alpha$_2$) are selective antagonists. Release of endogenous norepinephrine can be induced with tyramine. The source of norepinephrine can be important, since alpha$_1$ receptors may be more sensitive to neuronally released agonist, whereas alpha$_2$ receptors may be more responsive to circulation agonist (86).

a. *Cutaneous blood flow and pupillary responses*: Cutaneous blood flow is decreased by alpha$_1$-adrenergic stimulation. [133]Xenon washout, as described by Sejrsen (87), can be utilized to quantify alpha-receptor-mediated changes in blood flow. [133]Xenon, 100 nCi, is injected intracutaneously and the radioactivity is measured at 10-sec intervals. The time required for washout of radioactivity is proportional to the cutaneous blood flow (88). Intradermal injection of phenylephrine (0.01-200 ng) causes vasoconstriction. The dose of phenylephrine may be progressively increased until the xenon washout time is doubled (blood flow decreased by 50%) to provide a quantifiable response. The mean dose for this result in 38 normal subjects was determined to be 32 ± 7.5 ng (88). The phenylephrine response is antagonized by injection of phentolamine (10 ng) and potentiated by propranolol (10 ng) (88).

Pupillary response to conjunctival instillation of alpha-receptor agonist can also be used as a measure of receptor activation. Pupillary size is measured in the dark with a television camera sensitive to infrared lighting. The subject's gaze is focused upon a small red light at a fixed distance (typically 100 cm). Increasing concentrations of phenylephrine (0.5-3.5%) are then utilized until a mydriasis of 0.5 mm is induced. A mean concentration of 2.55 ± 0.08% was required in 57 normal subjects studied by Kaliner et al. (70).

Cutaneous blood flow and pupillary responses have been used to define alpha-receptor function in subjects with allergic diseases. The mean dose of phenylephrine required to decrease blood flow 50% was 32.1, 23.7, and 4.4 ng in normals, subjects with allergic rhinitis, and asthmatics, respectively (70). The mean concentration of phenylephrine required to increase the pupillary diameter by 0.5 mm was 2.55, 2.33, and 1.59% in normals, subjects with allergic rhinitis, and asthmatics (70). Asthmatics, and to a lesser degree subjects with allergic rhinitis, were therefore demonstrated to have alpha-receptor hyperresponsiveness. The results from two different means of evaluation of alpha-receptors correlated well.

    b. Testing for alpha-adrenergic responses in orthostatic hypo-
tension: The etiology of orthostatic hypotension can be differentiated
with the measurement of plasma catecholamines and pressor response
testing. Idiopathic orthostatic hypotension (IOH) (89) is a result
of peripheral sympathetic nerve degeneration which leads to decreased neuronal norepinephrine release, less synaptic spillover,
and low circulating norepinephrine. Multiple system atrophy
(MSA) (90), which also causes orthostatic hypotension, results from
a central nervous system defect in sympathetic nervous system activation. Although basal (supine) norepinephrine levels are only minimally
decreased, the normal increase in norepinephrine with standing does
not occur (Fig. 12). The decreased norepinephrine release in IOH
leads to receptor hypersensitivity. The dose-response curve for
infused norepinephrine is therefore shifted to the left such that the
initial blood pressure response is noted at a lower norepinephrine

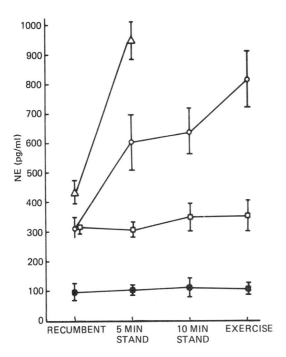

FIGURE 12 Plasma norepinephrine (pg/ml) with standing and exercise
in normal (o), idiopathic orthostatic hypotension (■), multisystem
atrophy (□), and subject with volume depletion (△). (From Ref. 93,
used with permission.)

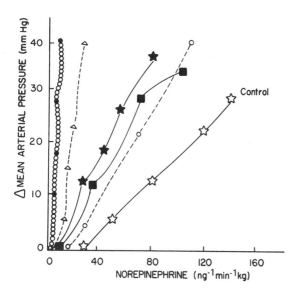

FIGURE 13 Norepinephrine dose response in subjects with idiopathic orthostatic hypotension compared with control. (Adapted from Ref. 94.)

concentration and incremental blood pressure increase (slope of dose-response curve) is greater than in normal subjects (Fig. 13) (91). A less prominent degree of receptor hypersensitivity occurs in MSA, so the dose-response curve, although still left of normal, is shifted less than with IOH. In one study, norepinephrine induced a detectable increment in mean blood pressure at plasma levels of 122, 98, and 32 pg/ml in controls, MSA, and IOH, respectively (92). Tyramine induced an increment in blood pressure at dose of 34, 22, and 9.4 µg/kg in controls, MSA, and IOH, respectively. Tyramine infusion in IOH leads to less than normal increase in norepinephrine, but because of the marked receptor hypersensitivity there is typically an enhanced pressor response compared to normal. Patients with MSA also have receptor hypersensitivity that causes an enhanced pressor response compared to normal. Norepinephrine release after tyramine administration is less severely affected in MSA than in IOH. Considerable interindividual variability occurs in pressor dose response and there is a marked spectrum of disease severity in IOH and MSA. Since tyramine pressor activity is dependent upon norepinephrine release, the tyramine response may be interpreted relative to the individual's norepinephrine sensitivity (assuming infused norepinephrine sensitivity reflects endogenous norepinephrine sensitivity). The tyramine/norepinephrine response ratio therefore provides a sensitive indication of tyramine-induced norepinephrine release. Dose-response curves for each agent are obtained. Normal responses to norepinephrine and tyramine reach the linear part of the dose-response curve before 200 and 40 µg/kg min, respectively (92). Using linear regression for the linear part of the response or parametric fit to a quadratic equation the pressure response to 100 ng/kg min norepinephrine and to 20 µg/kg min tyramine are extrapolated. In normal subjects, the pressor responses to these doses are equal (tyramine/norepinephrine ratio = 1). In IOH, the ratio is significantly decreased (92).

V. CONCLUSIONS

Investigation of adrenergic response using in vivo techniques provides clinically important information. As increasingly specific agonists and antagonists become available, organ system responses will become better defined. Results must be evaluated carefully and necessary assumptions clearly defined. Although clinical application of information may be straightforward, extrapolation of data to describe receptor characteristics is complicated by drug pharmacokinetics, local receptor milieu, and homeostatic responses.

REFERENCES

1. Powell, C. E., and Slater, I. H.: Blocking of inhibitory adrenergic receptors by a dichloro analog of isoproterenol. J. Pharmacol. Exp. Ther. 122:480-488, 1958.

2. Ahlquist, R. P.: A study of adrenotropic receptors. Am. J. Physiol. 153:586-600, 1948.

3. Clark, A. J.: *The Mode of Action of Drugs on Cells.* Edward Arnold, London, 1933.

4. Ariens, E. J., Simonis, A. M., and DeGroot, W. M.: Affinity and intrinsic activity in the theory of competitive and non-competitive inhibition and an analysis of some forms of dualism in action. Arch. Int. Pharmacodyn. Ther. 100:298-322, 1955.

5. Stephenson, R. P.: A modification of receptor theory. Br. J. Pharmacol. 11:379-393, 1956.

6. Langer, S. Z., and Shepperson, N. B.: Recent developments in vascular smooth muscle pharmacology: The post-synaptic alpha-2 adrenoceptor. Trends Pharmacol. Sci. 3:440-444, 1982.

7. Galant, S. P., Dureseti, L., Underwood, S., and Insel, P. A.: Decreased beta adrenergic receptors on polymorphonuclear leukocytes after adrenergic therapy. N. Engl. J. Med. 299:933-936. 1978.

8. Tashkin, D. P., Conolly, M. E., Deutsch, R. I., Hui, K. K., Littner, M., Scarpace, P. J., and Abrass, I. B.: Subsensitization of beta-adrenoceptors in airways and lymphocytes. Am. Rev. Respir. Dis. 125:185-193, 1982.

9. Conolly, M. E., and Greenacre, J. K.: The lymphocyte beta-receptor in normal subjects and patients with bronchial asthma. J. Clin. Invest. 58:1307-1316, 1976.

10. Fraser, J., Nadeau, J., Robertson, D., and Wood, A. J. J.: Regulation of human leukocyte beta-receptors by endogenous catecholamines. J. Clin. Invest. 67:1777-1784, 1981.

11. Krall, J. F., Connelly, M., and Tuck, M. L.: Acute regulation of beta-adrenergic catecholamine sensitivity in human lymphocytes. J. Pharmacol. Exp. Ther. 214:554-560, 1980.

12. Su, Y. F., Harden, T. K., Perkins, J. P.: Catecholamine-specific desensitization of adenylate cyclase. J. Biol. Chem. 255:7410-7419, 1980.

13. Chuang, D. M., and Costa, E.: Evidence for internalization of the recognition site of the beta-adrenergic receptor during receptor subsensitivity induced by (-)isoproterenol. Proc. Natl. Acad. Sci. U.S.A. 76:3024-3026, 1979.

14. Lands, A. M., Luduena, F. P., and Buzzo, H. J.: Differentiation of receptors responsive to isoproterenol. Life Sci. 6:2241-2249, 1967.

15. Minneman, K. P., Hegstrand, L. R., and Molinoff, P. B.: The pharmacologic specificity of beta-1 and beta-2 adrenergic receptors in rat heart and lung in vitro. Mol. Pharmacol. 16:21-33, 1979.

16. Goldstein, D. S., Feurstein, G., Isso, J. L., Koplin, I. J., and Deiser, H. R.: Validity and reliability of liquid chromatography with electrochemical detection for measuring plasma levels of norepinephrine and epinephrine in man. Life Sci. 28:467-475, 1981.

17. Silverberg, A. B., Shah, S. D., Haymond, M. W., and Cryer, P. E.: Norepinephrine: Hormone and neurotransmitter in man. Am. J. Physiol. 234:E252-E256, 1978.

18. Clutter, W., Bier, D., Shah, S., and Cryer, P.: Epinephrine plasma metabolic clearance rates and physiologic thresholds for

metabolic and hemodynamic actions in man. J. Clin. Invest. 66: 94-101, 1980.

19. Cryer, P. E., Rizza, R. A., Haymond, M. W., and Gerich, J. E.: Epinephrine and norepinephrine are cleared through beta-adrenergic, but not alpha-adrenergic, mechanisms in man. Metabolism 11:1114-1118, 1980.

20. Cryer, P. E.: Physiology and pathophysiology of the human sympathoadrenal neuroendocrine system. N. Engl. J. Med. 303: 436-444, 1980.

21. Wrenn, S., Homcy, C., and Haber, E.: Evidence for the beta-adrenergic receptor regulation of membrane bound catechol-O-methyltransferase activity in myocardium. J. Biol. Chem. 254: 5708-5712, 1979.

22. Halter, J. B., Pflug, A. E., and Tolas, A. G.: Arterial-venous differences of plasma catecholamines in man. Metabolism 29: 9-12, 1980.

23. Weinshilboum, R. M.: Biochemical genetics of catecholamines in human. Mayo Clin. Proc. 58:319-339, 1983.

24. O'Connor, D. T., and Bernstein, K. N.: Human chromagranin A, the major catecholamine storage vesicle soluble protein; quantitation by radioimmunoassay with application to human plasma as a probe of exocytotic sympathoadrenal activity and of pheochromocytoma. N. Engl. J. Med. 311:764-770, 1984.

25. Robertson, D., Johnson, G. A., Robertson, R. M., Nies, A. S., Shand, D. G., and Oates, J. A.: Comparative assessment of stimuli that release neuronal and adrenomedullary catecholamines in man. Circulation 59:637-642, 1979.

26. Robertson, D., Frölick, J. C., Carr, R. K., Watson, J. T., Hollifield, J. W., Shand, G. D., and Oates, J. A.: Effects of caffeine on plasma renin activity, catecholamines and blood pressure. N. Engl. J. Med. 298:181-186, 1979.

27. Vestal, R. E., Eiriksson, C. E., Jr., Musser, B., Ozaki, L. K., and Halter, J. B.: Effect of intravenous aminophylline on plasma levels of catecholamines and related cardiovascular and metabolic responses in man. Circulation 67:162-171, 1983.

28. Young, J. B., Rowe, J. W., Pallotta, J. A., Sparrow, D., and Landsberg, L.: Enhanced plasma norepinephrine response to upright posture and oral glucose administration in elderly human subjects. Metabolism 29:532-539, 1980.

29. Esler, M., Skews, H., Leonard, P., Jackman, G., Bobik, A., and Korner, P.: Age dependence of noradrenaline kinetics in normal subjects. Clin. Sci. 60:217-219, 1981.

30. Meier, A., Gubeliu, V., Weidmann, P., Grimm, M., Keusch, G., Gluck, Z., Minder, I., and Beretta-Piccoli, C.: Age related profile of cardiovascular reactivity to norepinephrine and angiotensin II in normal and hypertensive man. Klin. Wochenschr. 58:1183-1188, 1980.

31. Brown, H. C., Carruthers, S. G., Kelley, J. G., McDevitt, D. G., and Shanks, R. G.: Observations on the efficacy and pharmacokinetics of sotalol after oral administration. Eur. J. Clin. Pharmacol. 9:367-372, 1976.

32. Chamberlain, D. A., Turmer, P., and Sneddon, J. M.: Effects of atropine on heart rate in healthy man. Lancet 2:12-15, 1967.

33. Carruthers, S. G., Shanks, R. G., and McDevitt, D. G.: Intrinsic heart rate and exercise after beta-adrenoceptor blockade. Br. J. Clin. Pharmacol. 3:991-999, 1976.

34. Jose, A. D.: Effect of combined sympathetic and parasympathetic blockade on heart rate and cardiac function in man. Am. J. Cardiol. 18:476-478, 1966.

35. McDevitt, D. G.: The assessment of beta-adrenoceptor blocking drugs in man. Br. J. Clin. Pharmacol. 4:413-425, 1977.

36. Donald, D. E.: Myocardial performance after excision of the intrinsic cardiac nerves in the dog. Circ. Res. 34:417-424, 1974.

37. Conrad, K. A.: Comparison of inotropic and chronotropic effects of metoprolol and propranolol. J. Clin. Pharmacol. 21:213-218, 1981.

38. Sklar, J., Johnston, G. D., Overlie, P., Gerber, J. G., Brammell, H. L., Gal, J., and Nies, A. S.: The effect of a cardioselective (metoprolol) and a nonselective (propranolol) beta-adrenergic blocker on the response to dynamic exercise in normal men. Circulation 65:894-899, 1982.

39. Bobik, A., Jennings, G. L., Ashley, P., Koner, P. I.: Timolol pharmacokinetics and the effects on heart rate and blood pressure after acute and chronic administration. Eur. J. Clin. Pharmacol. 16:243-247, 1979.

40. Shand, D. G.: How should the proper dose of a beta blocker be determined? Circulation 67 (Suppl. I):86-88, 1983.

41. Oh, V. M. S., Kaye, C. M., Warrington, S. J., Taylor, E. A., and Wadsworth, J.: Studies of cardioselectivity and partial agonist

activity in beta adrenoceptor blockade comparing effects on heart
rate and peak expiratory flow rate during exercise. Br. J. Clin.
Pharmacol. 5:107-120, 1978.

42. Benson, M. K., Berrill, W. T., Cruikshank, J. M., and Sterling,
G. S.: A comparison of four adrenoceptor antagonists in patients
with asthma. Br. J. Clin. Pharmacol. 5:415-419, 1978.

43. Schluter, K. J., Aellig, W. H., Peerson, K. G., Rieband, H. C.,
Wehrli, A., and Kerp, L.: The influence of beta-adrenoceptor
blocking drugs with and without intrinsic sympathomimetic activity
on the hormonal responses to hypo and hyperglycemia. Br. J.
Clin. Pharmacol. 13:407S-417S, 1982.

44. Cleaveland, C. R., and Shand, D. G.: A standardized isoproterenol
sensitivity test. Arch. Intern. Med. 130:47-52, 1972.

45. George, C. F., Conolly, M. E., Fenyvesi, T., Briant, R., and
Dollery, C. T.: Intravenously administered isoproterenol sulphate
dose-response curves in man. Arch. Intern. Med. 130:361-364,
1972.

46. Arnold, J. M. O., and McDevitt, D. G.: Heart rate and blood
pressure response to intravenous bolus of isoproterenol in the
presence of propranolol, practolol and atropine. Br. J. Clin.
Pharmacol. 16:175-184, 1983.

47. Vestal, R. E., Wood, A. J. J., and Shand, D. G.: Reduced
beta-adrenoceptor sensitivity in the elderly. Clin. Pharmacol.
Ther. 26:181-186, 1979.

48. Lakatta, E. G.: Age related alterations in the cardiovascular
response to adrenergic mediated stress. Fed. Proc. 39:3173-3177,
1980.

49. Al-Jeboory, A. A., and Marshall, R. J.: Correlation between
the effects of salbutamol on contractions and on cyclic AMP
content of isolated fast and slow-contracting muscles of the guinea
pig. Naunyn Schmiedebergs Arch. Pharmacol. 305:201-206, 1978.

50. Cleaveland, C. R., and Shand, D. G.: Effect of route of administration on the relationship between beta-adrenergic blockade and
plasma propranolol level. Clin. Pharmacol. Ther. 13:181-185,
1972.

51. Stjarne, L., and Brundin, J.: Beta-2 adrenoceptors facilitating
noradrenaline secretion from human vasoconstrictor nerves. Acta
Physiol. Scand. 97:88-93, 1976.

52. Majewski, H.: Modulation of noradrenaline release through activation of presynaptic beta-adrenoceptors. J. Auton. Pharmacol. 3:
47-60, 1983.

53. Vincent, H. H., Man In't Veld, A. J., Boomsma, F., Wenting, G. J., Schalekamp, M. A. D. H.: Elevated plasma noradrenaline in response to beta adrenoceptor stimulation in man. Br. J. Clin. Pharmacol. 13:717-721, 1982.

54. Dunlop, D., and Shanks, R. G.: Selective blockade of adrenoceptor beta-receptors in the heart. Br. J. Pharmacol. 32:201-218, 1968.

55. Brick, I., Hutchinson, K. J., McDevitt, D. G., Roddie, I. C., and Shanks, R. G.: Comparison of the effects of ICI 50,172 and propranolol on the cardiovascular responses to adrenaline, isoprenaline and exercise. Br. J. Pharmacol. 36:35-45, 1968.

56. Stiles, G. L., Taylor, S., and Lefkowitz, R. J.: Human cardiac beta-adrenergic receptors: Subtype heterogeneity delineated by direct radioligand binding. Life Sci. 33:467-473, 1983.

57. Robberecht, P., Delhaye, M., Taton, G., De Neef, P., Waelbroeck, M., De Smet, J. M., Leclerc, J. L., Chatelain, P., and Christophe, J.: The human heart beta-adrenergic receptors. Mol. Pharmacol. 24:169-173, 1983.

58. Arnold, J. M. O., and McDevitt, D. G.: Interpretation of iso-prenaline dose-response curves in the presence of selective and non-selective beta-adrenoceptor blocking drugs. Br. J. Clin. Pharmacol. 13:581P, 1982.

59. Shand, D. G.: Pharmacokinetic properties of the beta-adrenergic receptor blocking drugs. Drugs 7:39-47, 1974.

60. Wagner, J. G.: A safe method for rapidly achieving plasma concentration plateaus. Clin. Pharmacol. Ther. 16:691-700, 1974.

61. Greenfield, A. D. M.: Venous occlusion plethysmography. Methods Med. Res. 8:293-301, 1960.

62. Arnold, J. M. O., Allen, J. A., Shanks, R. G., and McDevitt, D. G.: Changes in heart rate and forearm blood flow following intravenous boluses of isoprenaline in the presence of practolol and propranolol. Br. J. Clin. Pharmacol. 15:37-40, 1983.

63. McSorley, P. D., and Warren, D. J.: Effects of propranolol and metoprolol on the peripheral circulation. Br. Med. J. 2:1589-1600, 1978.

64. Marsden, C. W., and Bayliss, P. F. C.: Raynaud's phenomenon as a side effect of beta-blockers. Br. Med. J. 2:176, 1976.

65. Vale, J. A., and Jefferys, D. B.: Peripheral gangrene complicating beta-blockade. Lancet 1:1216-1217, 1978.

66. Arunlakshana, O., and Schild, H. O.: Some quantitative aspects of drug antagonists. Br. J. Pharmacol. 14:48-58, 1959.

67. Bowman, W. C., and Raper, C.: Adrenotropic receptors in skeletal muscle. Ann. N.Y. Acad. Sci. 139:741-753, 1967.

68. Arnold, J. M. O., and McDevitt, D. G.: An assessment of physiological finger tremor as an indicator of beta-adrenoceptor function. Br. J. Clin. Pharmacol. 16:167-174, 1983.

69. Larsson, S., Svedmyr, N.: Tremor caused by sympathomimetics is mediated by beta-2 adrenoceptors. Scand. J. Resp. Dis. 58: 5-10, 1977.

70. Kaliner, M., Shelhamer, J. H., Davis, P. B., Smith, L. J., and Venter, J. C.: Autonomic nervous system abnormalities and allergy. Ann. Intern. Med. 96:349-357, 1982.

71. Kendall, M. J., Dean, S., Bradley, D., Gibson, R., and Worthington, D. J.: Cardiovascular and metabolic effects of terbutaline. J. Clin. Hosp. Pharm. 7:31-36, 1982.

72. Phillips, P. J., Vedig, A. E., Jones, P. L., Chapman, M. G., Collins, M., Edwards, J. B., Smeaton, T. C., and Duncan, B. M.: Metabolic and cardiovascular side effects of the beta-2 adrenoceptor agonists salbutamol and rimiterol. Br. J. Clin. Pharmacol. 9:483-491, 1980.

73. Peart, W. S.: Renin 1978. Johns Hopkins Med. J. 143:193-206, 1978.

74. Staessen, J., Cattaert, A., Deschaepdryver, A., Fagard, R., Lijnen, P., Moerman, E., and Amery, A.: Effects of beta-1 adrenoceptor agonist on plasma renin activity in normal men. Br. J. Clin. Pharmacol. 16:553-556, 1983.

75. Meurer, K. A., Lang, R., Homback, V., and Helber, A.: Effects of a beta-1 selective adrenergic agonist in normal human volunteers. Klin. Wochenschr. 58:425-427, 1980.

76. Weber, M. A., Stokes, G. S., and Gain, J. M.: Comparison of the effects on renin release of beta-adrenergic antagonists with different properties. J. Clin. Invest. 54:1413-1419, 1974.

77. Lijnen, P. J., Amery, A. K., Fagard, R. H., Reybrouck, T. M., Moerman, E. J., and DeSchaepdryver, A. F.: The effects of beta adrenoceptor blockade on renin, angiotensin, aldosterone, and catecholamines at rest and during exercise. Br. J. Clin. Pharmacol. 7:175-181, 1979.

78. Derckx, F. H. M., Wenting, G. J., and Man In't Veld, A. J.: Inactive renin in human plasma. Lancet 2:496-499, 1976.

79. Loubatieres, A., Mariani, M. M., Sorel, G., and Savi, L.: The action of beta-adrenergic blocking and stimulating agents on

insulin secretion. Characterization of the type of beta-receptor. Diabetologia 7:127-129, 1971.

80. William-Olsson, T., Fellenius, E., Björntorp, P., and Smith, U.: Differences in metabolic responses to beta-adrenergic stimulation after propranolol or metoprolol administration. Acta Med. Scand. 205:201-206, 1979.

81. Smith, S. R., Kendall, M. J., Worthington, D. J., and Holder, R.: Can the biochemical responses to a beta-2 adrenoceptor stimulant be used to assess the selectivity of beta-adrenoceptor blockers? Br. J. Clin. Pharmacol. 16:557-560, 1983.

82. Lager, I., Blohme, G., and Smith, U.: Effect of cardioselective and noncardioselective beta-blockade on the hypoglycemic response in insulin-dependent diabetics. Lancet 1:458-462, 1979.

83. Petch, M. C., McKay, R., and Bethune, D. W.: The effect of beta-2 adrenergic blockade on serum potassium and glucose levels during open heart surgery. Eur. Heart J. 2:123-126, 1981.

84. Struthers, A. D., Reid, J. L., Whitesmith, R., and Rodger, J. C.: The effects of cardioselective and nonselective beta-adrenoceptor blockade on the hypokalemic and cardiovascular responses to adrenomedullary hormones in man. Clin. Sci. 65:143-147, 1983.

85. Sumner, D. J., Elliott, H. L., and Reid, J. L.: Analysis of the pressor dose response. Clin. Pharmacol. Ther. 32:450-458, 1982.

86. Ariens, E. J., and Simonis, A. M.: Physiological and pharmacological aspects of adrenergic receptor classification. Biochem. Pharmacol. 32:1539-1545, 1983.

87. Sejrsen, P.: Blood flow in cutaneous tissue in man studied by washout of radioactive xenon. Circ. Res. 25:215-229, 1969.

88. Henderson, W. R., Shelhamer, J. H., Reingold, D. B., Smith, L. J., Evans, R., and Kaliner, M.: Alpha-adrenergic hyperresponsiveness in asthma. N. Engl. J. Med. 300:642-647, 1979.

89. Bradbury, S., Eggleston, C.: Postural hypotension. A report of three cases. Am. Heart J. 1:73-86, 1926.

90. Shy, G. M., and Drager, G. A.: A neurological syndrome associated with orthostatic hypotension. Arch. Neurol. 25:511-527, 1960.

91. Polinsky, R. J., Kopin, I. J., Ebert, M. H., and Weise, V.: Pharmacologic distinction of different orthostatic hypotension syndromes. Neurology 31:1-7, 1981.

92. Demanet, J. C.: Usefulness of noradrenaline and tyramine infusion tests in diagnosis of orthostatic hypotension. Cardiology 61:213-224, 1976.

93. Ziegler, M. G.: Postural hypotension. Ann. Rev. Med. 31:239-245, 1980.

94. Chobanian, A. V., Tifft, C. P., Sachel, H., and Pitrezella, A.: Alpha and beta-adrenergic receptor activity in circulating blood cells of patients with idiopathic orthostatic hypotension and pheochromocytoma. Clin. Exp. Hypertens. A 4(4-5):793-806, 1982.

95. Gilman, A. G.: Pharmacodynamics: Mechanisms of drug action and the relationship between drug concentration and effect. In: *The Pharmacological Basis of Therapeutics* (6th ed.), A. G. Gilman and L. S. Goodman (eds.). New York, Macmillan, 1983, pp. 28-39.

96. Garber, A. J., Cryer, P. E., Santiago, J. V., Haymond, M. W., Aagliara, A. S., and Kipnis, P. M.: The role of adrenergic mechanisms in the substrate and humoral response to insulin induced hypoglycemia in man. J. Clin. Invest. 58:7-15, 1976.

2

Defining the Role of Adrenergic Receptors in Human Physiology

JOHN P. BILEZIKIAN   *College of Physicians and Surgeons of Columbia University, New York, New York*

I. INTRODUCTION

In this chapter, the actions of the catecholamines will be examined from a physiological point of view. The history leading to the development of the adrenergic receptor concept, and the subsequent refinement of the receptor concept to account for additional information that has accumulated over the years, will provide the background to a discussion of the diverse effects of the catecholamines upon a wide variety of tissues and organs. Not to be addressed in this chapter is information pertinent to the mechanisms by which the catecholamines act, data regarding their own metabolism or coverage of the dopaminergic system. With detailed knowledge of the physiological actions of the catecholamines and the opportunity to assign a specific receptor designation to most of these actions, the following chapters dealing with receptor identification, characterization, regulation, and their alterations in certain diseases may be better understood.

II. CONCEPT OF THE ADRENERGIC RECEPTOR

The notion that the catecholamines can be classified into two general groups, one often being a physiological counterpart of the other, was developed at the beginning of the twentieth century when Dale observed that the naturally occurring ergot-containing compounds were able to convert a predominant vasopressor response of epinephrine to a vasodepressor one (4). It was originally believed that this observation was

due to the elaboration of two mediators, an excitatory one "sympathin
E" and an inhibitory one, "sympathin I," both derived from a single
substance, "sympathin." The naturally occurring ergot alkaloids
were thought to block specifically the excitatory response. The
adrenergic receptor concept in the beginning, thus, was unitary:
One adrenergic receptor type was regarded to be the common mediator
of all adrenergic influences in the cell. In 1948, however, Ahlquist
(1) tested a hypothesis that revolutionized the notion of a single class
of adrenergic receptors. He considered the relative potency of five
sympathetic amines (norepinephrine [NE], epinephrine [E], isoproterenol [ISO], α-methylnorepinephrine [α-methyl NE], α-methylepinephrine
[α-methyl E]) upon four organ systems: vasculature, heart, uterus,
and intestine, in three animals: cats, dogs, and rabbits. The reason-
ing was simple. If the order of potency among the five catecholamines
was constant in all tissues and species, the differences in potency
would be best attributable to differences in their chemical structure—
and presumably their affinity for a single class of adrenergic receptors.
On the other hand, if the order of potency among the five catecholamines varied among the tissues studied, the properties of the target
cells themselves and presumably the receptors mediating the response
must be fundamentally different. Ahlquist (1) observed two general
orders of potency for these catecholamines among all the organs of
the species examined. For excitatory events in general (-) E > (±) -
E > (±) NE > α-methyl NE > α-methyl E > (±) ISO. For inhibitory
events, ISO > (-) - E > α-methyl E > (±) E > α-methyl NE > NE. The
results, therefore, suggested two receptor populations which became
designated alpha and beta. With some notable exceptions, the alpha-
adrenergic receptor was considered to mediate excitatory properties of
the catecholamines, whereas the beta-adrenergic receptor was considered
to mediate inhibitory properties of the catecholamines (26,33,36).

The original observations concerning the ergot alkaloids could
then be understood in terms of their inhibition of the alpha-adrenergic
response, allowing the beta-adrenergic actions to be expressed in
those tissues that possessed intrinsic beta-adrenergic activity (21).
The physiological distinction of the catecholamines into alpha and beta
components was strengthened further by the development of another
class of compounds that specifically inhibited the beta-adrenergic
response, allowing the contrasting alpha-adrenergic response to be
expressed in those tissues that possessed intrinsic alpha-adrenergic
activity (51,52). The prototypic beta-adrenergic inhibitors, dichloroiso-
proterenol and propranolol, were joined over the next 2 decades by an
encyclopedic array of other inhibitors. These inhibitors have strengthened markedly the concept of the dual adrenergic receptor and also,
notably, have become very important therapeutic agents for a wide
variety of clinical disorders.

The discovery of cyclic adenosine monophosphate (AMP) by Sutherland further helped to establish the dual receptor theory of the adrenergic catecholamines (40). Most of the activities characterized as beta-adrenergic could be shown to be mediated by the stimulation of adenylate cyclase and the cellular accumulation of cyclic AMP (41). In contrast, alpha-adrenergic responses were not, at that time, associated with major changes in the adenylate cyclase complex. Beta-adrenergic inhibitors were able to block selectively catecholamine-stimulated adenylate cyclase activity. Alpha-adrenergic inhibitors were without significant effects upon this enzyme. Thus it appeared reasonable to postulate the existence of two distinct populations of receptors, alpha and beta, each recognizing catecholamines in a different and predictable order of potency, each responsible for a specific set of physiological responses, and each mediated by different cellular mechanisms.

III. SUBTYPE SELECTIVITY OF THE PHYSIOLOGICAL RESPONSE

Further refinement of the hypothesis that the actions of the catecholamines were distinguishable into two different subclasses for each of the adrenergic receptors came from later studies that utilized the Ahlquist approach (22). In 1967, Lands (30) compared isoproterenol, norepinephrine, and epinephrine for their actions upon events already classified as beta-adrenergic: bronchodilatation, vasodepression, cardiac stimulation, and lipolysis. The order of potency of these compounds in heart and adipose tissue was ISO > NE $\geq$ E. The relative concentrations yielding responses that were 50% of the maximum (EC_{50}) displayed a ratio of 1:5-20:10-40. On the other hand, in the lungs and the vasculature, a different order of potency was appreciated: ISO > E > NE, the relative potencies being 1:3-15:60-400. It seemed likely therefore that within the general classification of the beta-adrenergic response, further subclassification into at least two subtypes, $beta_1$ and $beta_2$, was possible. $Beta_1$ activity was described by the behavior of heart and adipose tissue to the catecholamines (ISO > NE $\geq$ E); $beta_2$ activity was described by the behavior of the lungs and the vasculature to the catecholamines (ISO > E > NE).

Both $beta_1$- and $beta_2$-mediated activities utilized the cyclic AMP system, suggesting that subtype specificity was a property of the beta receptor itself and could not be better explained by postreceptor events. Since the early description by Lands (30), many additional physiological events have been similarly subclassified as $beta_1$ or $beta_2$ (16). A summary of this information is shown in Table 1. Along with subclassification of the beta-adrenergic response, there has been a major effort to develop agonists and antagonists with these same subtype selectivities, a partial listing of which is shown in Table 2.

TABLE 1 Subclassification of the Beta-Adrenergic Actions
of the Catecholamines

Beta$_1$	Beta$_2$
Increased heart rate	Smooth muscle relaxation (vasculature, trachea, bronchi)
Increased myocardial contractility	
Increased cardiac glycogenolysis	Small intestinal relaxation
Relaxation of coronary arteries	Skeletal muscle contraction
Relaxation of intestinal smooth muscle	Skeletal muscle glycogenesis
	Gallbladder relaxation
Calorigenesis (brown adipose tissue)	Diaphragm contraction
Lipolysis (adipose tissue)	Urinary bladder relaxation
Salivary gland secretion	Vas deferens relaxation
Renin release (kidney)	Fallopian tube relaxation
Stimulation of melatonin (pineal)	Uterus relaxation
Stimulation of glucagon (pancreas)	Glycogenolysis, gluconeogenesis (liver)
	Lymphocyte cyclic AMP accumulation
	Inhibition of lymphocyte mitogenesis
	Inhibition of histamine release from polymorphonuclear leukocytes
	ACTH release
	Thyroid hormone secretion
	Parathyroid hormone secretion
	Pancreatic insulin release
	Potassium uptake in muscle

The list of actions mediated by the subclasses of the beta-adrenergic
catecholamines includes only those tissues for which there is sufficient
evidence for subtype specificity.

TABLE 2 Subclassification of Beta-Adrenergic Agonists and Antagonists

Beta agonists	Beta antagonists
Beta$_1$	Beta$_1$
Dobutamine	Acebutolol
Tazolol	Atenolol
	Betaxolol
	Bunitrolol
	Metoprolol
	Practolol
	Tolamolol
Beta$_2$	Beta$_2$
Carbuterol	Butoxamine
Fenoterol	H35/25
Isoxsuprine	ICI 118.551
Orciprenaline	
Metaproterenol	
Ritodrine	
Salbutamol	
Soterenol	
Terbutaline	
Nonselective	Nonselective
Isoproterenol	Alprenolol
	Bunolol
	Nadolol
	Oxprenolol
	Pindolol
	Propranolol
	Sotalol
	Timolol

The same Ahlquist-like approach—namely, the study of the order of potency of a series of catecholamines in different tissues and species led to the subtype classification of the α-adrenergic catecholamines (8, 57). The subdivision of alpha-adrenergic actions into alpha$_1$ or alpha$_2$ was aided by the development of a great number of newer catecholamine analogs. In 1977, Berthelson and Pettinger (8) proposed that the subclassification of the alpha-adrenergic response be based upon functional criteria, alpha$_1$ agonists stimulating vascular smooth muscle; alpha$_2$ agonists inhibiting NE release from the postganglionic sympathetic neuron. The order of potency for the alpha$_1$ response was shown to be E > NE > oxymetazoline > naphazoline > phenylephrine > tramazoline > α-methyl NE > methoxamine. The order of potency of the alpha$_2$ response, on the other hand, was shown to be E > oxymetazoline > tramazoline > α-methyl NE > NE > naphazoline > phenylephrine > methoxamine. Although relatively weak compounds on the list of catecholamine analogs with alpha$_1$ activity, methoxamine and phenylephrine nevertheless are among the more selective alpha$_1$ agonists. Similarly, tramazoline, α-methyl NE, clonidine, and oxymetazoline are relatively selective alpha$_2$ agonists. A partial listing of alpha-adrenergic agonists and antagonists according to their subtype specificities is shown in Table 3. Physiological events for which subtype specificity is reasonably well established for alpha-adrenergic catecholamines is shown in Table 4. It should be noted that the majority of alpha-adrenergic actions have not yet been subclassified.

TABLE 3 Subclassification of Alpha-Adrenergic Agonists and Antagonists

Alpha agonists	Alpha antagonists
Alpha$_1$	Alpha$_1$
Phenylephrine	BE 2254
Metaraminol	Indoramin
Methoxamine	Labetalol
	Phenoxybenzamine
	Prazosin
	WB 4101
Alpha$_2$	Alpha$_2$
α-methylnorepinephrine	Clozapine
Clonidine	Piperoxan
Ergonovine	Rauwolscine
Guanabenz	Tolazoline
Oxymetazoline	Yohimbine
Tramazoline	
Xylazine	
Nonselective	Nonselective
Epinephrine	Ergocryptine, (dihydro-
Naphazoline	ergocryptine)
Norepinephrine	Phentolamine

TABLE 4 Subclassification of the Alpha-Adrenergic Actions
of the Catecholamines[a]

Alpha$_1$	Alpha$_2$
Contraction of smooth muscle	Inhibition of postganglionic norepinephrine release
Increased cardiac contractility	
Decreased cardiac automaticity[b]	Arterial constriction
Increased phosphatidylinositol turnover, intracellular and extracellular calcium fluxes	Reduction of blood pressure via central mechanisms
	Inhibition of hypothalamic heat conservation mechanisms
Increased secretion of ACTH (pituitary)	Stimulation of platelet aggregation
	Small intestinal relaxation and secretion
Inhibition of renin release	
	Inhibition of intestinal water and electrolyte secretion
	Inhibition of insulin release
	Inhibition of lipolysis
	Inhibition of vasopressin action in the renal tubules

[a]Many other alpha-adrenergic influences upon physiological events,
as described in this chapter, have not yet been subclassified and
thus do not appear in this table.
[b]Effects upon automaticity may be related to age, decreasing in the
elderly and increasing in the young.

IV. THE ROLE OF OTHER FACTORS IN INTERPRETING THE ADRENERGIC RESPONSE

Knowledge of the specific physiological events influenced by the
catecholamines in virtually all organs allows one to classify these actions
into alpha- or beta-mediated phenomena (2). As will be illustrated
in subsequent chapters, most of these physiological actions are concordant with the direct in vitro identification and subclassification
of the adrenergic receptors. The physiological description of catecholamine behavior, however, has to take into account a great number

of potential variables that may serve to complicate complete understanding of their actions. Some of these variables include the dose of the catecholamines employed (to distinguish between a physiological and pharmacological effect), the duration of exposure to the hormone(s), the preexisting state of the tissue, the particular combination of adrenergic agents employed, and the species investigated. In addition, a given tissue often will display several different adrenergic actions owing to the presence of several receptor classes. A small but significant effect may not be expressed until the predominant physiological effect is selectively inhibited. Some tissues demonstrate a heterogeneity of responses owing to a variable distribution of adrenergic receptors. Finally, it should be appreciated that the alpha- and beta-adrenergic response—when both are present—are not always opposite to each other but may be shared. These considerations have confounded complete understanding of this field. In the discussion to follow, a consensus is presented, although in some cases it has not been possible to be absolutely sure that the cautions noted above have been satisfactorily taken into account. Whenever possible, the information presented is derived from tissues of human subjects. In instances when the human data are not available, information based upon animal studies is presented. For some tissues, physiological responsiveness to the alpha- or beta-adrenergic catecholamines has not yet been classified into their subtypes. When this occurs, the text and the tables indicate only the major classifications.

V. SMOOTH MUSCLE

A. Nonvascular Smooth Muscle

In general, contraction of smooth muscle is induced by the alpha-adrenergic catecholamines, whereas relaxation of smooth muscle is a beta-adrenergic property (37). Many smooth muscles can be shown to have the dual capacity to contract or to relax after exposure to alpha- or beta-adrenergic catecholamines, respectively. However, the predominant physiological response, whether it be contraction (alpha) or relaxation (beta), appears to be a property of the particular smooth muscle. Presumably this observation implies a specific distribution of alpha and beta receptors in the particular tissue. A noteworthy exception to the contrasting actions of the alpha- and beta-adrenergic catecholamines in most smooth muscle preparations is found in the gastrointestinal tract, in which nonsphincteric smooth muscle responds to either alpha- or beta-adrenergic catecholamines with relaxation. The classification of smooth muscle according to tissue is shown in Table 5 and noted with respect to the physiological predominance of alpha- or beta-adrenergic behavior.

TABLE 5 Actions of the Catecholamines on Nonvascular Smooth Muscle

	Receptor type	Predominant response
Pulmonary (tracheal, bronchial)	$beta_2 \gg alpha$	Relaxation
Gastrointestinal tract		
motility	$alpha_1$ and $beta_2$	Decreased
sphincters	alpha > beta	Contraction
Gallbladder	$beta_2$	Relaxation
Spleen	alpha > beta	Contraction
Bladder		
detrusor	$beta_2$	Relaxation
trigone/sphincter	alpha	Contraction
Uterus	$beta_2$	Relaxation
	alpha	Contraction
Fallopian tubes	$beta_2$	Relaxation
Vas deferens	$alpha_2 > beta_2$	Contraction

B. Vascular Smooth Muscle

The smooth muscle of the vasculature represents a special situation
in which the catecholamines can indirectly alter organ activity by
influencing blood flow (6). In general, vasoconstriction due to smooth
muscle contraction is best described as alpha-adrenergic, whereas
vasodilatation due to smooth muscle relaxation is best described as
beta-adrenergic. The vasculature of many organs can respond to the
catecholamines with either vasoconstriction or vasodilatation and would
thus appear to contain both alpha and beta receptors (Table 6). The
predominant physiological response depends upon the degree to which
the particular organ is endowed with adrenergic receptors of either
type, its previous state of activity, and, most importantly, the
presence of other vasoactive principles that might augment, negate,
or otherwise influence the adrenergic receptor-mediated response.
For vasoconstriction, subtype specificity is fairly uniformly $alpha_1$
throughout all vascular beds. Recent evidence, however, in human
subjects has suggested that vasoconstriction in peripheral arteries
and veins may also be a property of $alpha_2$ receptors (24,27). For
vasodilatation, on the other hand, subtype designation may be $beta_1$
(coronary), $beta_2$ (skeletal muscle), or both $beta_1$ and $beta_2$ (kidney).

TABLE 6 Actions of the Catecholamines on Vascular Smooth Muscle

	Presence of alpha response (constriction)		Presence of beta response (relaxation)		Predominant response
Arterial					
Large vessels					
aorta	Yes	$(alpha_1)$	Yes	$(beta_2)$	Constriction
femoral	Yes	$(alpha_1)$			Constriction
brachial	Yes	$(alpha_1$ and $alpha_2)$			Constriction
Gastrointestinal (mesenteric)	Yes	$(alpha_1)$	Yes	$(beta_2)$	Relaxation (often followed by constriction)
Renal	Yes	$(alpha_1)$	Yes	$(beta_1$ and $beta_2)$	Constriction
Coronary	Yes	$(alpha_1)$	Yes	$(beta_1)$	Variable
Skeletal	Yes	$(alpha_1)$	Yes	$(beta_2)$	Variable
Pulmonary	Yes	$(alpha_1)$	Yes	$(beta_2)$	Relaxation
Skin and mucosa	Yes	$(alpha_1)$	Yes	$(beta_2)$	Constriction
Cerebral	Yes				Constriction (mild)
Ocular	Yes		Yes	$(beta_2)$	Variable
Salivary glands	Yes				Constriction
Venous					
Systemic	Yes	$(alpha_1$ and $alpha_2)$	Yes	$(beta_2)$	Variable
Portal	Yes		Yes		Constriction

It is evident that the physiological effect of the catecholamines in a given tissue is, to a variable extent, a function of how the vasculature of that tissue is influenced. In physiological terms, it is sometimes difficult to factor out this action on vascular smooth muscle when primary attention is directed toward the effects of the catecholamines upon the cellular activity itself whether it be metabolism, contractility, or electrolyte fluxes. Nevertheless, on the basis of extensive in vitro studies, in which the contribution of changes in blood flow to the physiological response has been minimized or negated, it would appear reasonably certain that most tissues respond to the catecholamines by virtue of functional adrenergic receptors in the tissue itself. The direct identification of adrenergic receptors in the cells of most of the organ systems covered in this chapter has validated this expectation.

VI. SKELETAL MUSCLE

The catecholamines increase the contractility of skeletal muscle. In "fast muscle," twitch tension and duration of contraction are increased; in "slow muscle," twitch tension and duration of contractions are decreased. The predominant physiological effect, to increase contractility, is mediated by a $beta_2$ mechanism. Associated with increased skeletal muscle contractility are increased glycogenolysis and potassium transport (Table 7). The accelerated uptake of potassium into skeletal muscle associated with the administration of catecholamines accounts for their well-established hypokalemic property. The hypokalemia is not due to renal potassium loss, and is independent of changes in insulin and aldosterone (13). The disposal of potassium into skeletal muscle could account for the hypokalemia seen in a variety of clinical disorders such as delirium tremens and other acute medical illnesses in which elevated catecholamine levels are commonly detected (17). In addition, the hypokalemia of familial periodic paralysis, thyrotoxicosis, and of asthma

TABLE 7 Actions of Catecholamines on Skeletal Muscle

	Alpha effect ($alpha_1$)	Beta effect ($beta_2$)
Contraction		Stimulation
Release of acetylcholine	? Increased	
Glycogenolysis		Stimulation
Vasculature	Constriction	Vasodilatation
Potassium uptake		Stimulation

therapy with beta agonists could be due to this mechanism. Similarly, propranolol may impair the extrarenal disposal of potassium by selectively inhibiting potassium uptake at the level of the skeletal muscle cell. It is of interest that catecholamine-stimulated uptake of potassium in muscle as well as vasodilatation of the vasculature, increased contractility, and increased glycogenolysis are all $beta_2$-adrenergic phenomena. It is likely, but not yet certain, that the mechanism by which potassium transport is stimulated in skeletal muscle by beta agonists involves stimulation of the ouabain-sensitive Na^+, K^+ pump. So far, except for alpha-agonist-mediated release of acetylcholine at the neuromuscular junction, there is little evidence for other actions mediated by the alpha-adrenergic catecholamines in skeletal muscle. Alpha-adrenergic agonists may counter the effects of beta-adrenergic agonists upon potassium uptake, but the evidence is not yet very compelling.

VII. HEART

The heart is a major target organ for the actions of the catecholamines (49,50). A variety of different cardiac functions has been shown to be specifically influenced by the beta-adrenergic catecholamines in an experimental literature that is secure and well documented (14). Specific effects upon the coronary blood vessels have already been reviewed. Heart rate and contractility, indices most commonly studied, are both clearly enhanced. The subtype specificity of the beta-adrenergic response is predominantly $beta_1$, although recent studies have suggested that to a limited extent, at least, $beta_2$ receptors might participate in these actions. With respect to the increase in contractility, the time required for peak tension development and relaxation is shortened. The effects upon automaticity are related to actions in the conduction systems of heart in several different sites. At the sinoatrial (SA) node, the rate of phase 4 depolarization is increased. Conduction through the SA node, the atrioventricular (AV) node, and the other specialized conducting fibers of the heart is accelerated. The duration of the action potential is shortened, including a reduction in the atrial and, perhaps, the ventricular refractory period.

Although most of the literature suggests that these physiological events are mediated rather exclusively by the beta-adrenergic receptor, it has recently been appreciated that there may also be an alpha-adrenergic component that contributes to these physiological responses (7,42). Recognition of the possibility that the catecholamines may alter cardiac function via an alpha receptor is in part owing to the selective use of beta-adrenergic inhibitors, thus permitting expression of the putative alpha response. Although it is not yet clear that the heart responds to alpha-adrenergic signals under physiological circumstances, the data

TABLE 8 Catecholamines in the Heart

	Alpha$_1$-adrenergic properties	Beta$_1$-adrenergic properties[a]
Contractility	Increased	Increased
Automaticity	↑, ↔ or ↓	Increased
Action potential (including specialized conducting tissue)	Prolonged	Shortened
Atrial and ventricular refractory period	Prolonged	Shortened
Coronary vasculature	Constriction	Relaxation
Metabolism		Glycogenolysis

[a]Although the overwhelming evidence is for a beta$_1$ effect on functions subserved by the beta-adrenergic catecholamines, recent demonstrations of beta$_2$-responsive physiological effects and adenylate cyclase activity in human heart raises the possibility of functions served by this subclass of receptors as well (12,55).

now suggest that this is a distinct possibility (56). It has been shown that contractility can be increased through an alpha$_1$-adrenergic mechanism similar to the known effect of beta$_1$-adrenergic catecholamines. In contrast to the beta-adrenergic effect, however, the time required for peak tension development is not shortened, the refractory period is prolonged, and the duration of systole may actually be extended. Electrophysiologically, the action potential is prolonged. The actions of the alpha$_1$-adrenergic catecholamines upon automaticity of the heart are not yet certain. Heart rate has been shown to be increased, slowed, or unchanged, depending upon the experimental protocol. Careful observations, however, would appear to favor a decrease in heart rate in older subjects, and an increase in heart rate in the young. A summary of the beta-adrenergic and putative alpha-adrenergic actions in the heart is shown in Table 8.

VIII. ADRENERGIC REGULATION OF BLOOD PRESSURE

The actions of epinephrine upon the blood pressure have been mentioned in connection with the historical development of the adrenergic receptor concept. The predominant physiological response to epinephrine and norepinephrine is an increase in the mean arterial pressure. Although both systolic and diastolic components are elevated, the pulse pressure rises because the increase in systolic pressure is greater than the increase in diastolic pressure. The overall effect of the catecholamines upon the blood pressure is due to increased strength of ventricular contractions, increased heart rate, and increased peripheral vascular resistance. The facts that inotropic and chronotropic influences are classically mediated by the beta-adrenergic receptor and that vaso-constriction is an alpha-mediated phenomenon suggest that catechola-mines administered in doses that demonstrate peripheral alpha and cardiac beta effects will be most effective in raising the blood pressure. In this regard, epinephrine and norepinephrine are both very potent. On the other hand, the beta agonist isoproterenol may not be a pressor agent—although it invariably increases cardiac output--because it is a peripheral vasodilator. Complicating the direct effects of the catechola-mines upon blood pressure are important reflex vagal discharges that may rapidly reduce the blood pressure and the chronotropic response. The secondary reflex effect has been used clinically in the treatment of paroxysmal atrial tachycardia with alpha-adrenergic catecholamines.

Finally, consideration of the effects of the catecholamines upon the blood pressure have to take into account their central actions (46). Via mechanisms and sites in the central nervous system that have yet to be elucidated, the alpha-adrenergic catecholamines will cause a re-duction in blood pressure. The hypotensive actions of α-methyldopamine and clonidine are attributed to this central effect. In the case of α-methyldopamine, metabolism in the central nervous system to the alpha-adrenergic agonist α-methylnorepinephrine occurs; clonidine appears to act directly at several central sites. There is some evidence that the antihypertensive properties of the beta-adrenergic inhibitors are due to separate opposing actions of beta agonists on the central regula-tion of blood pressure.

IX. RESPIRATORY TRACT

One of the more important physiological effects of the catecholamines is their property to relax the smooth muscles of the trachea and the bronchi. Administration of catecholamines or analogs with $beta_2$ proper-ties leads to relaxation of bronchospastic airways. The usefulness of beta-adrenergic catecholamines in asthma thus is well established (3). The $beta_2$ subtype of this response has led to the search for selective compounds that will serve this function without the undesirable effect of cardiac stimulation due to $beta_1$ activity. To a certain extent, it has been possible to achieve this goal, although no selective $beta_2$ agonist has yet been developed without some $beta_1$-agonist properties as well. The beta$_2$ response of bronchial and tracheal smooth muscle to epinephrine, which has both alpha and beta properties, illustrates the fact that the particular receptor composition of the tissue influences in a major way the physiological response. For example, the vasoconstricting actions of epinephrine due to its α-agonist properties predominate in most vascular beds, whereas its beta-adrenergic relaxant effect upon smooth muscle predominates in the lung. This observation implies that the distribution of alpha- and beta-adrenergic receptors accounts for these different actions. Another useful property of the catecholamines with respect to the therapy of asthma is that the release of chemical mediators of anaphylaxis such as histamine is inhibited. In addition to these major actions in the respiratory tract, other beta-adrenergic effects include vasodilatation of the nasal and pharyngeal blood vessels, augmented secretion of the bronchial glands, and increased mucociliary transport (Table 9). However, alpha-adrenergic effects, which are the converse of these beta effects (bronchial constriction, for example), can be observed under certain circumstances. This may be particularly troublesome in patients receiving therapy with beta-adrenergic blockers in whom the unopposed alpha-adrenergic actions of the endogenous catecholamines may induce bronchospasm. Conversely, alpha-vasoconstricting properties in the nasal and pharyngeal blood vessels as well as reduced bronchial gland secretions may be used advantageously in the clinical setting.

The overall effect of the catecholamines upon ventilation-perfusion relationships and upon blood oxygenation is related both to their effect

TABLE 9 Actions of Catecholamines on the Upper and Lower Respiratory Tracts

	Alpha		Beta
Nasal and pharyngeal vasculature	Constriction	>	Dilatation
Pulmonary arteries and veins	Constriction	<	Dilatation
Bronchial arteries and veins	Constriction	<	Dilatation
Bronchial smooth muscle	Constriction	<<	Relaxation
Bronchial gland secretion	Decreased	>	Increased
Mucociliary transport	Decreased	>	Increased
Ventilation-perfusion ratio			Decreased transiently

The symbols indicate the greater and lesser physiological responses under normal conditions.

upon minute ventilation and upon perfusion. Although both ventilation (via central mechanisms and peripheral chemoreceptors) and perfusion (via vasodilatation) are increased, the V/Q ratio may be reduced transiently owing to a somewhat greater effect upon pulmonary blood flow than upon ventilation.

X. THE GENITOURINARY SYSTEM

A. The Kidneys

The actions of the catecholamines upon the kidney account for major influences at the levels of hemodynamics, renin release, tubular electrolyte and water transport, and metabolism. With respect to the vasculature, the predominant physiological action is alpha-adrenergic constriction of the resistance vessels. Vasodilatation, which can also be demonstrated, is a mixture of $beta_1$ and $beta_2$ effects. Stimulation of renin release is a function of sympathetic nerve stimulation and is probably mediated by a $beta_1$ mechanism, although some evidence exists for $beta_2$ and nonselective effects (25). Alpha$_2$-mediated inhibition of renin release can also be demonstrated, but it is of minor physiological importance and may occur through indirect mechanisms. Direct alpha$_1$-mediated inhibition is also viewed as a minor physiological action. The beta-adrenergic catecholamines decrease urine flow, free water clearance, and electrolyte secretion by means that are independent of changes in renal hemodynamics. The antidiuretic properties appear to be dependent upon the presence of an intact neurohypophysis, suggesting that release of arginine vasopressin (AVP) is augmented (44,45). Current evidence suggests that augmented AVP release via a beta effect occurs indirectly through the baroreceptors. There is very little evidence for a direct action of beta-agonists upon renal tubular function. In contrast, the alpha$_2$-adrenergic catecholamines lead to an increase in urine flow and free water clearance. The diuretic properties of the alpha$_2$-adrenergic catecholamines occur via their inhibitory effects upon AVP action at the level of the renal tubule (29). A contributing factor to alpha$_2$-adrenergic-mediated diuresis is inhibition of AVP via an effect upon baroreceptor function. Finally, gluconeogenesis is enhanced by the catecholamines in a manner that has not yet been demonstrated to be clearly alpha or beta (Table 10).

B. Lower Urinary Tract

The ureters are stimulated with respect to motility and tone by alpha-adrenergic agonists. The bladder is under dual control. The detrusor muscle is relaxed through a $beta_2$ mechanism; the trigone and sphincter are contracted via an alpha-adrenergic effect.

TABLE 10 Actions of the Catecholamines on the Genitourinary Tract

	Alpha		Beta	
Kidney				
Vasculature	Constriction	$(alpha_1)$ >>	Dilatation	$(beta_1$ and $beta_2)$
Renin release	Inhibition	$(alpha_1)$ <<	Stimulation	$(beta_1$ and nonselective)
Erythropoietin release			Stimulation	$(beta_2)$
Tubular electrolyte and water transport	Increased	$(alpha_2)$ <	Decreased	
Free water clearance	Increased	$(alpha_2)$ <	Decreased	
Gluconeogenesis	Increased	$\approx$	Increased	
Lower Urinary Tract				
Ureters	Increased motility and tone			
Bladder				
detrusor				
trigone and sphincter	Contraction		Relaxation	$(beta_2)$
Genital Tract				
Vas deferens	Contraction	>>	Relaxation	$(beta_2)$
Ejaculation	Stimulation			
Fallopian tubes			Contractility inhibited	$(beta_2)$
Uterus	Contraction	>	Relaxation	$(beta_2)$

The symbols indicate the greater and lesser physiological responses under normal conditions.

C. Genital Tracts

In the male, the vas deferens is physiologically stimulated by alpha agonists. To a much lesser and questionably physiological extent, a $beta_2$-relaxing action upon the vas deferens has been demonstrated. Ejaculation is fostered by the alpha-adrenergic catecholamines. In the female, the contractile state of the fallopian tubes is inhibited by $beta_2$ agonists. The responsiveness of the uterus to the catecholamines is influenced to a large extent by the levels of estrogen and progesterone. In the pregnant state, the uterus becomes more sensitive to the contracting influence of alpha agonists. In this setting, however, it can be demonstrated that the $beta_2$ agonists will relax the uterus, an observation that has led to the widespread use of $beta_2$ agonists in the control of premature labor.

XI. THE GASTROINTESTINAL TRACT

The gastrointestinal tract, like so many organs, is affected in multiple ways by the catecholamines. It is of great interest that one of the major actions of the catecholamines, inhibition of nonsphincteric smooth muscle tone, is mediated both by alpha and beta agonists (54). Delineation of this shared property of the alpha- and beta-adrenergic catecholamines

TABLE 11   Actions of Catecholamines on the Gastrointestinal Tract

	Alpha		Beta	
Nonsphincteric smooth muscle	Relaxation	$(alpha_1$ and $alpha_2) \cong$	Relaxation	$(beta_1)$
Sphincteric smooth muscle	Contraction	$(alpha_1) >>$	Relaxation	$(beta_1)$
Vasculature	Constriction		Vasodilatation	$(beta_2)$
Gastric and small intestinal secretion	Inhibition	$(alpha_2)$	Inhibition	$(beta_1)$
Acetylcholine release	Inhibition			
Gallbladder and ducts			Relaxation	$(beta_2)$

was in part due to the observation that neither alpha- nor beta-adrenergic inhibitors alone could completely negate the inhibitory effect of the catecholamines upon motility. When both alpha- and beta-adrenergic inhibitors were present, however, complete inhibition of the catecholamine effect could be achieved. Concordant actions upon inhibiting gastric and small intestinal secretions are also best described as both alpha and beta adrenergic. In contrast to these shared actions, sphincteric smooth muscle is contracted by alpha-adrenergic catecholamines and inhibited, although to a minor physiological extent, by the beta-adrenergic catecholamines. Acetylcholine release from enteric neurons is inhibited by alpha-adrenergic catecholamines. The direct effect of catecholamines to stimulate water and electrolyte absorption is predominantly alpha$_2$ (53). The vasculature of the gastrointestinal tract (arterial and venous) appears to respond to both alpha and beta agents. Physiologically, an initial constriction (alpha) is often followed quickly thereafter by relaxation (beta) ("autoregulatory escape"). Finally, the gallbladder demonstrates a beta$_2$-mediated relaxant effect following exposure to the catecholamines (Table 11).

XII. METABOLISM

A. Liver

The breakdown of glycogen in the liver is one of the classic physiological actions of the catecholamines. Indeed, it was the study of this particular biochemical event that led to the discovery of cyclic AMP. Typically, therefore, glycogenolysis has been regarded to be a beta-adrenergic activity (18). Additionally, the beta-adrenergic catecholamines stimulate gluconeogenesis, pyruvate kinase activity, glycogen synthase, potassium outflux, mitochondrial pyruvate carboxylation, and amino acid transport in the liver. Confusing to many investigators, however, have been the persistently inconclusive results regarding the exclusivity of these beta-adrenergic-mediated processes. Indeed, very recently it has been appreciated that these same actions may also be mediated by the alpha-adrenergic catecholamines (43). In the presence of total beta-adrenergic blockade, for example, epinephrine-induced glycogenolysis will proceed only to be inhibited completely by the alpha-adrenergic inhibitor phenoxybenzamine. It would appear in man that the alpha-adrenergic effect predominates at physiological concentrations of the catecholamines. Beta-adrenergic actions are of smaller magnitude and require higher agonist concentrations. The concerted actions of the alpha- and beta-adrenergic catecholamines upon these features of hepatic metabolism further illustrate the fact that alpha and beta actions are not always in opposition to each other.

B. Adipose Tissue

The white fat cell responds to the catecholamines with lipolysis. Lipolysis is regarded along with the beta-adrenergic-mediated responses of the
heart to be classic prototypes of $beta_1$ subtype selectivity (19). Nevertheless, even with these two classic $beta_1$-specific organs there is a
subtle but noteworthy difference: The $beta_1$ inhibitor practolol is not
as potent in the fat cell as it is in the heart. Although $beta_1$-adrenergic-
mediated lipolysis is the predominant response in all animals, $alpha_2$-
adrenergic inhibition of lipolysis can also be demonstrated in human
subjects.

Despite the minimal antilipolytic actions of alpha-adrenergic
catecholamines in the adipocyte, a set of important metabolic processes—
not yet clearly related to lipolysis—are rather exclusively mediated by
$alpha_1$ agonists. These actions include accelerated phosphatidylinositol
(PI) turnover via increased synthesis of PI and increased metabolism
of PI to PI-4, PI-5, bisphosphate, and the subsequent formation of
inositol trisphosphate, cyclic inositol phosphate, and diacylglycerol.
In addition, $alpha_1$-adrenergic agents stimulate the entry of calcium
into the fat cell and release bound, intracellular calcium into the cytosol
compartment of the cell (Table 12). These $alpha_1$-adrenergic properties
of the white fat cell illustrate yet another principle when considering
the actions of the catecholamines; namely, that in some cases a rather
exclusive subtype response for both alpha and beta agonists can be
demonstrated with no significant opposing or shared effect by the other.

C. Thermogenesis

Thermogenesis is an enormously complicated subject owing to the influences of the catecholamines upon a great variety of cells and cellular
processes that ultimately play a role in heat production, heat conservation, and heat loss (48). One might focus upon the brown fat cell in this
regard because it is a thermogenic organ in which fatty acids, released
by lipolysis, are oxidized. This action is best described as $beta_1$.
Whether thermogenesis via this cell is a major determinant of heat
production is problematical because of the paucity of brown fat in
human subjects. Other features that must be taken into account when
considering the thermogenic properties of the catecholamines are central
nervous system locations of heat conservation mechanisms, sedation,
and appetite all influenced by alpha agonists. In addition, important
peripheral actions upon skeletal muscle, skin, adipose tissue, and
other hormones that affect metabolism serve to influence thermogenesis.
In general, the peripheral actions lead to calorigenesis, whereas the
central actions lead to heat dissipation. It is, however, difficult to
predict with any degree of accuracy in a given situation how the
catecholamines alter thermogenesis or to what extent the net effect
is due to a particular distribution of individual responses.

TABLE 12 Actions of the Catecholamines on Adipose Tissue

	Alpha		Beta	
Lipolysis	Decreased	$(alpha_2)$ <<	Increased	$(beta_1)$
Phosphatidylinositol turnover	Increased	$(alpha_1)$		
Calcium entry	Increased	$(alpha_1)$		
Released of bound intracellular calcium	Increased	$(alpha_1)$		

XIII. ENDOCRINE SECRETIONS

Complicating our complete understanding of the direct physiological actions of the catecholamines is the fact that they influence the secretion of a great number of other hormones. In some cases, therefore, the catecholamines may alter cellular events indirectly through the stimulation or inhibition of endocrine secretions. The antidiuretic actions attributed to the beta-adrenergic catecholamines, for example, are in part due to stimulation of AVP release from the posterior pituitary gland (47). The opposing actions of the alpha-adrenergic catecholamines are in part due to inhibition of AVP release.

In the anterior pituitary gland, most of the established adrenergic actions are stimulatory and are due to an alpha-adrenergic influence. Alpha-adrenergic stimulation of the anterior pituitary leads to the release of growth hormone and the gonadotropins, especially luteinizing hormone (5). $Alpha_1$-stimulation leads to release of ACTH via the precursor molecule pro-opiomelanocortin (POMC) (28). Adrenocorticotropic hormone (ACTH) release from POMC is also stimulated by $beta_2$ agonists (34). The other hormones of the pituitary, prolactin and thyroid-stimulating hormone (TSH), do not appear to be influenced significantly by alpha- or beta-adrenergic catecholamines.

The release of thyroid hormone from the thyroid gland is stimulated by both alpha and $beta_2$ agonists by mechanisms independent of TSH (31). Earlier observations suggesting an inhibitory alpha-adrenergic effect upon TSH-mediated thyroid hormone release were most likely due to vasoconstriction and reduced delivery of TSH to the thyroid follicle. Of interest is a second observation that propranolol inhibits the peripheral conversion of thyroxine (T_4) to triiodothyronine (T_3), suggesting that another action of the beta-adrenergic catecholamines is to facilitate peripheral conversion of T_4 to T_3. Similar to thyroid hormone, parathyroid hormone (PTH) release from human parathyroid glands is stimulated by $beta_2$ agonists (11). In contrast to thyroid hor-

mone, however, PTH release appears to be inhibited by alpha$_2$ agonists
(10). This latter effect has been demonstrated primarily by observing
greater beta-mediated release of PTH in the presence of alpha-adrenergic
inhibition. It is still not absolutely clear, however, that alpha-
adrenergic mechanisms directly inhibit PTH release.

The endocrine pancreas responds to the catecholamines with a rapid
and physiologically significant inhibition of insulin release via alpha$_2$
receptors (32,58). It can also be demonstrated, however, that insulin
secretion is stimulated by beta$_2$-adrenergic agonists. The physiological
significance of this latter, slower effect is not well understood. Pancreatic glucagon is stimulated by beta$_2$ agonists and may be inhibited
slightly by alpha-adrenergic agonists. Somatostatin release from the
delta cells of the pancreas is also stimulated by beta-adrenergic agonists.

Several other hormone secretions are influenced by the catecholamines. The pineal gland secretes melatonin in response to beta$_1$
agonists. The release of gastrin is mediated both by alpha and beta
agonists. As mentioned previously, renin is released from the kidney
following beta$_1$ stimulation (Table 13).

XIV. EXOCRINE SECRETIONS

The flow rate of pancreatic exocrine secretions is reduced by alpha-
adrenergic catecholamines. Salivary glands are stimulated to secrete
water and potassium by an alpha effect and stimulated to secrete amylase
via a beta mechanism. The secretion of milk from mammary glands is
inhibited by beta agonists. Finally, sweat gland secretions are stimulated by alpha-adrenergic catecholamines (Table 14).

XV. CIRCULATING BLOOD ELEMENTS

Platelets aggregate in response to alpha$_2$-adrenergic agents. Heightened
aggregatability of platelets appears to be mediated by an alpha$_2$-mediated
decrease in cyclic AMP. There is some evidence that the rate of production of factor VIII might also be facilitated by alpha-adrenergic agonists.
Several reports have suggested an adenylate cyclase responsiveness to
beta$_2$-adrenergic catecholamines in platelets, leading to increased
platelet cyclic AMP and decreased aggregation. Although under carefully defined experimental conditions (i.e., in the presence of the
alpha$_2$ inhibitor yohimbine) it has been possible to demonstrate decreased
platelet aggregatability in response to epinephrine via a beta$_2$ effect,
the physiological significance of this observation is not yet certain (38).

Human lymphocytes can be shown to have a cyclic AMP system that
is stimulated by beta$_2$-adrenergic catecholamines (9). The physiological

TABLE 13 Actions of Catecholamines on Endocrine Secretions

	Alpha		Beta	
Pituitary				
arginine vasopressin[a]	Inhibited		Released	
ACTH	Released	(alpha$_1$)	Released	(beta$_2$)
growth hormone	Released			
luteinizing hormone	Released			
Thyroid hormone	Released		Released	(beta$_2$)
Parathyroid hormone	Inhibited	(alpha$_2$)	Released	(beta$_2$)
Pancreas				
insulin	Inhibited	(alpha$_2$) >>	Released	(beta$_2$)
glucagon	Slight in-		Released	(beta$_2$)
somatostatin	hibition		Released	
Pineal				
melatonin			Released	(beta$_2$)
Gastrin	Stimulated		Released	
Kidney				
renin	Inhibited	(alpha$_1$) <<	Released	(beta$_1$ and nonselective)
erythropoietin			Released	(beta$_2$)[b]

[a]The effects of the catecholamines on vasopressin release would appear to occur indirectly via baroreceptor mechanisms.
[b]Through a beta$_2$ mechanism, catecholamines also appear to enhance the actions of erythropoietin on the red cell production in the bone marrow (35).

TABLE 14 Actions of the Catecholamines on Exocrine Glands

	Alpha	Beta
Pancreas	Reduced secretions	
Salivary glands		
water and potassium	Stimulated	
amylase		Stimulated
Mammary		Inhibited
Sweat glands	Stimulated	

significance of this observation has not yet been elucidated (15), although mitogenesis appears to be inhibited as a result. A recent effect of catecholamine-associated cyclic AMP levels to be attenuated in lymphocytes of older subjects is of interest and may be correlated with changes in agonist affinity for the lymphocyte beta receptor (20).

Polymorphonuclear leukocytes can also be shown to respond to beta$_2$ agonists with an increase in cyclic AMP. As a result, antigen-induced release of histamine and hydrolytic enzymes is inhibited. The antianaphylactic properties of the catecholamines could well be accounted for, at least in part, by this property. There is very little evidence for alpha-adrenergic responses in lymphocytes or granulocytes.

Human erythrocytes do not appear to respond to the catecholamines, either with respect to cyclic AMP levels or to any physiological indices. This observation is in marked contrast to the erythrocytes of birds (turkeys, chickens, ducks) in which furosemide-sensitive monovalent cation transport is markedly stimulated by beta$_1$-adrenergic catecholamines (23). The differences in red cell responsiveness among species could well be due to the fact that the nonnucleated human erythrocyte has lost this function in the course of evolving into an exclusive oxygen-carrying element. Supporting this possibility is the fact that the human erythrocyte does contain two of the regulatory proteins of the adenylate cyclase complex.

XVI. EYE

The eye represents another organ system that is influenced by a multiplicity of physiological effects owing to the catecholamines (39). In general, alpha-adrenergic actions predominate in the contraction of the radial muscle of the iris leading to mydriasis and contraction of extraocular skeletal muscles (Table 15). A beta-adrenergic activity in the eye is relaxation of the ciliary muscle. The catecholamines reduce intraocular pressure, but the mechanisms involved are complex owing to the multiple factors that impinge upon the hydrodynamics of aqueous humor formation and removal. Aqueous humor is produced in the ciliary body and by ultrafiltration of the ciliary body vessels. It is drained through outflow channels (trabecular network, Schlemm's canal, and the uveoscleral pathway). Intraocular pressure is modulated by the rate of formation and outflow of aqueous humor as well as by central nervous system mechanisms. Alpha agonists thus may reduce intraocular pressure by decreasing the rate of aqueous humor formation, via decreased ciliary body secretion, and by reduced ultrafiltration due to vasoconstriction. Beta agonists also are associated with reduced intraocular pressure because of their greater effect to increase aqueous humor drainage than to increase the rate of its formation. Very interestingly,

TABLE 15 Actions of Catecholamines on the Eye

	Alpha	Beta
Vasculature		
conjunctivae	Constriction	
uvea	Constriction	Relaxation (beta$_2$)
ciliary body	Constriction (reduced ultrafiltration)	Relaxation (beta$_2$)
Ciliary epithelium	Secretion inhibited (alpha$_2$)	Secretion stimulated (beta$_2$)
Outflow of aqueous humor (trabecular meshwork)		Stimulation (beta$_2$)
Radial muscle dilator	Contraction (mydriasis)	
Intraocular pressure[a]	Decreased (alpha$_2$)	Decreased
Extraocular muscles	Contraction	
Ciliary muscle		Relaxation

[a]The adrenergic effects on intraocular pressure are complex due to a variety of actions at multiple sites. Hence, both β agonists and antagonists may decrease intraocular pressure (see text).

beta antagonists also reduce intraocular pressure by decreasing the rate of aqueous humor formation, by permitting greater alpha-adrenergic vessel tone to decrease ultrafiltration of the ciliary vasculature, and, in the case of timolol, perhaps by a separate mechanism in the central nervous system.

XVII. CENTRAL NERVOUS SYSTEM

There is abundant evidence to document the presence of norepinephrine in the central nervous system. The distribution of noradrenergic-containing cell bodies is extensive and includes the following discrete areas and tracts: medulla, pons, hypothalamus, nucleus tractus solitarii, nucleus dorsalis motorius nervi vagi, sympathetic intermediolateral columns, reticular formation, and nucleus commissuralis. Many of these systems are organized into ascending and descending tracts that receive and transmit noradrenergic signals. Of interest is the recent demonstration that epinephrine-containing cell bodies and tracts are also found in the central nervous system. It would appear that the catecholamines are formed centrally and not derived from the periphery because the catecholamines do not cross the blood-brain barrier.

Consideration of the wide spectrum of adrenergic influences in the body has to take into account, therefore, the presence of a central component that might contribute to the overall physiological effect. The cardiovascular system is a noteworthy example. In addition, other features of central adrenergic activity suggest primary roles for the catecholamines in areas such as the control of wakefulness, psychomotor activity, memory, learning, thermoregulation, respiration, and apetite. Most of these potentially important areas of adrenergic control are not yet well understood.

XVIII. CONCLUSIONS

The hypothesis that the actions of the catecholamines can be understood best by the existence of subpopulations of adrenergic receptors is based upon and supported by the wealth of physiological information covered in this chapter. Classification of the adrenergic response, as monitored by physiological endpoints in a wide variety of cells and organ systems, permits an accurate prediction of the existence of adrenergic receptors and their subtypes in these tissues. With the advent of radioligand-binding techniques by which the identification of alpha- and beta-adrenergic receptors can be made directly these physiological expectations have been confirmed and extended to a greater understanding of altered receptor function in human diseases.

REFERENCES

1. Ahlquist, R. P.: Study of adrenotropic receptors. Am. J. Physiol. 153:586-600, 1948.

2. Arnold, A.: Sympathomimetic amine-induced responses of effector organs subserved by alpha-, beta$_1$-, and beta$_2$-adrenoceptors. L. Szakeres (ed.). In: *Handbook of Experimental Pharmacology*. Heidelberg, Springer-Verlag, 54(I):63-88, 1980.

3. Atkins, P. C., and Zweiman, B.: Pharmacologic therapy of asthma. In: *Update: Pulmonary Diseases and Disorders*. A. P. Fishman (ed.). New York, McGraw-Hill, 1982, pp. 336-348.

4. Barger, G., and Dale, H. H.: Chemical structure and sympathomimetic actions of amines. J. Physiol. (Lond.) 91:19-59, 1910.

5. Barraclough, C. A., and Wise, P. M.: The role of catecholamines in the regulation of pituitary luteinizing hormone and follicle-stimulating hormone. Endocr. Rev. 3:91-119, 1983.

6. Baum, T.: Fundamental principles governing the regulation of circulatory function. In: *Cardiovascular Pharmacology*. M. Antonaccio (ed.). New York, Raven, 1972, pp. 1-43.

7. Benfey, B. G.: Cardiac α-adrenoceptors. Can. J. Physiol. Pharmacol. 58:1145-1157, 1980.

8. Berthelson, S., and Pettinger, W. A.: A functional basis for classification of α-adrenergic receptors. Life Sci. 21:595-606, 1977.

9. Bourne, H. R., Lichtenstein, L. M., Melmon, K. L., Henney, C. S., Weinstein, Y., and Shearer, G. M.: Modulation of inflammation and immunity by cyclic AMP. Science 184:19-28, 1974.

10. Brown, E. M., Hurwitz, S. H., and Aurbach, G. D.: α-Adrenergic inhibition of adenosine 3'5'-monophosphate accumulation and parathyroid hormone release. Endocrinology 103:893-899, 1978.

11. Brown, E. M., Gardner, D. G., Windeck, R. A., Hurwitz, S., Brennan, M. F., and Aurbach, G. D.: β-adrenergically stimulated adenosine 3'5'-monophosphate accumulation and parathyroid hormone release from dispersed human parathyroid cells. J. Clin. Endocrinol. Metab. 48:618-626, 1979.

12. Brown, J. E., McLeod, A. A., and Shand, D. G.: Evidence for cardiac β_2-adrenoceptors in man. Clin. Pharmacol. Exp. Ther. 33:424, 1983.

13. Brown, M. J., Brown, P. C., and Murphy, M. B.: Hypokalemia from beta$_2$ receptor stimulation by circulating epinephrine. N. Engl. J. Med. 309:1414-1419, 1983.

14. Carlsson, E., Hedberg, A., and Mattsen, H.: Classification and function of adrenoceptors. In: *Catecholamines and the Heart*. W. Delius, E. Gerlach, H. Crobecker, and W. Kubler (eds.). Heidelberg, Springer-Verlag, 1981, pp. 19-28.

15. Crary, B., Hauser, S. L. Borysenko, M., Kutz, I., Hoban, C., Ault, K. A., Weiner, H. L., and Benson, H.: Epinephrine-induced changes in the distribution of lymphocyte subsets in peripheral blood of humans. J. Immunol. 131:1178-1181, 1983.

16. Daly, M. J., and Levy, G. P.: The subclassification of β-adrenoceptors: Evidence in support of the dual β-adrenoceptor hypothesis. In: *Trends in Autonomic Pharmacology I*. Kalsner, S. (ed.). Baltimore, Urban and Schwarzenberg, 1978, pp. 347-385.

17. Epstein, F. H., and Rosa, R. M.: Adrenergic control of serum potassium. N. Engl. J. Med. 309:1450-1451, 1983.

18. Exton, J. H.: Mechanisms involved in effects of catecholamines on liver carbohydrate metabolism. Biochem. Pharmacol. 28:2237-2240, 1979.

19. Fain, J. N., and Garcia-Sainz, J. A.: Adrenergic regulation of adipocyte metabolism. J. Lipid. Res. 24:945-966, 1983.

20. Feldman, R. D., Limbird, L. E., Nadeau, J., Robertson, D., and Wood, A. J. J.: Alterations in leukocyte β-receptor affinity with aging. A potential explanation for altered β-adrenergic sensitivity in the elderly. N. Engl. J. Med. 310:815-819, 1984.

21. Furchgott, R. F.: Pharmacological characteristics of adrenergic receptors. Fed. Proc. 29:1352-1361, 1970.

22. Furchgott, R. F.: The classification of adrenoceptors. An evaluation from the standpoint of receptor therapy. In: *Handbook of Experimental Pharmacology*. H. Blaschko and E. Muscholl (eds.). New York, Springer-Verlag, 1972, pp. 283-335.

23. Furukawa, H., Bilezikian, J. P., and Loeb, J. N.: Effects of ouabain and isoproterenol on potassium influx in the turkey erythrocyte: Quantitative relation to ligand binding and cyclic AMP generation. Biochem. Biophys. Acta 598:345-356, 1980.

24. Glusa, E., and Markwardt, F.: Characterisation of postjunctional α-adrenoceptors in isolated human femoral veins and arteries. Naunyn Schmiedebergs Arch. Pharmacol. 323:101-105, 1983.

25. Insel, P. A., and Snavely, M. D.: Catecholamines and the kidney: Receptors and renal function. Ann. Rev. Physiol. 43:625-636, 1981.

26. Jenkinson, D. H.: Classification and properties of peripheral adrenergic receptors. Br. Med. Bull. 29:142-147, 1973.

27. Klowski, W., Hulthen, U. L., Ritz, R., and Buhler, F. R.: α₂ Adrenoceptor-mediated vasoconstriction of arteries. Clin. Pharmacol. Ther. 34:565-569, 1983.

28. Krieger, D. T.: Physiopathology of Cushing's disease. Endocr. Rev. 4:22-43, 1983.

29. Krothapalli, R. K., and Suki, W. N.: Functional characterization of the alpha adrenergic receptor modulating the hydroosmotic effect of vasopressin on the rabbit cortical collecting tubule. J. Clin. Invest. 73:740-749, 1984.

30. Lands, A. M., Luduena, F. P., and Buzzo, H. J.: Differentiation of receptor responsiveness to isoproterenol. Life Sci. 6:2241-2249, 1967.

31. Landsberg, L.: Catecholamines and the sympathoadrenal system. In: *The Thyroid*, 4th ed. S. C. Werner and S. H. Ingbar (eds.). New York, Harper and Row, 1978, pp. 791-799.

32. Langer, J., Panten, U., and Zielmann, S.: Effects of α-adrenergic antagonists on clonidine-induced inhibition of insulin secretion by isolated pancreatic islets. Br. J. Pharmacol. 79:415-420, 1983.

33. Weiner, N., and Taylor, P.: Drugs acting at synaptic and neuro-effector junctional sites. In: *The Pharmacological Basis of Therapeutics*, 6th ed. A. G. Gilman, L. S. Goodman, T. W. Rall, and F. Murad (eds.). New York, Macmillan, 1985, pp. 66-222.

34. Mezey, E., Reisine, T. D., Palkovits, M., Brownstein, M. J., and Axelrod, J.: Direct stimulation of β_2-adrenergic receptors in rat anterior pituitary induces the release of adrenocorticotropin in vivo. Proc. Natl. Acad. Sci. U.S.A. 80:6726-6731, 1983.

35. Mladenovic, J., and Adamson, J. W.: Adrenergic modulation of erhythopoiesis: *In vitro* studies of colony-forming cells in normal and polycythaemic man. Br. J. Haematol. 56:323-332, 1984.

36. Moran, N. C.: Adrenergic receptors. In: *Handbook of Physiology*. Vol. 6, Sect. 7. H. Blascho, G. Sayers, and A. D. Smith (eds.). Washington, D.C., American Physiological Society, 1975, pp. 447-472.

37. Morton, I. K. M., and Halliday, J.: Adrenoceptors in smooth muscle. In: *Adrenoceptors and Catecholamine Action*. G. Kunos (eds.). New York, Wiley, 1981, pp. 1-68.

38. Motulsky, H. J., and Insel, P. A.: Adrenergic receptors in man. N. Engl. J. Med. 307:18-29, 1982.

39. Potter, D. E.: Adrenergic pharmacology of aqueous humor dynamics. Pharmacol. Rev. 33:133-153, 1981.

40. Robison, G. A., Butcher, R. W., and Sutherland, E. W.: *Cyclic AMP*. New York, Academic Press, 1971.

41. Robison, G. A., Butcher, R. W., and Sutherland, E. W. The catecholamines. In: *Biochemical Actions of Hormones*. G. Litwack (ed.). New York, Academic Press, 1972, pp. 81-111.

42. Rosen, M. R., Rabine, L. M., Danilo, P., Jr., and Hordof, A. J.: Alpha and beta-adrenergic effects on cardiac arrhythmias due to automaticity. In: *β-Adrenergic Blockade: A New Era in Cardiovascular Medicine*. E. Braunwald (ed.). Amsterdam, Excerpta Medica, 1978, pp. 179-189.

43. Rosen, S. G., Clutter, W. E., Shah, S. D., Miller, J. P., Bier, D. M., and Cryer, P. E.: Direct α-adrenergic stimulation of

hepatic glucose production in human subjects. Am. J. Physiol. 245:E616-E626, 1983.

44. Schrier, R. W., Lubuman, R., and Ufferman, R. C.: Mechanism of antidiuretic effect of beta-adrenergic stimulation. J. Clin. Invest. 51:97-111, 1972.

45. Schrier, R. W.: Effect of adrenergic nervous system and catecholamines on systemic and renal hemodynamics, sodium and water excretion and renin excretion. Kidney Int. 6:291-306, 1974.

46. Scriabine, A., Clineschmidt, B. V., and Sweet, C. S.: Central noradrenergic control of blood pressure. Ann. Rev. Pharmacol. Toxicol. 16:113-122, 1976.

47. Sklar, A. H., and Schrier, R. W. Central nervous system mediators of vasopressin release. Physiol. Rev. 63:1243-1280, 1983.

48. Smith, U. (ed.): Adrenergic control of metabolic functions. Acta Med. Scand. (Suppl.) 672:5-126, 1983.

49. Starke, K., and Majewski, H.: Role pf presynaptic and postsynaptic adrenoceptors in cardiac function. In: *Catecholamines and the Heart*. W. Delius, E. Gerlach, H. Crobecker, and W. Kubler (eds.). Heidelberg, Springer-Verlag, 1981, pp. 29-37.

50. Stull, J. P., and Mayer, S. E.: Adrenergic and cholinergic mechanisms of modulation of myocardial contractility. In: *The Cardiovascular Systems. Handbook of Physiology*. R. M. Berne (ed.). American Physiological Society, 1979, pp. 741-774.

51. Szekeres, L. (ed.): Adrenergic activators and inhibitors. In: *Handbook of Experimental Pharmacology*. Heidelberg, Springer-Verlag, 54(II):3-36, 129-288, 345-362, 569-578, 1980.

52. Szekeres, L. (ed.): Adrenergic activators and inhibitors. In: *Handbook of Experimental Pharmacology*. Heidelberg, Springer-Verlag. 54(I):522-594, 1980.

53. Tapper, E. J.: Local modulation of intestinal ion transport by enteric neurons. Am. J. Physiol. 7:G457-G468, 1983.

54. Vizi, E. S.: The role of α-adrenoceptors situated in Auerbach's plexus in the inhibition of gastrointestinal motility. In: *Physiology of Smooth Muscles*. E. Bulbring and M. F. Shuter (eds.). New York, Raven, 1976, pp. 357-367.

55. Waelbroeck, M., Taton, G., Delhaye, M., Chatelain, P., Camus, J. C., Pochet, R., LeClerc, J. L., DeSmet, J. M., Robberecht, P., and Christophe, J.: The human heart beta-adrenergic receptors. Coupling of beta2-adrenergic receptors with the adenylate cyclase system. Mol. Pharmacol. 24:174-182, 1983.

56. Wagner, J., Schumann, H. J., Knorr, A., Rohm, N., and Reidemeister, J. C.: Stimulation by adrenaline and dopamine but not by noradrenaline of myocardial α-adrenoceptors indicating positive inotropic effects in human atrial preparations. Naunyn Schmiedebergs Archiv. Pharmacol. 312:99-102, 1980.

57. Wikberg, J. E. S.: The pharmacological classification of adrenergic α_1 and α_2 receptors and their mechanism of action. Acta Physiol. Scand. (Suppl.) 468:11-99, 1979.

58. Yamazaki, S., Katada, T., and Vi, M.: Alpha$_2$-adrenergic inhibition of insulin secretion via interference with cyclic AMP generation in rat pancreatic islets. Mol. Pharmacol. 21:648-653, 1982.

3

Adrenergic Receptors as Pharmacological Targets: The Beta-Adrenergic Blocking Drugs

WILLIAM H. FRISHMAN   *The Albert Einstein College of Medicine,*
Bronx, New York

I. INTRODUCTION

The discovery of β-adrenoceptor blocking drugs has emerged as one of the most important advances in clinical medicine. The versatility of these drugs in medical practice has been demonstrated beyond question. Originally prescribed for patients with angina pectoris, the β-blockers are used to treat an every-growing list of disorders in a list that ranges from hypertension to gastrointestinal bleeding (1,2). Among the most established uses of the drugs are in the treatment of angina pectoris, hypertension, arrhythmia, thyrotoxicosis, hypertrophic cardiomyopathy, migraine, and glaucoma, and for reducing the risk of death and nonfatal reinfarction in survivors of a myocardial infarction (1-7).

More than 15 β-adrenoceptor blocking drugs are now available worldwide. As of 1986, with the introduction of labetalol, seven orally active β-blockers are marketed for approved uses in the United States. These are propranolol for angina pectoris, arrhythmia, systemic hypertension, prevention of migraine headache, and for reducing the risk of mortality in survivors of acute myocardial infarction; atenolol and nadolol for angina pectoris and hypertension; acebutolol for hypertension and ventricular arrhythmias; timolol for hypertension, open-angle glaucoma and for reducing the risk of mortality and reinfarction in survivors of acute myocardial infarction; metoprolol for hypertension and for reducing the risk of mortality in survivors of acute myocardial infarction; pindolol for hypertension; and labetalol for hypertension and hypertensive emergencies (1,2,8-12).

Five β-blockers (bevantolol, bisoprolol, celiprolol, esmolol, and penbutolol) are now under consideration by the United States Food and Drug Administration for marketing approval.

To assess the clinical value of the different β-adrenoceptor blockers, one must compare the pharmacodynamic and pharmacokinetic properties of these drugs. To date, experience with beta-blockers suggests that no one drug is significantly more effective than the others in treating cardiovascular disease (1-3,9); in other words, any beta-blocker in proper dose is equally effective in treating patients with arrhythmia, hypertension, and angina pectoris of effort (1,2,8-13). However, one drug may be more effective in that it reduces the chance of adverse reactions or side effects in a specific patient, or reduces the incidence of undesirable drug interaction, or allows for greater ease in dosing (1,2,8-13). All of these possibilities are important to the clinician. Therefore, in this chapter, comparative pharmacodynamics and pharmacokinetics of the beta-blockers are discussed before the therapeutic actions of these drugs are reviewed.

II. CLINICAL PHARMACOLOGY OF BETA-ADRENOCEPTOR BLOCKING DRUGS

The beta-adrenoceptor blocking drugs differ from one another in respect to several pharmacodynamic and pharmacokinetic properties (2,14-16). Some of these properties, such as selectivity or presence of partial agonist activity, are important to the clinician. Others, such as potency and membrane-stabilizing activity, seem less important. In still other cases, the clinical relevance of a drug's property is unclear, a matter for controversy or research. The properties to be considered here are potency, chemical structure, and its relationship to drug action, membrane-stabilizing activity, selectivity, intrinsic sympathomimetic activity (partial agonist activity), and, finally, pharmacokinetic properties.

A. Potency

As noted earlier, beta-adrenoceptor blocking drugs are competitive inhibitors of catecholamine binding at beta-adrenoceptor sites. They act to reduce the effect of the catecholamine agonist on a sensitive tissue (1). In the presence of the drug, the dose-response curve of the agonist is shifted to the right; that is, a higher concentration of the agonist is required to provoke the response (1,3,7).

The potency of a beta-blocker tells us how much of the drug must be administered in order to inhibit the effects of an adrenergic agonist. One way of assessing potency is by noting the dose of the drug which is needed to inhibit tachycardia produced by the agonist isoproterenol (or alternately by exercise). Potency differs from drug to drug, with pindolol being the most potent and esmolol the least potent (Table 1) (1,2). While differences in potency explain the different dosages needed to achieve effective beta-adrenergic blockade, they have no therapeutic relevance, except when switching patients from one drug to another (1,14).

B. Structure–Activity Relationships

The chemical structures of most beta-adrenergic blockers have several features in common with the agonist isoproterenol (Fig. 1), which consists of an aromatic ring with a substituted ethanolamine side chain linked to it by an $-OCH_2$ group (1,2,15).

Among the beta-blockers, timolol is unique in that it has a catecholamine-mimicking side chain; this is attached to a five-membered heterocyclic ring containing nitrogen and sulfur (a thiadiazole), which is in turn attached to another heterocyclic ring containing nitrogen and oxygen (a morpholino compound). Whether or not this thiadiazole-morpholino structure confers on timolol properties not possessed by other beta-blockers remains to be determined.

Most beta-blockers exist as pairs of optical isomers, and are marketed as racemic mixtures (1,15,16). Almost all of the beta-blocking activity is found in the negative (-) levorotatory stereoisomer, which can be up to 100 times more active than the positive (+) dextrorotatory isomer. The two stereoisomers of beta-adrenergic blockers are useful to investigators seeking to differentiate between the pharmacological effects of the beta-blockade and other unrelated effects. If the effect is produced when only the d-isomer is present, it can be assumed to be unrelated to catecholamine inhibitory actions. d-Isomers of beta-blocking drugs have, of themselves, no apparent clinical value (1,15,16).

C. Membrane–Stabilizing Activity

At very high concentrations, certain beta-blockers (i.e., propranolol) have a quinidine or local anesthetic effect on the cardiac action potential (1,15,16). The property is unrelated to beta-adrenergic blockade (a finding suggested by the fact that it is exhibited equally by both stereoisomers) (1,15). There is no evidence that membrane-stabilizing activity is responsible for any direct negative inotropic effect of the beta-blockers, since drugs with and without this property are equally effective in depressing left ventricular function (1,16). In therapeutic situations, the concentrations of the beta-blockers are probably too small to produce the membrane-stabilizing activity. Only during massive beta-blocker intoxication is the activity manifested (1,17).

TABLE 1 Pharmacodynamic Properties of β-Adrenoceptor
Blocking Drugs

Drug	β_1-Blockade potency ratio (propranolol=1.0)	Relative β_1 selectivity	Intrinsic sympatho-mimetic activity	Membrane-stabilizing activity
Acebutolol	0.3	+	+	+
Atenolol	1.0	++	0	0
Bevantolol	0.3	++	0	0
Bisoprolol	10.0	++	0	0
Carteolol	10.0	0	+	0
Celiprolol[a]	0.4	+	+?	0
Esmolol	0.02	++	0	0
Labetalol[b]	0.3	0	+?	0
Metoprolol	1.0	++	0	0
Nadolol	1.0	0	0	0
Oxprenolol	0.5-1.0	0	+	+
Penbutolol	1.0	0	+	0
Pindolol	6.0	0	++	+
Propranolol	1.0	0	0	++
Sotalol[c]	0.3	0	0	0
Timolol	6.0	0	0	0
Isomer-propranolol	—	—	—	++

[a]Celiprolol may have additional peripheral α_2-adrenergic blocking
activity at high doses.
[b]Labetalol has additional α_1-adrenergic blocking activity and direct
vasodilatory actions (β_2-agonism).
[c]Sotalol has additional type of antiarrhythmic activity.

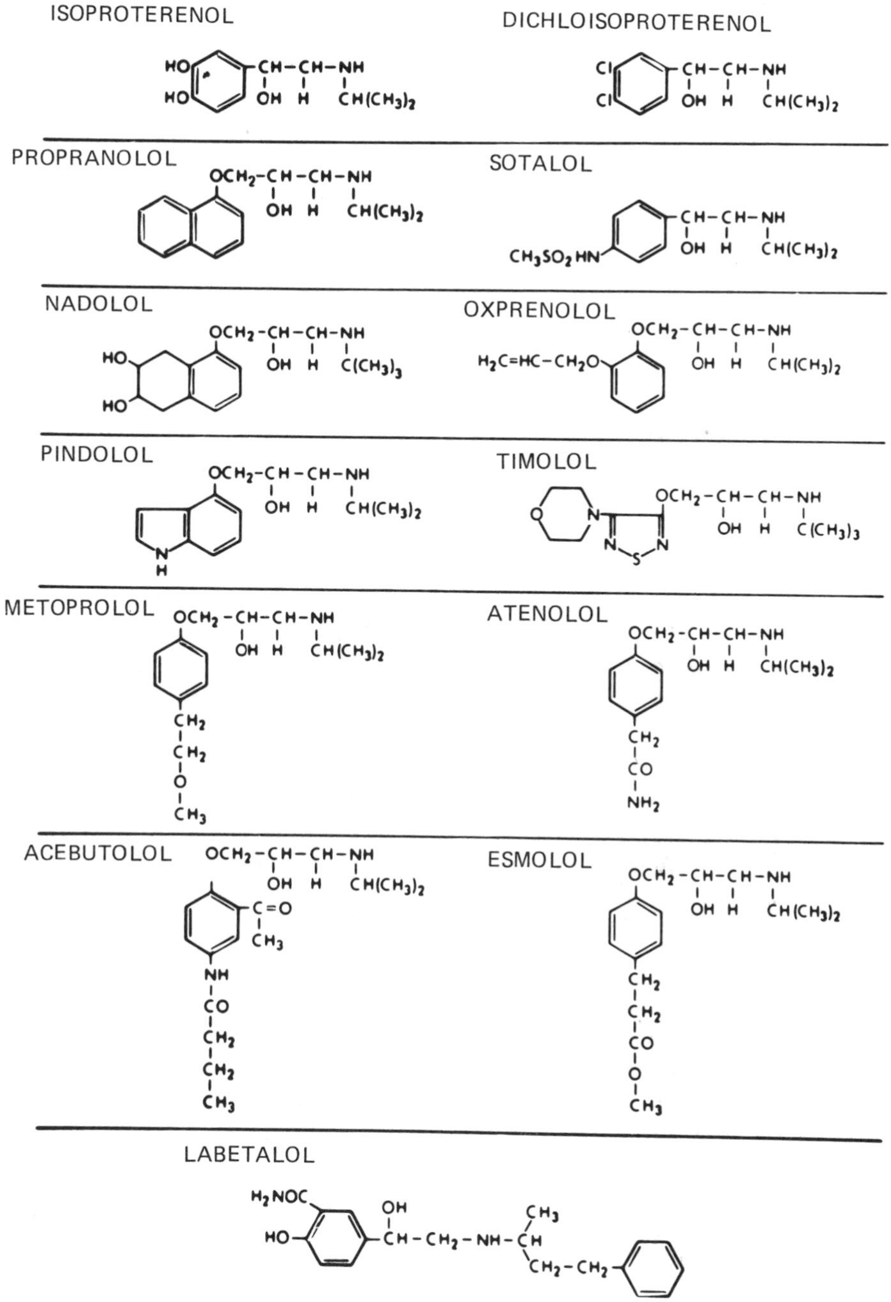

FIGURE 1 Molecular structure of the beta-adrenergic agonist isoproterenol and some beta-adrenergic blocking drugs. (From Ref. 2, used
with permission.)

D. Selectivity

The beta-adrenoceptor blockers can be classified as selective or non-selective, according to their relative abilities to antagonize the actions of sympathomimetic amines in some tissues at lower doses than those required in other tissues (1,9). Drugs have been developed with a degree of selectivity for two subgroups of the beta-adrenoceptor population: Beta$_1$ receptors such as those in the heart, and beta$_2$ receptors such as those found in the peripheral circulation and bronchi (1,2,9). For some time, it has been known that selective beta$_1$-blockers, such as atenolol and metoprolol, inhibit cardiac beta$_1$ receptors but have less influence on bronchial and vascular beta-adrenoceptors (beta$_2$) (9,10).

Because selective beta$_1$-blockers have less of an inhibitory effect on beta$_2$ receptors, they offer two theoretical advantages. The first is that selective beta$_1$-blockers may be safer than nonselective drugs when beta$_2$ blockade is clearly undesirable. In patients with asthma or obstructive pulmonary disease, there beta$_2$ receptors must remain available to mediate adrenergic bronchodilatation (1,2,9,18), relatively low doses of beta$_1$-selective drugs have been shown to cause a lower incidence of bronchial side effects than similar doses of a nonselective drug, such as propranolol (1,2). It should be noted that even selective beta$_1$-blockers may aggravate bronchospasm in some patients, and so these drugs are not generally recommended for patients with asthma and other bronchospastic disease (1,9,10).

A second theoretical advantage of using the selective beta$_1$-blockers relates to the fact that they have less of an inhibitory effect on the beta$_2$ receptors that mediate dilation of arterioles, and are thus less likely to impair peripheral blood flow (1). In the presence of epinephrine, nonselective beta-blockers can cause a pressor response by blocking beta$_2$-receptor-mediated vasodilation (since alpha-adrenergic vasoconstriction receptors remain operative). Selective beta$_1$-blockers may not induce this effect (1,2,9,10).

The theoretical advantages associated with the use of selective beta$_1$-blockers may translate into clinical advantages. For example, because selective beta$_1$-blockers in low doses do not block the beta$_2$ receptors that mediate dilation of arterioles, these drugs may offer advantages in the treatment of hypertension (a possibility that has yet to be clearly demonstrated). In general, leaving beta$_2$ receptors unblocked and responsive to epinephrine may be functionally important in treating patients who have asthma, hypoglycemia, hypertension, or peripheral vascular disease (1,2,9,10).

E. Intrinsic Sympathomimetic Activity
(Partial Agonist Activity)

Certain beta-adrenoceptor blockers (acebutolol, pindolol, and possibly labetalol) possess partial agonist activity (3). These drugs cause a slight to moderate activation of the beta receptor (1,2,12); even as they prevent the access of natural and synthetic catecholamines to the receptor sites (Fig. 2). The result is a weak stimulation of the receptor.

The quantitative assessment of the partial agonist activity of a beta-blocker is gained from observing the actions of the drug in animals whose resting sympathetic tone has been abolished by adrenalectomy and pretreatment with reserpine or syringoserpine (8,12). If the beta-blocker increases the heart rate or force of myocardial contraction, the drug has partial agonist activity (12). The effects are known to be mediated through beta-adrenergic stimulation because they can be antagonized by propranolol (12). The pharmacological effects of beta-adrenoceptor blocking drugs are, of course, much weaker than those of the agonists epinephrine and isoproterenol (1,12). In laboratory animals, pindolol, for example, may have as much as 50% of the agonist activity of isoproterenol; however, the activity is probably lower in humans (2,12).

The assessment of the partial agonist activity of beta-blockers in humans is complicated by the need to study the intact subject. However, the significance in partial agonist activity can be evaluated in clinical trials in which equivalent pharmacological doses of beta-blockers with and without this property are compared. In such trials, drugs with partial agonist activity, such as pindolol, have been shown

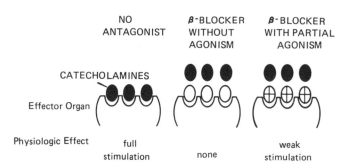

FIGURE 2 Physiological effects of beta-adrenergic blocking drugs with and without partial agonist activity in the presence of circulating catecholamines. When circulating catecholamines (●) combine with beta-adrenergic receptors, they produce a full physiological response. When these receptors are occupied by a beta-blocker lacking partial agonist activity (○), no physiological effects from catecholamine stimulation can occur. A beta-blocking drug with partial agonist activity (⊕) also blocks the binding of catecholamines to beta-adrenergic receptors, but in addition the drug also causes a relative weak stimulation of the receptor. (From Ref. 12, used with permission.)

to cause less slowing of the resting heart rate than do drugs without
partial agonist activity, such as propranolol, atenolol, or metoprolol
(12,18). It is important to note, by contrast, that both types of beta-
blockers similarly reduce the increases in heart rate that occur with
exercise or isoproterenol (1,12). An explanation for these findings
is that the importance of the partial agonist effect of pindolol, relative
to its beta-blocker action, is greatest when sympathetic tone is low,
as it is in resting subjects (12). During exercise, when sympathetic
tone is high, the beta-blocking effect of pindolol predominates over
its partial agonist activity. It is for this reason that all beta-blockers
have been found to be equally effective in reducing the increases in
heart rate and blood pressure that occur with exercise (2,12).

Whether or not partial agonist activity in a beta-blocker offers
an overall advantage in cardiac therapy remains a matter of controversy (12). Some investigators suggest that drugs with this property
may reduce peripheral vascular resistance, and may depress atrioventricular conduction less than other beta-blockers (12). Other
investigators claim that partial agonist activity in a beta-blocker
protects against myocardial depression, bronchial asthma, and peripheral vascular complications in patients receiving therapy (12,18).
However, these claims have not yet been substantiated by definitive
clinical trials.

F. Alpha-Adrenergic Activity

Labetalol is the first of a group of beta-blockers that act as comparative
pharmacological antagonists at both alpha and beta adrenoceptors (2,19).
Labetalol is four to 16 times more potent at beta than at alpha adrenoceptors. In a series of tests, the drug has been shown to be six to
10 times less potent than phentolamine at alpha adrenoceptors and one
and a half to four times less potent than propranolol at beta adrenoceptors (2,19).

Whether or not concomitant alpha-adrenergic activity is generally
advantageous in a beta-blocker has not yet been determined. In the
case of labetalol, the additional alpha-adrenergic blocking action does
result in a reduction of peripheral vascular resistance which may be
useful in treatment of hypertensive emergencies, and unlike most
other beta-blockers, the drug may maintain cardiac output in patients
(2,19). Labetalol, like other beta-blockers, has been shown to be
effective in the treatment of arrhythmias, hypertension, and angina
pectoris (2,19,20).

G. Pharmacokinetic Properties

Although the beta-adrenergic blocking drugs have similar therapeutic
effects, their pharmacokinetic properties differ significantly (Tables 2
and 3) (14,16,21,22); that is, in ways that may influence their clinical

TABLE 2 Pharmacokinetic Properties of β-Adrenoceptor Blocking Drugs

Drug	Extent of absorption (% of dose)	Extent of bioavailability (% of dose)	Dose-dependent bioavailability (major first-pass hepatic metabolism)	Interpatient variations in plasma levels	β-Blocking plasma concentrations	Protein binding (%)	Lipid solubility[a]
Acebutolol	≈70	≈40	No	seven-fold	0.2-2.0 μg/ml	25	Moderate
Atenolol	≈50	≈40	No	four-fold	0.2-5.0 μg/ml	<5	Weak
Bevantolol	≈40	≈55	No	four-fold	0.1-3.0 μg/ml	95	Moderate
Bisoprolol	>90	≈90	No	two-fold	5-80 ng/ml	30	Moderate
Carteolol	≈90	≈40	No	two-fold	40-160 μg/ml	20-30	Weak
Celiprolol	≈30	≈30	No	three-fold	–	≈30	Weak
Esmolol[b]	NA[c]	NA	NA	five-fold	0.15-1.0 μg/ml	55	Weak
Labetalol	>90	≈33	Yes	ten-fold	0.7-3.0 μg/ml	≈50	Moderate
Metoprolol	>90	≈50	No	seven-fold	50-100 ng/ml	12	Moderate
Nadolol	30	≈30	No	seven-fold	50-100 ng/ml	≈30	Weak
Oxprenolol	90	≈40	No	five-fold	80-100 ng/ml	80	Moderate
Penbutolol	>90	≈90	No	four-fold		98	High
Pindolol	>90	≈90	No	four-fold	5-15 ng/ml	57	Moderate
Propranolol	>90	≈30	Yes	twenty-fold	50-100 ng/ml	93	High
Long-acting propranolol	>90	≈20	Yes	ten-twenty-fold	20-100 ng/ml	93	High
Sotalol	70	≈60	No	four-fold	0.5-4.0 μg/ml	0	Weak
Timolol	>90	≈75	No	seven-fold	5-10 ng/ml	≈10	Weak

[a]Determined by the distribution ratio between octanol and water.
[b]Ultrashort-acting β-blocker only available in intravenous form.
[c]NA indicates not applicable.
Source: From Ref. 2, used with permission.

TABLE 3 Elimination Characteristics of Orally Active β-Adrenoceptor Blocking Drugs

Drug	Elimination half-life (H)	Total body clearance (ml/min)	Urinary recovery of unchanged drug (% of dose)	Total urinary recovery (% of dose)	Predominant route of elimination[a]	Active metabolites	Drug accumulation in renal disease
Acebutolol	3-4[b]	6-15	≈ 40	> 90	RE (≈40% unchanged and HM)	Yes	Yes
Atenolol	6-9	130	≈ 40	> 95	RE	No	Yes
Bevantolol	2-4	960	1	74	HM	No	No
Bisoprolol	10-12	300	≈ 50	95	RE (≈ 50% unchanged and HM)	No	Yes
Carteolol	5-6	497	40-68	90	RE	Yes	Yes
Celiprolol	5	500	≈ 90	≈ 30	RE (≈50% unchanged and HM)	Yes	No
Esmolol[c]	9 min	27,000	<2	70-90	RE[d]	No	No
Labetalol	3-4	2,700	<1	> 90	HM	No	No
Metoprolol	3-4	1,100	≈ 3	> 95	HM	No	No

Nadolol	14-24	200	70	70	RE	No	Yes
Oxprenolol	2-3	380	2-5	70-95	HM	No	No
Penbutolol	27	350	50-70	90	RE	No	No
Pindolol	3-4	400	≈40	>90	RE (≈40% un-changed and HM)	No	No
Propranolol	3-4	1,000	<1	>90	HM	Yes	No
Long-acting propranolol	10	1,000	<1	>90	HM	Yes	No
Sotalol	8-10	150	≈60	>90	RE	No	Yes
Timolol	4-5	660	20	65	RE (≈20% un-changed and HM)	No	No

a = RE denotes renal excretion; HM denotes hepatic metabolism.
b = Acebutolol has an active metabolite with elimination half-life of 8-13 hours.
c = Ultrashort-acting β-blocker only available in intravenous form.
d = Metabolized by blood esterases.
Source: From Ref. 2, used with permission.

usefulness in some patients. Among the individual drugs, there are differences in completeness of gastrointestinal absorption, amount of first-pass hepatic metabolism, lipid solubility, protein binding, extent of distribution in the body, penetration into the brain, concentration in the heart, rate of hepatic biotransformation, pharmacological activity of metabolites, and renal clearance of a drug and its metabolites (1, 14,16,22).

On the basis of their pharmacokinetic properties, the beta-blockers can be classified into two broad categories: those eliminated by hepatic metabolism, and those eliminated unchanged by the kidney (1). Drugs in the first group, for example, propranolol and metoprolol, are lipid soluble, are almost completely absorbed by the small intestine, and are largely metabolized by the liver (1,9,14). They tend to have highly variable bioavailability and relatively short plasma half-lives (1,9,14). In contrast, drugs in the second category are more water soluble, are incompletely absorbed through the gut, and are eliminated unchanged by the kidney (11,16). They show less variable bioavailability and have longer half-lives (10,11).

Many of the beta-blockers, including those with short plasma half-lives, can be administered as infrequently as once or twice daily (1). Of course, the longer the half-life, the more useful the drug is likely to be for patients who experience difficulty in complying with beta-blocker therapy (10). A recent addition to the list of available beta-blockers is a long-acting sustained-release preparation of propranolol which provides beta-blockade for 24 hr (Tables 2 and 3) (23-26). Studies have shown that this compound provides a much smoother curve of daily plasma levels than comparable divided doses of conventional propranolol, and that it has fewer side effects (2,23-26).

Ultrashort-acting beta-blockers, with a half-life of no more than 10 min, also offer advantages to the clinician (e.g., in patients with questionable congestive heart failure in which beta-blockers may be harmful). Such drugs are now being tested. Among the most promising are esmolol (ASL-8052), a selective beta-blocker (Tables 1-3) that has been shown to be effective in treating supraventricular tachy-arrhythmias. The short half-life of esmolol relates to the rapid metabolism of the drug by blood tissue and hepatic esterases (27-29).

In medical practice, the pharmacokinetic properties of the different beta-adrenergic blockers are important. The clinician who is administering a drug must be knowledgeable about the extent of its first-pass metabolism, its active metabolites, and its lipid solubility. The dose of the drug, for example, depends on its first-pass metabolism; if the first-pass effect is extensive, only a fraction of an orally administered drug will reach the systemic circulation (1,14,21), and so dosage will have to be larger than the intravenous dose would be (9,14,21). Knowing if the drug is transformed into active metabolites as opposed to inactive metabolites is important in gauging the total pharmacological effect (1,2). Finally, for some beta-blockers, lipid solubility has been associated with the entry of these drugs into the brain, resulting in side effects that are probably unrelated to beta blockade, such as lethargy, mental depression, and even hallucinations (1,11). Whether or not drugs that are less lipid soluble cause fewer of these adverse reactions remains to be determined (10,11).

III. CLINICAL EFFECTS AND THERAPEUTIC APPLICATIONS

The therapeutic efficacy and safety of beta-adrenoceptor blocking drugs has been well established in patients with angina pectoris, cardiac arrhythmias, hypertension, and for reducing the risk of mortality and nonfatal reinfarction in survivors of acute myocardial infarction (1,2,6). The drugs are also used for a multitude of other cardiac (Table 4) (7,31-41) and noncardiac (Table 5) (2,42-47) indications.

TABLE 4 Reported Cardiovascular Indications for β-Adrenoceptor Blocking Drugs

1. Hypertension

2. Angina pectoris

3. Supraventricular arrhythmias

4. Ventricular arrhythmias

5. Reducing the risk of mortality and reinfarction in survivors of acute myocardial infarction

6. Hyperacute phase of myocardial infarction

7. Dissection of the aorta

8. Hypertrophic cardiomyopathy

9. Digitalis intoxication

10. Mitral valve prolapse

11. "QT interval" prolongation syndrome

12. Tetralogy of Fallot

13. Mitral stenosis

14. Congestive cardiomyopathy

15. Fetal tachycardia

16. Neurocirculatory asthenia

TABLE 5 Reported Noncardiovascular Indications
for β-Adrenoceptor Blocking Drugs

Neuropsychiatric
1. Prophylaxis against migraine
2. Essential tremor
3. Anxiety
4. Alcohol withdrawal (delirium tremens)

Endocrine
5. Thyrotoxicosis
6. Hyperparathyroidism

Other
7. Glaucoma
8. Portal hypertension and gastrointestinal
   bleeding

A. Cardiovascular Effects

1. Effects on Elevated Systemic Blood Pressure (Table 6)

It is now well recognized that beta-adrenergic blockers are effective in reducing the blood pressure of many patients with systemic hypertension. Black patients appear to be less responsive than white patients to monotherapy with beta-blockers. At the present time, there is no consensus as to the mechanism(s) whereby these drugs lower blood pressure. It is probable that some or all of the following proposed mechanisms play a part.

a. *Negative chronotropic and inotropic effects:* Slowing of the heart rate and some decrease in myocardial contractility with beta-blockers lead to a decrease in cardiac output, which in the short- and long-term may lead to a reduction in blood pressure (1). It might be expected that these factors would be of particular importance in the treatment of hypertension related to high cardiac output (5, 48) and increased sympathetic tone.

b. *A central nervous system effect:* There is now good clinical and experimental evidence to suggest that beta-blockers cross the blood-brain barrier and enter the central nervous system (49). Although there is little doubt that beta-blockers with high lipophilicity (e.g., metoprolol, propranolol) enter the central nervous system in high concentrations, a direct antihypertensive effect mediated by their presence is not well defined. As noted earlier, those beta-blockers which are less lipid soluble, and thus less likely to concentrate in the brain, appear to be as effective in lowering blood pressure as propranolol (10,11).

TABLE 6 Proposed Mechanisms to Explain the Antihypertensive
Actions of Beta-Blockers

---

1. Reduction in cardiac output

2. Inhibition of renin

3. Reduction in plasma volume

4. Central nervous system effects

5. Reduction in peripheral vascular resistance

6. Resetting of baroreceptor levels

7. Reduction in venomotor tone

8. Effects on prejunctional beta-receptors—reductions in
   norepinephrine release

Most Important Effect of Beta-Blockers
   Prevents the pressor response to catecholamines with exercise
   and stress

Source: From Ref. 2, used with permission.

   c. Differences in effects on plasma renin: The relationship
between the hypotensive action of beta-blocking drugs and their
ability to reduce plasma renin activity remains one of the more controversial areas in hypertension research. There is no doubt that
beta-blocking drugs can antagonize sympathetically mediated renin
release (50). Adrenergic activity is not the only mechanism whereby
renin release is mediated, however. Other major determinants are
sodium balance, posture, and renal perfusion pressure.
   The important question is whether or not there is a clinical correlation between the beta-blocker effect on plasma renin activity and
the lowering of blood pressure. Some investigators (50) have found
that "high-renin" patients respond well to propranolol; that "low-renin"
patients do not respond or may even show a rise in blood pressure;
and that "normal-renin" patients do not respond in a predictable way.
Other investigators have been unable to confirm these relationships
either for propranolol or other beta-blockers (10-12). In the high-
renin hypertensive patient, it has been suggested that renin may not
be the only factor maintaining the high blood pressure state. At
present, the exact role of renin reduction in blood pressure control
is not well defined.
   d. Venous tone and plasma volume: Reduced plasma volume and
venous return are thought to play a role in the control of blood pressure
by beta-blockers. Few studies involving patients in whom heart failure

was not present have demonstrated these actions of beta-blockers in
both acute and long-term clinical trials. The effect is important in
that one would ordinarily expect an impaired cardiac output with beta
blockade to cause a reflex increase instead of a decrease in plasma
volume. These results, though not yet fully substantiated, are of
interest (2).

    e. Peripheral resistance: Nonselective beta-blockers have no
primary action in lowering peripheral resistance, and indeed may
cause it to rise by leading unopposed the alpha-stimulatory mechanisms
(51). The vasodilating effect of catecholamines on skeletal muscle
blood vessels is $beta_2$ mediated; this suggests that there are therapeutic advantages in using $beta_1$-selective blockers, agents with partial
agonist activity, and drugs with alpha-blocking activity when blood
pressure control is desired. Since $beta_1$ selectivity diminishes as the
drug dosage is raised, and since hypertensive patients generally have
to be given far larger doses than are required to block the $beta_1$
receptors alone, $beta_1$ selectivity (52) offers the clinician little, if
any, real advantage in antihypertensive treatment (9,52).

    f. "Quinidine effect" (membrane-stabilizing activity): Some
early clinical investigations (53) indicated that the antihypertensive
effect of propranolol paralleled the antihypertensive effect of quinidine,
suggesting that the "membrane-stabilizing" effect in the beta-blocker
might be important. Subsequent studies refuted these early findings
(54). All beta-blockers appear to reduce blood pressure, regardless
of membrane effects (10). It is worth noting that d-propranolol, with
predominant membrane effects, does not affect blood pressure; it is
only when l-propranolol is added to it in the racemic mixture that
the drug has an antihypertensive effect.

    g. Resetting of baroreceptors: In patients with long-standing
hypertension, the baroreceptors may reset less strongly to a reduction
in blood pressure than they would in a normal subject. It may be that
beta-blockers achieve some of this antihypertensive effect by "resetting"
or increasing the sensitivity of the baroreceptor (55). The clinical
significance of this proposed mechanism is unknown.

    h. Effects on prejunctional receptors: Blockade of the prejunc-
tional beta receptors apart from its effects on postjunctional tissue
beta receptors is believed to be involved in the hemodynamic actions
of beta-blocking drugs. The stimulation of prejunctional $alpha_2$
receptors leads to a reduction in the quantity of norepinephrine
released by the postganglionic sympathetic fibers (56,57). Conversely,
stimulation of prejunctional beta receptors is followed by an increase
in the quantity of norepinephrine released by postganglionic sympathetic
fibers (58-60). Blockade of the prejunctional beta receptors should,

therefore, diminish the amount of norepinephrine released, leading to a weaker stimulation of postjunctional alpha receptors, an effect which would produce less vasoconstriction and lower blood pressure. Opinions differ, however, both on the contributions and presynaptic beta blockade to a reduction in the peripheral vascular resistance, and on its role in the antihypertensive effects of beta-blocking drugs.

2. Effects in Angina Pectoris

Sympathetic innervation of the heart causes the release of norepinephrine, thereby activating beta adrenoreceptors in myocardial cells. This adrenergic stimulation causes an increment in heart rate, isometric contractile force, and maximal velocity of muscle fiber shortening, all of which lead to an increase in cardiac work and myocardial oxygen consumption (61). The decrease in intraventricular pressure and volume caused by the sympathetic-mediated enhancement of cardiac contractility tends, on the other hand, to reduce myocardial oxygen consumption by reducing myocardial wall tension (law of LaPlace) (62). Although there is a net increase in myocardial oxygen demand, this is normally balanced by an increase in coronary blood flow. Angina pectoris is felt to occur when oxygen demand exceeds supply; i.e., when coronary blood flow is restricted by coronary atherosclerosis. Since the conditions which precipitate anginal attacks (exercise, emotional stress, food, and so forth) cause an increase in cardiac sympathetic activity, it might be expected that blockade of cardiac beta-adrenoreceptors would relieve the symptoms of the anginal syndrome. It is on this basis that the early clinical studies with beta-blocking drugs in angina were initiated (63).

To understand how beta-blockers relieve angina, it is necessary to examine the four main factors which influence myocardial oxygen consumption—heart rate, ventricular systolic pressure, rate of rise of left ventricular pressure, and the size of the left ventricle. Of these, heart rate and systolic pressure appear to be the most important (the product of heart rate × systolic blood pressure is a reliable index for predicting the precipitation of angina in a given patient) (64,65).

Beta blockade produces a reduction in heart rate that has two favorable consequences: (1) a decrease in cardiac work, thereby reducing myocardial oxygen needs, and (2) a longer diastolic filling time associated with a slower heart rate, allowing for increased coronary perfusion. Beta-blockade also reduces exercise-induced blood pressure increments, the velocity of cardiac contraction, and oxygen consumption at a given workload (64,65).

Some investigators have reported a decrease in coronary blood flow. For example, with beta-blockers studies in dogs have shown that propranolol causes such a decrease in coronary blood flow (66). However, subsequent animal studies have demonstrated that beta-blocker-induced shunting occurs in the coronary circulation, maintaining blood flow to ischemic areas, especially in the subendocardial region (67). In humans, concomitant with the decrease in myocardial oxygen consumption, beta-blockers can cause a reduction in coronary blood flow and a rise in coronary vascular resistance (68). In patients taking beta-blockers, the reduction in myocardial oxygen demand may be sufficient cause for this decrease in coronary blood flow (64,65).

Virtually all beta-blockers—whether or not they have partial agonist activity, alpha-blocking effects, membrane-stabilizing activity, or general or selective beta-blocking properties—produce some degree of increased work capacity without pain in patients with angina pectoris. Therefore, it must be concluded that this results from their common property: blockade of cardiac beta receptors (64).

Although exercise tolerance and work capacity improves with beta blockade in patients, the increments in heart rate and blood pressure with exercise are blunted, and the rate-pressure product (systolic blood pressure × heart rate) achieved when pain occurs is less than that reached during a control run (69). The depressed pressure-rate produce at the onset of pain (about 20% reduction from control) is reported to occur with various beta-blocking drugs. It accounts for the fact that even with the increased exercise tolerance made possible by beta blockade patients exercise less than might be expected. This effect probably relates to the action of beta-blockers to increase left ventricular size, causing increased left ventricular wall tension and an increase in oxygen consumption at a given blood pressure (70).

The therapeutic benefit of beta-blockers in angina pectoris is now established beyond question. Many double-blind studies of beta-blockers in patients have demonstrated a significant reduction in the frequency of angina attacks (2,64). Observed improvement is dose related and dosage must be titrated for each individual patient.

3.   Combined Use of Beta-Blockers with Other Antianginal Therapies in Stable Angina

a. *Nitrates*: Combined therapy with nitrates and beta-blockers may be more efficacious for the treatment of angina pectoris than either drug alone (64,71). The primary effect of beta-blockers are to cause a reduction in both resting and exercise-induced heart rate. Nitrates produce a reflex increase in heart rate (owing to a reduction in arterial pressure), and concomitant beta-blocker therapy will be extremely effective simply because it will block this increment in reflex heart rate. Similarly, the preservation of diastolic coronary flow with a reduced heart rate will be beneficial (64). In patients with a propensity for myocardial failure who might have a slight increase in heart size with the beta-blockers, the nitrates will counteract this tendency:

The peripheral venodilatory effects of these drugs reduce the left ventricular volume. When nitrates are administered, the reflex increase in contractility that is mediated through the sympathetic nervous system will be checked by the presence of beta-blockers. Similarly, the increase in coronary resistance associated with beta-blockers can be ameliorated by the administration of nitrates (64).

b. *Calcium-entry blockers*: Calcium-entry blockers are a group of antianginal drugs that block transmembrane calcium currents in vascular smooth muscle to cause arterial vasodilation. Some calcium-entry blockers also will slow the heart rate and reduce atrioventricular conduction. Combined therapy with beta-adrenergic and calcium-entry blockers can provide substantial clinical benefits for patients with angina pectoris who remain symptomatic with either agent used alone (64). Because adverse effects can occur (heart block, heart failure), however, patients being considered for such treatment need to be carefully selected and observed (64).

4. Angina at Rest and Vasospastic Angina

Although beta-blockers are effective in treatment of patients with angina of effort, their use in angina at rest is not so well established. To date, clinical studies in patients with angina at rest have been based largely on uncontrolled observations and have not been conclusive. The rationale for therapy with beta-blockers was based on the assumption that the pathogenesis of chest pain at rest was similar to that in patients with exertional symptoms. Recent studies, however, emphasize that angina pectoris can be caused by multiple mechanisms and that coronary vasospasm is responsible for ischemia in a significant proportion of patients with angina at rest (64,72). Therefore, drugs such as propranolol and other beta-blockers that primarily reduce myocardial oxygen consumption, but fail to exert vasodilating effects on the coronary vasculature, may not be as effective in patients in whom angina is caused by dynamic alterations in coronary luminal diameter (64,73). Despite their theoretical dangers in rest and vasospastic angina, beta-blockers have been successfully used alone and in combination with vasodilating agents in many patients (64).

B. Antiarrhythmic Effects

1. Electrophysiological Effects

Beta-adrenoreceptor blocking drugs have two main effects on the electrophysiological properties of specialized cardiac tissue (Table 7) (2). The first effect of beta-blockers results from the specific blockade

TABLE 7 Antiarrhythmic Mechanisms for Beta-Blockers

---

1. Beta-Blockade
   electrophysiology: depress excitability; depress conduction
   prevention of ischemia: decrease automaticity; inhibit reentrant
       mechanisms

2. Membrane-stabilizing effects
   Local anesthetic, "quinidinelike" properties: depress excitability;
       prolong refractory period; delay conduction
   Clinically: probably not significant

3. Special pharmacological properties (beta-selectivity, intrinsic
   sympathomimetic activity) do not appear to contribute to increased
   antiarrhythmic effectiveness.

Source: From Ref. 2, used with permission.

of adrenergic stimulation of cardiac pacemaker potentials. This is
undoubtedly important in the control of arrhythmias caused by enhanced
automaticity. In concentrations causing significant inhibition of adrenergic receptors, the beta-blockers produce little change in the transmembrane potentials of cardiac muscle. However, by competitively
inhibiting adrenergic stimulation, beta-blockers decrease the slope of
phase 4 depolarization and the spontaneous firing rate of sinus or
ectopic pacemakers, and thus decrease automaticity. Arrhythmias
occurring in the setting of enhanced automaticity as seen in myocardial
infarction, digitalis toxicity, hyperthyroidism, and pheochromocytoma
would therefore be expected to respond well to beta blockade (74,75).

The second electrophysiological effect of beta-blockers is one of
membrane-stabilizing action, also known as the "quinidinelike" or
"local anesthetic" action. Characteristic of this effect is a reduction
in the rate of rise of the intracardial action potential without any effect
on the spike duration of the resting potential (75). Associated features
include an elevated electrical threshold of excitability, delay in conduction velocity, and a significant increase in the effective refractory
period. This effect and its attendant changes has been explained by
an inhibition of the depolarizing inward sodium current (75).

Among the beta-blockers, sotalol is unique in that it possesses
class III antiarrhythmic properties, causing prolongation of the action
potential period and thereby delaying repolarization (2). Clinical
studies have verified the efficacy of sotalol in the control of arrhythmias (76,77), but additional investigation will be required to determine
whether its class III antiarrhythmic properties contribute significantly
to its efficacy as an antiarrhythmic agent.

The most important mechanism underlying the antiarrhythmic effect of beta-blockers (with the possible exclusion of sotalol) is felt to be beta blockade with resultant inhibition of pacemaker potentials. If this view is accurate, then one would expect *all* beta-blockers to be similarly effective at a comparable level of beta blockade. In fact, this appears to be the case. No superiority of one beta-blocking agent over another in the therapy of arrhythmias has been convincingly demonstrated (2). Differences in overall clinical usefulness are related to their other associated pharmacological properties (2).

2. Therapeutic Uses in Cardiac Arrhythmias

Beta-adrenergic blocking drugs have become an important treatment modality for various cardiac arrhythmias (Table 8) (2). While it has long been acknowledged that beta-blockers are more effective in treating supraventricular than ventricular arrhythmias, it has only been recently appreciated that these agents can be quite useful in the treatment of ventricular tachyarrhythmias, especially in the setting of myocardial ischemia (2, 78-80).

a. *Supraventricular arrhythmias*: These arrhythmias have a variable response to beta-blockade. Beta-blockers are not only therapeutically useful, but diagnostically important; by slowing a very rapid heart rate, the drug may permit an accurate EKG diagnosis of an otherwise puzzling arrhythmia.

Sinus tachycardia: This arrhythmia usually has an obvious cause (e.g., fever, hyperthyroidism, congestive heart failure) and therapy should be addressed to correction of the underlying condition. However, if the rapid heart rate itself is compromising the patient—for example, causing recurrent angina in a patient with coronary artery disease, then direct intervention with a beta-blocker is effective and indicated therapy. Patients with heart failure, however, should not be treated with beta-blockers unless they have been placed on diuretic therapy and cardiac glycosides, and even then the beta-blockers should be administered with extreme caution.

Supraventricular ectopic beats: As with sinus tachycardia, specific treatment of these extrasystoles is seldom required and therapy should be directed to the underlying cause. Although supraventricular ectopic beats are often the precursors to atrial fibrillation (especially in acute myocardial infarction, thyrotoxicosis, and mitral stenosis), there is no evidence that prophylactic administration of beta-blockers can prevent the development of atrial fibrillation. Supraventricular ectopic beats due to digitalis toxicity generally respond well to beta blockade. Beta-blockers can be useful for those patients in whom supraventricular ectopic activity causes discomforting palpitations.

TABLE 8 Effects of Beta-Blockers in Various Arrhythmias

Arrhythmia	Comment
Supraventricular sinus tachycardia	Treat underlying disorder; excellent response to beta-blocker if need to control rate (e.g., ischemia).
atrial fibrillation	Beta-blockers reduce rate, rarely restore sinus rhythm. May be useful in combination with digoxin and/or verapamil.
atrial flutter	Beta-blockers reduce rate, sometimes restore sinus rhythm.
atrial tachycardia	Effective in slowing ventricular rate, may restore sinus rhythm. Useful in prophylaxis.

Ventricular
  premature ven-
  tricular contrac-
  tions
	Good response to beta-blockers, especially when caused by digitalis, exercise (ischemia), mitral valve prolapse, or hypertrophic cardiomyopathy.

ventricular tachy- cardia	Usually not effective, except in digitalis toxicity or exercise (ischemia)-induced.
ventricular fibril- lation	Electrical defibrillation is treatment of choice. Beta-blockers can be used to prevent recurrence in cases of excess digitalis or sympathomimetic amines. Appear to be effective in reducing the incidence of ventricular fibrillation and sudden death postmyocardial infarction.

Source: From Ref. 2, used with permission.

Paroxysmal supraventricular tachycardia (SVT): These may be divided into two groups: (1) those related to abnormal conduction (e.g., reciprocating AV nodal tachycardia, the reentrant tachycardias, as in the Wolff-Parkinson-White syndrome, in which there is abnormal conduction through an AV nodal bypass tract); and (2) those caused by ectopic atrial activity, as in digitalis toxicity. Since beta blockade delays AV conduction (e.g., increased A-H interval in the bundle of His electrocardiograms) and prolongs the refractory period of the reentrant pathways, it is not surprising that many cases of paroxysmal supraventricular tachycardia respond to beta-blockers. In acute episodes, vagal maneuvers after beta blockade may effectively terminate an arrhythmia when they may have been previously unsuccessful without beta blockade. Even when beta-blockers do not convert an arrhythmia to sinus rhythm, by increasing atrioventricular nodal refractoriness, they will often slow the ventricular rate. Additionally, the use of beta-blocking drugs still allows the option of direct current countershock cardioversion (which would be more hazardous if digitalis in high doses was initially used).

Atrial flutter: Beta blockade can be used to slow the ventricular rate (by increasing AV block) and may restore sinus rhythm in a large percentage of patients. This is a situation in which beta blockade may be of diagnostic value: Given intravenously, beta-blockers slow the ventricular response and permit the differentiation of flutter waves, ectopic P waves, or sinus mechanism.

Atrial fibrillation: The major action of beta-blockers in rapid atrial fibrillation is the reduction in the ventricular response by increasing the refractory period of the AV node. While all beta-blocking drugs have been effective in slowing ventricular rates in patients with atrial fibrillation, they are less effective than quinidine or DC cardioversion in the reversion of atrial fibrillation to sinus rhythm (although this can occur, especially when the atrial fibrillation is of recent onset).

Beta-blockers must be used cautiously when atrial fibrillation occurs in the setting of a severely diseased heart which is dependent on high levels of adrenergic tone so as to avoid myocardial failure. These drugs may be particularly useful in controlling the ventricular rate in situations in which this is difficult to achieve with maximally tolerated doses of digitalis (e.g., thyrotoxicosis, hypertrophic cardiomyopathy, mitral stenosis, and so forth).

Many patients with paroxysmal atrial fibrillation or flutter may have "sick sinus" or "tachy-brady" syndrome, and administration of beta-blockers may precipitate severe bradycardic episodes. These patients often require both antiarrhythmic therapy and a pacemaker.

b. Ventricular Arrhythmias: Beta-adrenoreceptor blocking drugs can decrease the frequency or abolish ventricular ectopic beats in various conditions. They are particularly useful if these arrhythmias are related to excessive catecholamines (e.g., exercise, halothane anesthesia, pheochromocytoma, exogenous catecholamines), myocardial ischemia, or digitalis.

Premature ventricular contractions: The response of these arrhythmias to beta blockade is as good as that seen with quinidine. The best response can be expected to occur in ischemic heart disease, particularly when the arrhythmia is secondary to an ischemic event. Since beta-blockers are effective in preventing ischemic episodes, arrhythmias generated by these episodes may be prevented.

Beta-blockers are also quite effective in controlling the frequency of premature ventricular contractions in hypertrophic cardiomyopathy and in mitral valve prolapse. In these situations, a beta-blocker is generally the antiarrhythmic drug of first choice.

Ventricular tachycardia: Beta-blocking drugs should not be considered agents of choice in the treatment of acute ventricular tachycardia. Cardioversion or other antiarrhythmic drugs (e.g., lidocaine, quinidine, procainamide) should be the initial mode of therapy. Beta-blockers have, however, been shown to be of benefit for prophylaxis against recurrent ventricular tachycardia, particularly if sympathetic stimulation appears to be a precipitating cause. There have been several reported studies showing the prevention of exercise-induced ventricular tachycardia by beta-blockers; in many previous cases there had been a poor response to digitalis or quinidine (2).

Prevention of ventricular fibrillation: Beta-blocking agents can attenuate cardiac stimulation by the sympathetic nervous system, and perhaps the potential for reentrant ventricular arrhythmias and sudden death (78). Experimental studies have shown that beta-blockers raise the ventricular fibrillation threshold in the ischemic myocardium (78). Placebo-controlled clinical trials have shown that beta-blockers reduce the number of episodes of ventricular fibrillation and cardiac arrest during the acute phase of myocardial infarction (79). The long-term beta-blocker postmyocardial infarction trials and other clinical studies with beta-blockers have demonstrated that there is a significant reduction of complex ventricular arrhythmias (80).

TABLE 9   Possible Mechanisms by which Beta-Blockers Protect the Ischemic Myocardium

---

1. Reduction in myocardial oxygen consumption
    a. Reduction in heart rate, blood pressure, and myocardial contractility

2. Augmentation of coronary blood flow
    a. Increase in diastolic perfusion time by reducing heart rate
    b. Augmentation of collateral blood flow
    c. Redistribution of blood flow to ischemic areas

3. Alterations in myocardial substrate utilization

4. Decrease in microvascular damage

5. Stabilization of cell and lysosomal membranes

6. Shift of oxyhemoglobin dissociation curve to the right

7. Inhibition of platelet aggregation

Source: From Refs. 2 and 81, used with permission.

C. Effects in Survivors of Acute Myocardial Infarction

Beta-adrenergic blockers have beneficial effects on many determinants of myocardial ischemia (Table 9) (2,65,81). The results of placebo-controlled long-term trials with some beta-adrenergic blocking drugs in survivors of acute myocardial infarction have demonstrated a favorable effect on total mortality, on cardiovascular mortality (including sudden and nonsudden cardiac deaths), and on the incidence of non-fatal reinfarction (6). These beneficial results with beta-blocker therapy can be explained by both the antiarrhythmic (Table 9) and anti-ischemic effects of these drugs (65,78,82). Two nonselective beta-blockers, propranolol and timolol, are approved by the FDA for reducing the risk of mortality in infarct survivors when started 5-28 days postinfarction. Metoprolol, a beta$_1$-selective blocker was recently approved for this same use in both intravenous and oral form. Beta-blockers have also been suggested as a treatment for reducing the extent of myocardial injury and mortality during the hyperacute phase of myocardial infarction (7,83), but their role in this situation still remains unclear (6,84).

D. Other Cardiovascular Applications

Although beta-blockers have been studied most extensively in patients with angina pectoris, arrhythmia, and hypertension, they have also been shown to be safe and effective for other cardiovascular conditions (Table 4). Some of these conditions are described below.

1. Hypertrophic Cardiomyopathy

Beta-adrenoceptor blocking drugs have been proven to be efficacious in the therapy of patients with hypertrophic cardiomyopathy or idiopathic hypertrophic subaortic stenosis (IHSS) (32,85). These drugs are useful in controlling the symptoms of this disease—specifically, dyspnea, angina, and syncope (2). Beta-blockers have also been shown to lower the intraventricular pressure gradient both at rest and with exercise.

Of course, outflow pressure gradient is not the only abnormality in hypertrophic cardiomyopathy; more important is the loss of ventricular compliance which impedes normal left ventricular functioning. It has been shown by invasive and noninvasive methods that propranolol can improve left ventricular function in patients with this condition (86). The drug also produces favorable changes in ventricular compliance while it relieves symptoms. Propranolol is approved for treatment of hypertrophic cardiomyopathy and may be combined with the calcium-channel blocker verapamil in patients not responding to the beta-blocker alone.

The salutary hemodynamic and symptomatic effects produced by propranolol derive from its inhibition of sympathetic stimulation to the heart (87). However, there is no evidence that the drug alters the primary cardiomyopathic process; many patients remain in or return to their severely symptomatic state, and some die despite the administration of the drug (32,85).

2. Mitral Valve Prolapse

This auscultatory complex characterized by a nonejection systolic click, a late systolic murmur, or a midsystolic click followed by a late systolic murmur has been studied extensively over the last 15 years (88). Atypical chest pain, malignant arrhythmias, and nonspecific ST and T wave abnormalities have been observed with mitral valve prolapse. Beta-adrenergic blockers, by decreasing sympathetic tone, have been shown to be useful for relieving the chest pains and palpitations that many of these patients experience, and for reducing the incidence of life-threatening arrhythmias and other electrocardiographic (ECG) abnormalities (34).

3. Dissecting Aneurysms

Beta-adrenergic blockade plays a major role in the treatment of patients with acute aortic dissection. During the hyperacute phase, the administration of beta-blocking drugs is mandatory to reduce the force and velocity of myocardial contraction (dp/dt) and, hence, to arrest the progress of the dissecting hematoma (31). However, such beta-blocker therapy must be initiated simultaneously with other antihypertensive therapy that may cause reflex tachycardia and increases in cardiac output—factors which could aggravate the dissection process. Initially, propranolol is administered intravenously to reduce the heart rate below 60 beats/min. Once a patient is stabilized and long-term medical management is contemplated, the patient should be maintained on oral beta-blocker therapy to prevent the recurrence of dissection (89).

Recently, it has been demonstrated that long-term beta-blocker therapy might also reduce the risk of dissection in patients prone to this complication (e.g., in patients with Marfan's syndrome). Systolic time intervals are used to assess the adequacy of beta blockade in children with Marfan's syndrome.

4. Tetralogy of Fallot

By reducing the effects of increased adrenergic tone on the right ventricular infundibulum in tetralogy of Fallot, beta-blockers have been shown to be useful for the treatment of severe hypoxic spells and hypercyanotic attacks (36). With chronic use, the drugs have also been shown to prevent prolonged hypoxic spells (36). These drugs should only be looked at as palliative, and definitive surgical repair of this condition is usually required.

5. Q-T Interval Prolongation Syndromes

The syndrome of ECG Q-T interval prolongation is usually a congenital condition associated with deafness, syncope, and sudden death (35). Abnormalities in sympathetic nervous system functioning in the heart have been proposed as an explanation for the electrophysiological aberrations seen in patients with this syndrome (35). Propranolol appears to be the most effective drug for treatment; it reduces the frequency of syncopal episodes in the majority of patients, and may prevent sudden death (35). The drug will reduce the ECG Q-T interval.

E. Noncardiovascular Applications

Beta-adrenergic receptors are ubiquitous in the human body and their blockade affects a variety of organ and metabolic systems, causing them to be effective in treating many noncardiovascular disorders (glaucoma and migraine prophylaxis are FDA-approved indications). Noncardiovascular conditions for which beta-blockers have been considered are listed in Table 5 (42-47) and are described below.

1. Thyrotoxicosis

Many of the symptomatic and physical manifestations of thyrotoxicosis resemble those produced by the sympathetic nervous system or by the administration of catecholamines (46,90-93). The physiological basis for these sympathomimetic features of thyroid hormone excess is obscure. Possible mechanisms include an enhanced tissue sensitivity to catecholamines, owing to increased numbers of beta receptors (94), to more efficient coupling of catecholamine-binding to activation of adenylate cyclase, or to inhibition of tissue phosphodiesterase activity (95); increased delivery of circulating catecholamines, owing to increased tissue perfusion; and the occurrence of similar but separate and additive effects of thyroid hormones and catecholamines (46,96). Despite the inability to define precisely the relationship between catecholamines and hyperthyroidism, certain antiadrenergic agents are capable of alleviating many of the sympathomimetic manifestations of the thyrotoxic state (97,98). Since these drugs act within the peripheral tissues, their symptomatic effect is much more prompt than that of traditional approaches to treating hyperthyroidism, which achieve their effects by decreasing thyroid hormone synthesis or release (46). Therefore, antiadrenergic agents like reserpine, guanethidine, and beta-blockers have particular value in treating severely thyrotoxic patients (97, 98). Because of their relative freedom from side effects, ease of administration, and rapid onset of action, beta-blockers are the agents of choice (46). Although the largest experience has been garnered with propranolol, other beta-blockers with and without beta$_1$ selectivity have also proven useful (99-103).

The exact mechanism of beta-blocker benefit in hyperthyroidism is not fully defined. It is not resolved whether the effects of beta blockade are mediated by adrenergic blockade or by blocking the peripheral conversion of T$_4$ (thyroxine) to T$_3$ (triiodothyronine) (104).

Particular benefit has been obtained with beta-blocking drugs in the management of thyrotoxic excess (thyrotoxic storm) (105). In this situation, beta blockade produces a rapid reduction in fever, heart rate, and adverse central nervous system effects, such as restlessness and disorientation. Most of the experience with beta-blockers to date in "thyroid storm" has been reported with propranolol, although other beta-blockers may also be effective (106). Beta-blockers have also been used preoperatively in thyrotoxic patients undergoing partial thyroidectomy and other surgical procedures (107).

As part of routine medical management of hyperthyroidism, beta-blocking drugs are of less certain value. All are capable of reducing the heart rate, although drugs with partial agonist activity are probably less effective (108). Other manifestations of thyrotoxicosis—tremor, hyperreflexia, agitation, hemodynamic changes, hyperkinesia, and those eye signs attributable to sympathetically innervated smooth muscle—may be reduced by beta$_1$-selective and nonselective beta-blockers (46, 107-111).

When employed chronically as the sole therapeutic agent, beta-blockers alleviate but do not eliminate the symptomatic and physiological manifestations of thyrotoxicosis (46). The drugs have no effect on thyroid hormone secretion, the peripheral disposal of the hormone, or the thyrotropic or prolactin responses to thyrotropin-releasing hormone (112). Patients fail to gain weight satisfactorily, and evidence for an increased metabolic rate persist (113). Consequently, beta-blockers cannot be considered a substitute for specific antithyroid therapies.

2. Prophylaxis of Migraine

The use of beta-adrenergic blocking drugs to prevent migraine headache was first suggested in 1966 (114). Rabkin et al. and others reported a beneficial effect of propranolol on migraine headaches in patients being treated for angina pectoris or arrhythmia. These early observations led to clinical trials that confirmed the safety and efficacy of propranolol for the prophylaxis of common migraine (42, 115, 116); the FDA approved the drug for this indication in 1979. Propranolol is not approved for the treatment of migraine headache or for the prevention and treatment of cluster headaches.

The causes of vascular headache syndromes, including common migraine, are not well defined (116). Therefore, the exact mechanisms by which propranolol prevents migraine are not known. Other beta-blockers may also be effective in migraine, but they need more intensive study. The use of propranolol for migraine is based on the fact that the drug concentrates in the brain and presumably inhibits beta-adrenoceptor-mediated vasodilatation. Dilatation of branches of the external carotid artery is assumed to be one source of pain during an episode of migraine. Propranolol may also prevent the uptake of serotonin, which is then available for vasotonic actions on cerebral blood flow (117).

Propranolol decreases the frequency of common migraine and can completely suppress headaches in some patients. One-third of patients with common migraine have an excellent response to propranolol, with more than a 50% reduction in the number of attacks and a markedly reduced need for ergotamine and analgesic medication; another third have a smaller reduction in the number of attacks; and the remaining third either have no response or become worse (118). In a comparative trial, propranolol was demonstrated to be as effective as methysergide in reducing the frequency and severity of migraine headaches (119). However, fewer adverse reactions were seen during propranolol treatment (119). Direct comparisons with other prophylactic regimens for migraine (cyproheptadine, tricyclic antidepressants, papaverine, and monoamine oxidase inhibitors) have not been made (120).

Several reports have appeared on the combined use of propranolol and ergot preparations; this combination apparently had no untoward effects (119). Severe migraine attacks have been reported to follow abrupt withdrawal from propranolol. It is recommended that the drug be gradually withdrawn over a 2-week period if the maximal dosage has not produced a satisfactory response within 4-6 weeks. Adverse reactions in patients receiving propranolol for migraine are similar to those in patients given this drug for hypertension or angina pectoris (1).

3. Open-Angle Glaucoma

In treating systemic hypertension with beta-adrenoceptor blocking drugs, it was fortuitously discovered that these agents reduced intra-ocular pressure in patients with concomitant glaucoma (120). As early as 1968, topical application of propranolol was shown to reduce intra-ocular pressure (121); however, its mild local anesthetic properties made investigators reluctant to use it for treatment of glaucoma. Topical application of timolol, a nonselective beta-blocker without this local anesthetic property or partial agonist activity also reduced intraocular pressure (122). The mechanism of its ocular hypotensive effect has not been firmly established, but it may reduce the pressure by decreasing the production of aqueous humor (122). Timolol maleate

(Timoptic) was approved by the FDA in 1978 for the topical treatment
of increased intraocular pressure in patients with chronic open-angle
glaucoma. It is also approved for patients with aphakia and glaucoma,
for some patients with secondary glaucoma, and for patients with elevated intraocular pressure who are at sufficient risk to require lowering
of the intraocular pressure.

Timolol ophthalmic solution is usually well tolerated. Mild eye
irritation occurs occasionally, and a few patients have reported blurred
vision after initial doses. Objective measurements of ophthalmic status
during topical timolol treatment have shown few changes (122). The
oral dose of timolol maleate for the treatment of systemic hypertension
is 20-60 mg/day. In contrast, the amount of timolol in four drops of
0.5% ophthalmic solution (the maximal daily dose) is only about 1 mg.
Plasma levels of the drugs after ocular administration are far below
those of the cardiovascular therapeutic dose range. However, aggrava-
tion or precipitation of certain cardiovascular and pulmonary disorders
has been reported and is presumably related to the systemic effects
of beta-adrenoceptor blockade (122). These effects include bradycardia, hypotension, syncope, confusion, and bronchospasm (predominantly in patients with bronchospastic disease). Caution is
recommended in prescribing timolol eye drops when a systemic beta-
adrenergic blocking drug may be contraindicated, as in patients with
preexisting asthmatic conditions, heart block, or heart failure. Patients
who are taking an oral beta-adrenergic drug and are given topical
timolol should be observed for a potential additive effect on intraocular
pressure, and on the known systemic effects of beta-blockade.

Recently two new beta-blocker ophthalmic solutions were approved
for clinical use in open-angle glaucoma. Betaxolol is a beta$_1$-selective
drug for twice daily use. Levobunolol is a non-selective beta blocker
for once or twice daily use. These new ophthalmic solutions appear
to be comparable in efficacy and safety to timolol.

IV. ADVERSE EFFECTS OF BETA-BLOCKERS

Comparing and tabulating adverse effects from different studies of
beta-adrenergic blockers is very difficult because of a number of
factors. These include differences in the definitions of side effects,
the kinds of patients studied, and features of study design. Methods
of ascertaining and reporting adverse side effects also differ significantly from study to study (123). When these differences are taken
into account and the results are analyzed, it appears that the types
and frequencies of adverse effects attributed to various beta-blocker
compounds are similar (124). The profile of side effects of beta-blockers
is also remarkably close to that seen with concurrent placebo treatments,
attesting to the remarkable safety margin of the beta-blockers as a
group (4,123).

The adverse effects of beta-adrenoceptor blockers can be divided into two categories: (1) those reactions that result from known pharmacological consequences of beta-adrenoceptor blockade; and (2) those that do not appear to result from beta-adrenoreceptor blockade.

Side effects of the first type are widespread because of the ubiquitous nature of the sympathetic nervous system in the control of physiological and metabolic function. They include asthma, heart failure, hypoglycemia, bradycardia and heart block, intermittent claudication, and Raynaud's phenomenon. The incidence of these adverse effects varies with the type of beta-blocker used (2,123).

A. Adverse Cardiac Effects Related to Beta-Adrenoceptor Blockade

1. Myocardial Failure

There are several circumstances by which blockade of beta-receptors may cause congestive heart failure: (1) in an enlarged heart with impaired myocardial function in which excessive sympathetic drive is essential to maintain the myocardium on a compensated Starling curve; and (2) in hearts in which the left ventricular stroke volume is restricted and tachycardia is needed to maintain cardiac output.

Another important component of heart failure may be the increases in peripheral vascular resistance produced by nonselective agents (e.g., propranolol, timolol, sotalol) (125). It has been claimed that beta-blockers with partial agonist activity are better in preserving left ventricular function and less likely to precipitate heart failure (126), but there have been few in vivo studies in humans to support this contention (18).

In patients with impaired myocardial function who require beta-blocking agents, digitalis and diuretics can be used, preferably with drugs having intrinsic sympathomimetic activity or alpha-adrenergic blocking properties.

2. Sinus Node Dysfunction and Atrioventricular Conduction Delay

Slowing of the resting heart rate is a normal response to treatment with beta-blocking drugs with and without partial agonist activity. In most cases, this does not present a problem: Healthy individuals can sustain a heart rate of 40-50 without disability, unless there is clinical evidence of heart failure (15). Drugs with partial agonist activity do not lower the resting heart rate to the same degree as propranolol (127). However, all beta-blocking drugs are contraindicated (unless an artificial pacemaker is present) in patients with "sick sinus" syndrome (75).

If there is a partial or complete atrioventricular conduction defect, use of a beta-blocking drug may lead to a serious bradyarrhythmia (15). The risk of atrioventricular impairment may be less with beta-blockers having partial agonist activity (128).

3. Overdosage

Suicide attempts and accidental overdosing with beta-blockers are being described with increasing frequency. Since beta-adrenergic blockers are competitive pharmacological antagonists, their life-threatening effects (bradycardia, myocardial and ventilatory failure) can be overcome with an immediate infusion of beta-agonist agents like isoproterenol and dobutamine. In situations in which catecholamines are not effective, i.v. amrinone and glucagon have been used (17).

Close monitoring of cardiorespiratory function is necessary for at least 24 hr after the patient responds to therapy. Patients who recover will usually have no long-term sequelae; however, they should be observed for signs of the beta-blocker withdrawal phenomenon (17).

4. Beta-Adrenoceptor Blocker Withdrawal

Following abrupt cessation of chronic beta-blocker therapy, exacerbation of angina pectoris and, in some cases, acute myocardial infarction and death have been reported (129-131).

Observations made in multiple double-blind randomized trials have confirmed the reality of a propranolol withdrawal reaction (129-132). The exact mechanism for this reaction is unclear. There is some evidence that the withdrawal phenomenon may be due to the generation of additional beta-adrenoceptors during the period of beta-adrenoceptor blockade. When the beta-adrenoceptor blocker is subsequently withdrawn, the increased beta-receptor population results in excessive beta-receptor stimulation, which will be clinically important when the delivery and use of oxygen is finely balanced, as in ischemic heart disease. Other suggested mechanisms for the withdrawal reaction include heightened platelet aggregability (131), an elevation in thyroid hormone activity (133), and an increase in circulating catecholamines (134).

B. Noncardiac Adverse Side Effects Related to Beta-Adrenoreceptor Blockade

1. Effect on Ventilatory Function

The bronchodilator effects of catecholamines on the bronchial beta-adrenoreceptors (beta$_2$) are inhibited by nonselective beta-blockers (e.g., propranolol, nadolol) (135). Comparative studies have shown that beta-blocking compounds with partial agonist activity (12), beta$_1$ selectivity (9,10), and alpha-adrenergic blocking actions (136) are less likely to increase airways resistance in asthmatics than propranolol. Beta$_1$-selectivity, however, is not absolute, and may be lost with high therapeutic doses as shown with atenolol and metoprolol. It is possible in asthma to use a beta$_2$-selective agonist (such as albuterol) in certain patients with concomitant low-dose $beta_1$-selective blocker treatment (137). As noted earlier, all beta-blockers should be avoided in patients with bronchospastic disease.

2. Peripheral Vascular Effects (Raynaud's Phenomenon)

Cold extremities and absent pulses have been reported to occur more frequently in patients receiving beta-blockers for hypertension than in those receiving methyldopa (21). Among the beta-blockers, the incidence was highest with propranolol and lower with drugs having $beta_1$ selectivity or intrinsic sympathomimetic activity. In some instances, peripheral vascular compromise has been severe enough to cause cyanosis and impending gangrene (138). This is probably due to the reduction in cardiac output and blockade of $beta_2$-adrenoceptor-mediated vasocilation, resulting in unopposed alpha-adrenoceptor vasoconstriction (139). Beta-blocking drugs with $beta_1$ selectivity or partial agonist activity will not affect peripheral vessels to the same degree as propranolol.

Raynaud's phenomenon is one of the more common side effects of propranolol treatment (140). It is more troublesome with propranolol than metoprolol, atenolol, or pindolol, probably because of the $beta_2$-blocking properties of propranolol.

Patients with peripheral vascular disease who suffer from intermittent claudication often report worsening of the claudication when treated with beta-blocking drugs (141). Whether drugs with $beta_1$ selectivity or partial agonist activity can protect against this adverse reaction has yet to be determined.

3. Hypoglycemia and Hyperglycemia

Several authors have described severe hypoglycemic reactions during therapy with beta-adrenergic blocking drugs (142). Some of the patients affected were insulin-dependent diabetics, while others were nondiabetic. Studies of resting normal volunteers have demonstrated that propranolol produces no alteration in blood glucose values (143), although the hyperglycemic response to exercise is blunted.

The enhancement of insulin-induced hypoglycemia and its hemodynamic consequences may be less with $beta_1$-selective agents (in which there is no blocking effect on $beta_2$ receptors) and agents with intrinsic sympathomimetic activity (which may stimulate $beta_2$ receptors) (144).

There is also a marked diminution in the clinical manifestations of the catecholamine discharge induced by hypoglycemia (tachycardia) (145). These findings suggest that beta-blockers interfere with compensatory responses to hypoglycemia and can mask certain "warning signs" of this condition. Other hypoglycemic reactions, such as diaphoresis, are not affected by beta-adrenergic blockade.

4. Central Nervous System Effects

Dreams, hallucinations, insomnia, and depression can occur during therapy with beta-blockers (140,146). These symptoms are evidence of drug entry into the central nervous system (CNS), and we have noted them to be especially common with the highly lipid-soluble beta-blockers (propranolol, metoprolol). It has been claimed that beta-blockers with less lipid solubility (atenolol, nadolol) will cause fewer CNS side effects (10,11). This claim is intriguing, but to be useful, it must be corroborated with more extensive clinical experiences.

5. Miscellaneous Side Effects

Diarrhea, nausea, gastric pain, constipation, and flatulence have been seen occasionally with all beta-blockers (in 2-11% of patients (147).
Hematological reactions are rare: rare cases of purpura (148) and agranulocytosis (149) have been described with propranolol.
A devastating blood pressure rebound effect has been described in patients who discontinued clonidine while being treated with non-selective beta-blocking agents. The mechanism for this may be related to an increase in circulating catecholamines and an increase in peripheral vascular resistance (150). Whether $beta_1$-selective or partial agonist beta-blockers have similar effects following clonidine withdrawal, remains to be determined. It is worth noting that blood pressure rebound effect has not been a problem with labetalol (151).

C. Adverse Effects Unrelated to
   Beta-Adrenoceptor Blockade

1. Oculomucocutaneous Syndrome

A characteristic immune reaction, the oculomucocutaneous syndrome, affecting singly or in combination the eyes, mucous and serous membranes, and the skin (often in association with a positive antinuclear factor), has been reported in patients treated with practolol and has led to the curtailment of this drug in clinical practice (152). Close attention has been focused on this syndrome because of fears that other beta-adrenoreceptor blocking drugs may be associated with this syndrome. In 19 patients who experienced such a reaction with practolol, the lesions healed after switching to atenolol treatment (10).

V. DRUG INTERACTIONS

The wide diversity of diseases for which beta-blockers are employed raises the likelihood of their concurrent administration with other drugs. It is imperative, therefore, that clinicians become familiar with the

TABLE 10 Drug Interactions that May Occur with Beta-Adrenoceptor Blocking Drugs

Drug	Possible effects	Precautions
Aluminum hydroxide gel	Decreases beta-blocker absorption and therapeutic effect	Avoid beta-blocker-aluminum hydroxide combination.
Aminophylline	Mutual inhibition	Observe patient's response.
Antidiabetic agents	Enhanced hypoglycemia: hypertension	Monitor for altered diabetic response.
Calcium channel inhibitors (e.g., verapamil, diltiazem)	Potentiation of bradycardia, myocardial depression, and hypotension	Avoid use, although few patients show ill effects.
Cimetidine	Prolongs half-life of propranolol	Combination should be used with caution.
Clonidine	Hypertension during clonidine withdrawal	Monitor for hypertensive response; withdraw beta-blocker before withdrawing clonidine.
Digitalis glycosides	Potentiation of bradycardia	Observe patient's response; interactions may benefit angina patients with abnormal ventricular function.
Epinephrine	Hypertension; bradycardia	Administer epinephrine cautiously; cardioselective beta-blocker may be safer.
Ergot alkaloids	Excessive vasoconstriction	Observe patient's response; few patients show ill effects.
Glucagon	Inhibition of hyperglycemic effect	Monitor for reduced response.

(continued)

TABLE 10 *(Continued)*

Drug	Possible effects	Precautions
Halofenate	Reduced beta-blocking activity; production of propranolol withdrawal rebound syndrome	Observe for impaired response to beta-blockade.
Indomethacin	Inhibition of antihypertensive response to beta-blockade	Observe patient's response.
Isoproterenol	Mutual inhibition	Avoid concurrent use or choose cardiac-selective beta-blocker.
Levodopa	Antagonism of levodopa's hypotensive and positive inotropic effects	Monitor for altered response; interaction may have favorable results.
Lidocaine	Propranolol pretreatment increases lidocaine blood levels and potential toxicity	Combination should be used with caution; use lower doses of lidocaine.
Methyldopa	Hypertension during stress	Monitor for hypertensive episodes.
Monoamine oxidase inhibitors	Uncertain, theoretical	Manufacturer of propranolol considers concurrent use contraindicated.
Phenothiazines	Additive hypotensive effects	Monitor for altered response especially with high doses of phenothiazine.
Phenylpropranolamine	Severe hypertensive reaction	Avoid use, especially in hypertension controlled by both methyldopa and beta-blockers.

TABLE 10 *(Continued)*

Drug	Possible effects	Precautions
Phenytoin	Additive cardiac depressant effects	Administer i.v. phenytoin with great caution.
Quinidine	Additive cardiac depressant effects	Observe patient's response; few patients show ill effects.
Reserpine	Excessive sympathetic blockade	Observe patient's response.
Tricyclic antidepressants	Inhibits negative inotropic and chronotropic effects of beta-blockers	Observe patient's response.
Tubocurarine	Enhanced neuromuscular blockade	Observe response in surgical patients, especially after high doses of propranolol.

Source: From Refs. 2, 153, 154, used with permission.

interactions of beta-blockers with other pharmacological agents. The list of commonly used drugs with which beta-blockers interact is extensive (Table 10) (2,153,154). The majority of the reported interactions have been associated with propranolol, the best studied of the beta-blockers, and may not apply to other drugs in this class.

VI. CONCLUSIONS

A. How to Choose a Beta-Blocker

The various beta-blocking compounds given in adequate dosage appear to have comparable, antihypertensive, antiarrhythmic, and antianginal effects. Therefore, the beta-blocking drug of choice in an individual patient is determined by the pharmacodynamic and pharmacokinetic differences between the drugs, in conjunction with the patient's other medical conditions (Table 11) (2,8).

TABLE 11 Clinical Situations that Would Influence the Choice of a Beta-Blocking Drug

Condition	Choice of beta-blocker
Asthma, chronic bronchitis with bronchospasm	Avoid all beta-blockers if possible; however small doses of $beta_1$-selective blockers (e.g., acebutolol, atenolol, metoprolol) can be used. $Beta_1$-selectivity is lost with higher doses. Drugs with partial agonist activity (e.g., pindolol, oxprenolol) and labetalol with alpha-adrenergic blocking properties can also be used.
Congestive heart failure	Drugs with partial agonist activity and labetalol might have an advantage, although beta-blockers are usually contraindicated.
Angina	In patients with angina at low heart rates, drugs with partial agonist activity probably contraindicated. Patients with angina at high heart rates but who have resting bradycardia might benefit from a drug with partial agonist activity. In vasospastic angina, labetalol may be useful; other beta-blockers should be used with caution.
Atrioventricular conduction defects	Beta-blockers generally contraindicated but drugs with partial agonist activity and labetalol can be tried with caution.
Bradycardia	Beta-blockers with partial agonist activity and labetalol have less pulse-slowing effect and are preferable.
Raynaud's phenomenon, intermittent claudication, cold extremities	$Beta_1$-selective blocking agents, labetalol, and those with partial agonist activity might have an advantage.

Depression	Avoid propranolol. Substitute a beta-blocker with partial agonist activity.
Diabetes mellitus	Beta$_1$-selective agents and partial agonist drugs are preferable.
Thyrotoxicosis	All agents will control symptoms but agents without partial agonist activity are preferred.
Pheochromocytoma	Avoid all beta-blockers unless alpha-blocker is given. Labetalol may be used as a treatment of choice.
Renal failure	Use reduced doses of compounds largely eliminated by renal mechanisms (nadolol, sotalol, atenolol) and of those drugs whose bioavailability is increased in uremia (propranolol, alprenolol). Also consider possible accumulation of active metabolites (alprenolol, propranolol).
Insulin and sulfonylurea use	Danger of hypoglycemia. Possibly less using drugs with beta$_1$ selectivity.
Clonidine	Avoid sotalol (other nonselective beta-blockers). Severe rebound effect with clonidine withdrawal.
Oculomucocutaneous syndrome	Stop drug. Substitute any other beta-blocker.
Hyperlipidemia	Avoid nonselective beta-blockers; use agents with partial agonism, beta$_1$-selectivity or labetalol.

Source: From Refs. 2 and 9, used with permission.

REFERENCES

1. Frishman, W. H.: β-Adrenoceptor antagonists new drugs and new indications. N. Engl. J. Med. 305:500, 1981.

2. Frishman, W. H.: Clinical Pharmacology of the β-Adrenoceptor Blocking Drugs, 2nd ed. Norwalk, Connecticut, Appleton-Century-Crofts, 1984.

3. The Norwegian Multicenter Study Group: Timolol induced reduction in mortality and reinfarction in patients surviving acute myocardial infarction. N. Engl. J. Med. 304:801, 1981.

4. Beta-Blocker Heart Attack Trial Research Group: A randomized trial of propranolol in patients with acute myocardial infarction. I. Mortality results. J.A.M.A. 247:1707, 1982.

5. Braunwald, E.: Treatment of the patient after myocardial infarction. N. Engl. J. Med. 302:290, 1980.

6. Frishman, W. H., Furberg, C. D., and Friedewald, W. T.: β-Adrenergic blockade for survivors of acute myocardial infarction. N. Engl. J. Med. 310:830, 1984.

7. Hjalmarson, Å., Elmfeldt, D., Herlitz, J., et al.: Effect of mortality of metoprolol in acute myocardial infarction. Lancet 2:823, 1981.

8. Frishman, W. H.: The beta-adrenoceptor blocking drugs. Int. J. Cardiol. 2:165, 1982.

9. Koch-Weser, J.: Metoprolol. N. Engl. J. Med. 301:698, 1979.

10. Frishman, W. H.: Atenolol and timolol, two new systemic β-adrenoceptor antagonists. N. Engl. J. Med. 306:1456, 1982.

11. Frishman, W. H.: Nadolol: A new β-adrenoceptor antagonist. N. Engl. J. Med. 305:678, 1981.

12. Frishman, W. H.: Pindolol: A new β-adrenoceptor antagonist with partial agonist activity. N. Engl. J. Med. 308:940, 1983.

13. Frishman, W., and Silverman, R.: Clinical pharmacology of the new beta-adrenergic blocking drugs. Part 3. Comparative clinical experience and new therapeutic applications. Am. Heart J. 98: 119, 1979.

14. Frishman, W.: Clinical pharmacology of the new beta-adrenergic blocking drugs. Part 1. Pharmacokinetic and pharmacodynamic properties. Am. Heart J. 98:663, 1979.

15. Conolly, M. E., Kersting, F., and Dollery, C. T.: The clinical pharmacology of beta-adrenoceptor blocking drugs. Prog. Cardiovasc. Dis. 19:203, 1976.

16. Opie, L. H.: Drugs and the heart. 1. Beta-blocking agents. Lancet 1:693, 1980.

17. Frishman, W., Jacob, H., Eisenberg, E., and Ribner, H.: Clinical pharmacology of the new beta-adrenergic blocking drugs. Part 8. Self-poisoning with beta-adrenoceptor blocking drugs: Recognition and management. Am. Heart J. 98:798, 1979.

18. Taylor, S. H., Silke, B., and Lee, P. S.: Intravenous beta-blockade in coronary heart disease: Is cardioselectivity or intrinsic sympathomimetic activity hemodynamically useful? N. Engl. J. Med. 306:631, 1982.

19. Frishman, W., and Halprin, S.: Clinical pharmacology of the new beta-adrenergic blocking drugs. Part 7. New horizons in beta-adrenoceptor blocking therapy: Labetalol. Am. Heart J. 98:660, 1979.

20. Frishman, W. H., Strom, J., Kirschner, M., et al.: Labetalol therapy in patients with systemic hypertension and angina pectoris: Effects of combined alpha- and beta-adrenergic blockade. Am. J. Cardiol. 48:917, 1981.

21. Waal-Manning, H. J.: Hypertension: Which beta-blocker? Drugs 12:412, 1976.

22. Johnsson, G., and Regardh, C. G.: Clinical pharmacokinetics of β-adrenoceptor blocking drugs. Clin. Pharmacokinet. 1:233, 1976.

23. Frishman, W. H., and Teicher, M.: Long-acting propranolol. Cardiovasc. Rev. Rep. 4:1100, 1983.

24. Halkin, H., Vered, I., Saginer, A., and Rabinowitz, B.: Once-daily administration of sustained release propranolol capsules in the treatment of angina pectoris. Eur. J. Clin. Pharmacol. 16:387, 1979.

25. Parker, J. O., Porter, A., and Parker, J. D.: Propranolol in angina pectoris: Comparison of long-acting and standard formulation propranolol. Circulation 65:1351, 1982.

26. Mishriki, A. A., and Weidler, D. J.: Long-acting propranolol (Inderal LA): Pharmacokinetics, pharmacodynamics and therapeutic use. Pharmacotherapy 3:334, 1983.

27. Zaroslinski, J., Borgman, R. J., O'Donnel, J. P., et al.: Ultra-short acting beta-blockers: A proposal for the treatment of the critically ill patient. Life Sci. 31:899, 1982.

28. Gorczyski, R. J., Shaffer, J. E., Lee, R. J., and Vuong, A.: Pharmacology of ASL-8052, a novel beta-adrenergic receptor

antagonist with an ultra-short duration of action. J. Cardiovasc. Pharmacol. 5:668, 1983.

29. Murthy, V. S., Hwang, T. F., Zagar, M. E., et al.: Cardiovascular pharmacology of ASL-8052. An ultra-short acting beta-blocker. Eur. J. Pharmacol. 94:43, 1983.

30. Cohn, J. N.: Nitroprusside and dissecting aneurysms of aorta. N. Engl. J. Med. 295:567, 1976.

31. Wheat, M. W., Jr.: Treatment of dissecting aneurysms of the aorta: Current status. Prog. Cardiovasc. Dis. 16:87, 1973.

32. Cohen, L. S., and Braunwald, E.: Amelioration of angina pectoris in idiopathic hypertrophic subaortic stenosis with beta-adrenergic blockade. Circulation 35:847, 1967.

33. Turner, J. R. B.: Propranolol in the treatment of digitalis-induced and digitalis-resistant tachycardia. Am. J. Cardiol. 18:450, 1966.

34. Winkle, R. A., Lopes, M. G., Goodman, D. S., et al.: Propranolol for patients with mitral valve prolapse. Am. Heart J. 93:422, 1970.

35. Vincent, G. M., Abildskov, J. A., and Burgess, M. J.: Q-T interval syndromes. Prog. Cardiovasc. Dis. 16:523, 1974.

36. Shah, P. M., and Kidd, L.: Circulatory effects of propranolol in children with Fallot's tetralogy. Observations with isoproterenol infusion, exercise and crying. Am. J. Cardiol. 19:653, 1967.

37. Meister, S. G., Engel, T. R., Feitosa, G. S., et al.: Propranolol in mitral stenosis during sinus rhythm. Am. Heart J. 94:685, 1977.

38. Bhatia, M. L., Shrivastava, S., and Roy, S. G.: Immediate haemodynamic effects of a beta-adrenergic blocking agent— propranolol—in mitral stenosis at fixed heart rates. Br. Heart J. 34:638, 1972.

39. Svedberg, K., Hjalmarson, Å., and Waagstein, F.: Beneficial effects of long-term beta-blockade in congestive cardiomyopathy. Br. Heart J. 44:117, 1980.

40. Teuscher, A., Bossi, E., Imhof, P., et al.: Effect of propranolol on fetal tachycardia in diabetic pregnancy. Am. J. Cardiol. 42:304, 1978.

41. Furberg, C., and Morsing, C.: Adrenergic beta-receptor blockade in neurocirculatory asthenia. Pharmacol. Clin. 1:168, 1969.

42. Weber, R. B., and Reinmuth, O. M.: The treatment of migraine with propranolol. Neurology 22:366, 1972.

43. Young, R. R., Growdon, J. H., and Shahani, B. T.: Beta-adrenergic mechanisms in action tremor. N. Engl. J. Med. 293:950, 1975.

44. Granville-Grossman, K. L., and Turner, P.: The effect of propranolol on anxiety. Lancet 1:788, 1966.

45. Sellers, E. M., Degani, N. C., Silm, D. H., and MacLeod, S. M.: Propranolol decreased noradrenaline secretion and alcohol withdrawal. Lancet 1:94, 1976.

46. Ingbar, S. H.: The role of antiadrenergic agents in the management of thyrotoxicosis. Cardiovasc. Rev. Rep. 2:683, 1981.

47. Caro, J. F., Castro, J. H., and Glennon, J. A.: Effect of long-term propranolol administration on parathyroid hormone and calcium concentration in primary hyperparathyroidism. Ann. Intern. Med. 91:740, 1979.

48. Frohlich, E. D.: Hyperdynamic circulation and hypertension. Postgrad. Med. 5:64, 1972.

49. Myers, M. G., Lewis, P. J., Reid, J. L., and Dollery, C. T.: Brain concentration of propranolol in relation to hypotension effects in the rabbit with observations on brain propranolol levels in man. J. Pharmacol. Exp. Ther. 192:327, 1975.

50. Laragh, J. H.: Vasoconstriction-volume analysis for understanding and treating hypertension: The use of renin and aldosterone profiles. Am. J. Med. 55:261, 1973.

51. Prichard, B. N. C.: Propranolol as an antihypertensive agent. Am. Heart J. 79:128, 1970.

52. Imhof, P. R.: Characterization of beta-blockers as antihypertensive agents in the light of human pharmacology studies. In: *Beta-Blockers—Present Status and Future Prospects*, W. Schweizer (ed.). Bern, Huber, 1974, pp. 40-50.

53. Waal, H. J.: Hypotensive action of propranolol. Clin. Pharmacol. Ther. 7:558, 1966.

54. Rahn, K. H., Hawlina, A., Kersting, F., and Planz, G.: Studies on the antihypertensive action of the optical isomers of propranolol in man. Naunyn Schmiedebergs Arch. Pharmacol. 286:319, 1974.

55. Pickering, T. G., Gribbin, B., Petersen, E. S., et al.: Effects of autonomic blockade on the baroreflex in man at rest and during exercise. Circ. Res. 30:177, 1972.

56. Langer, S. Z.: Presynaptic receptors and their role in the regulation of transmitter release. Br. J. Pharmacol. 60:481, 1977.

57. Berthelsen, S., and Pettinger, W. A.: A functional basis for classification of α-adrenergic receptors. Life Sci. 21:77, 1977.

58. Yamaguchi, N., de Champlain, J., and Nadeau, R. L.: Regulation of norepinephrine release from cardiac sympathetic fibers in the dog by presynaptic α- and β-receptors. Circ. Res. 41:108, 1977.

59. Stjarne, L., and Brundin, J.: β-Adrenoceptors facilitate noradrenaline secretion from human vasoconstrictor nerves. Acta. Physiol. Scand. 97:88, 1976.

60. Majewski, H. J., McCulloch, M. W., Rand, M. J., and Story, D. F.: Adrenaline activation of pre-junctional β-adrenoceptors in guinea pig atria. Br. J. Pharmacol. 71:435, 1980.

61. Sonnenblick, E. H., Ross, J., Jr., and Braunwald, E.: Oxygen consumption of the heart. Newer concepts of its multifactorial determination. Am. J. Cardiol. 22:328, 1968.

62. Sonnenblick, E. H., and Skelton, C. L.: Myocardial energetics: Basic principles and clinical implications. N. Engl. J. Med. 285:668, 1971.

63. Black, J. W., and Stephenson, J. S.: Pharmacology of a new adrenergic beta-receptor blocking compound (Nethalide). Lancet 2:311, 1962.

64. Frishman, W. H.: Beta-adrenergic blockade in the treatment of coronary artery disease. In: *Clinical Essays on the Heart*. Vol. 2. J. W. Hurst (ed.). New York, McGraw-Hill, 1983, p. 25.

65. Frishman, W. H.: Multifactorial actions of β-adrenergic blocking drugs in ischemic heart disease: Current concepts. Circulation (Suppl. 1) 67, 11, 1983.

66. Parratt, J. R., and Grayson, J.: Myocardial vascular reactivity after β-adrenergic blockade. Lancet 1:388, 1966.

67. Becker, L. C., Fortuin, N. J., and Pitt, B.: Effects of ischemia and antianginal drugs on the distribution of radioactive microspheres in the canine left ventricle. Circ. Res. 28:263, 1971.

68. Wolfson, S., and Gorlin, R.: Cardiovascular pharmacology of propranolol in man. Circulation 40:501, 1969.

69. Gianelly, R. S., Goldman, R. H., Treister, B., and Harrison, D. C.: Propranolol in patients with angina pectoris. Ann. Intern. Med. 57:1216, 1967.

70. Robinson, B. F.: The mode of action of beta-antagonists in angina pectoris. Postgrad. Med. J. (Suppl. 2) 47:451, 1971.

71. Parmley, W. W.: The combination of beta-adrenergic blocking agents and nitrates in the treatment of stable angina pectoris. Cardiol. Rev. Rep. 3:1425, 1982.

72. Maseri, A., L'Abbate, A., Ballestra, A. M., et al.: Coronary vasospasm in angina pectoris. Lancet 1:713, 1977.

73. Parodi, O., Simonetti, I., L'Abbate, A., and Maseri, A.: Vera-pamil versus propranolol for angina at rest. Am. J. Cardiol. 50: 923, 1982.

74. Opie, L. H., and Lubbe, W. F.: Catecholamine-mediated arrhythmia in acute myocardial infarction: Experimental evidence and role of beta-adrenoceptor blockade. S. Afr. Med. J. 56:871, 1979.

75. Singh, B. N., and Jewitt, D. E.: β-Adrenoceptor blocking drugs in cardiac arrhythmias. In: *Cardiovascular Drugs*. Vol. 2. G. Avery (ed.). Baltimore, University Park Press, 1977, pp. 141-142.

76. Latour, Y., Dumont, G., Brosseau, A., and LeLorie, J.: Effects of sotalol in twenty patients with cardiac arrhythmias. Int. J. Clin. Pharmacol. 15:275, 1977.

77. Fogelman, F., Lightman, S. L., Sillett, R. W., and McNicol, M. W.: The treatment of cardiac arrhythmias with sotalol. Eur. J. Clin. Pharm. 5:72, 1972.

78. Pratt, C., and Lichstein, E.: Ventricular anti-arrhythmic effects of beta-adrenergic blocking drugs: A review of mechanism and clinical studies. J. Clin. Pharmacol. 22:335, 1982.

79. Ryden, L., Ariniego, R., Arnman, K., et al.: A double-blind trial of metoprolol in acute myocardial infarction: Effects on ventricular tachyarrhythmias. N. Engl. J. Med. 308:614, 1983.

80. Lichstein, E., Morganroth, J., Harriet, R., and Hubble, E.: Effect of propranolol on ventricular arrhythmias—The beta-blockers: Preliminary data from the heart attack trial experience. Circulation (Suppl. 1) 67:I-32, 1983.

81. Braunwald, E., Muller, J. E., Kloner, R. A., and Maroko, P. R.: Role of beta-adrenergic blockade in the therapy of patients with myocardial infarction. Am. J. Med. 74:113, 1983.

82. Furberg, C. D., Hawkins, C. M., and Lichstein, E.: Effect of propranolol in post-infarction patients with mechanical or electrical complications. Circulation 69:761, 1984.

83. International Collaborative Study Group. Reduction of infarct size with the early use of timolol in acute myocardial infarction. N. Engl. J. Med. 310:9, 1984.

84. Muller, J., Roberts, R., Stone, P., et al.: Failure of propranolol administration to limit infarct size in patients with acute myocardial infarction. Circulation (Suppl. III) 68:294 (abstract).

85. Swan, D. A., Bell, B., Oakley, C. M., and Goodwin, J.: Analysis of symptomatic course and prognosis and treatment of hypertrophic obstructive cardiomyopathy. Br. Heart J. 33:671, 1971.

86. Hubner, P. J. B., Ziady, G. M., Lane, G. K., et al.: Double-blind trial of propranolol and practolol in hypertrophic cardio-myopathy. Br. Heart J. 35:1116, 1973.

87. Epstein, S. E., Henry, W. L., Clark, C. E., et al.: Assymmetric septal hypertrophy. Ann. Intern. Med. 81:650, 1974.

88. Jeresaty, R. M.: Mitral valve prolapse syndrome. Prog. Cardio-vasc. Dis. 15:623, 1973.

89. Slater, E. E., and DeSanctis, R.: Dissection of the aorta. Med. Clin. N. Am. 63:141, 1979.

90. Levey, G. S.: Catecholamine hypersensitivity thyroid hormone and the heart—a reevaluation. Am. J. Med. 50:413, 1971.

91. Brewster, W. R., Isaacs, J. P., Osgood, P. F., and King, T. L.: The hemodynamic and metabolic interrelationships in the activity of epinephrine, norepinephrine, and the thyroid hormones. Circulation 13:1, 1956.

92. Ramsay, I.: Adrenergic blockade in hyperthyroidism. Br. J. Clin. Pharmacol. 2:385, 1975.

93. Landsberg, L.: Catecholamines and hyperthyroidism. Clin. Endocrinol. Metab. 3:697, 1977.

94. Williams, L. T., Leibowitz, R. J., Watanabe, A. M., et al.: Thyroid hormone regulation of beta-adrenergic receptor numbers. J. Biol. Chem. 252:2787, 1977.

95. Malbon, C. C., Moreno, F. J., Cabelli, R. J., and Fain, J. N.: Fat cell adenylate cyclase and beta-adrenergic receptors in altered thyroid states. J. Biol. Chem. 253:671, 1978.

96. Levey, G. S., and Epstein, S. E.: Myocardial adenyl cyclase: Activation by thyroid hormones and evidence for two adenyl cyclase systems. J. Clin. Invest. 48:1663, 1969.

97. Canary, J. J., Schaff, M., Duffy, B. J., and Kyle, L. H.: Effects of oral and intramuscular administration of reserpine in thyrotoxicosis. N. Engl. J. Med. 257:435, 1957.

98. Wilson, W. R., Theilin, F. O., and Fletcher, F. W.: Pharmaco-dynamic effects of beta-adrenergic blockade in patients with hyperthyroidism. J. Clin. Invest. 41:1697, 1967.

99. Jones, M. K., John, R., and Jones, G. R.: The effect of ox-prenolol, acebutolol, and propranolol on thyroid hormones in hyperthyroid subjects. Clin. Endocrinol. 13:343, 1980.

100. How, J., Khir, A. S. M., and Bewsher, P. D.: The effect of atenolol on serum thyroid hormones in hyperthyroid patients. Clin. Endocrinol. 13:299, 1980.

101. Peden, N. R., Isles, T. E., Stevenson, I. H., and Crooks, J.: Nadolol in thyrotoxicosis. Br. J. Clin. Pharmacol. 13(6):835, 1982.

102. Nilsson, O. R., Karlberg, B. E., Kagedal, B., et al.: Nonselective and selective β_1-adrenoceptor blocking agents in the treatment of hyperthyroidism. Acta. Med. Scand. 206:21, 1979.

103. Wahlberg, P., and Carlsson, S. A.: Long-term control of thyrotoxicosis with the beta-blocker sotalol—A model of untreated hyperthyroidism. Acta. Endocrinol. 87:734, 1978.

104. Wiersinga, W. M., and Touber, J. L.: The influence of β-adrenoceptor blocking drugs on plasma thyroxine and triiodothyronine. J. Clin. Endocrinol. Metab. 45:293, 1977.

105. Mackin, J. F., Canary, J. J., and Pittman, C. S.: Thyroid storm and its management. N. Engl. J. Med. 291:1396, 1974.

106. Caswell, H. T. R., Marks, A. D., and Channick, B. J.: Propranolol for the preoperative preparation of patients with thyrotoxicosis. Surg. Gynecol. Obstet. 146(6):908, 1978.

107. Lee, T. C., Coffey, R. J., Currier, B. M., Ma, X. P., and Canary, J. J.: Propranolol and thyroidectomy in the treatment of thyrotoxicosis. Ann. Surg. 195(6):766, 1982.

108. Turner, P., and Hill, R. C.: A comparison of three beta-adrenergic blocking drugs in thyrotoxic tachycardia. J. Clin. Pharmacol. 8:268, 1968.

109. Schelling, J. L., Scazziga, B., Dufour, R. J., et al.: Effect of pindolol, a beta-receptor antagonist in hyperthyroidism. Clin. Pharmacol. Ther. 14:158, 1973.

110. Shanks, R. G., Hadden, D. R., Lowe, D. C., and McDevitt, D. G.: Controlled trial of propranolol in thyrotoxicosis. Lancet 1:993, 1969.

111. Grossman, W., Robin, N. I., Johnson, L. W., et al.: The effect of beta-blockade on the peripheral manifestations of thyrotoxicosis. Ann. Intern. Med. 74:875, 1971.

112. Wartofsky, L., Dimond, R. C., Noel, G. L., et al.: Failure of propranolol to alter thyroid iodine release, thyroxine turnover, or the TSH and PRL responses to thyrotropin-releasing hormone in patients with thyrotoxicosis. J. Endocrinol. Metab. 41:485, 1975.

113. Georges, L. P., Santangelo, R. P., Machin, J. F., and Canary, J. J.: Metabolic effects of propranolol in thyrotoxicosis. 1. Nitrogen, calcium and hydroxyproline. Metabolism 24:11, 1975.

114. Rabkin, R., Stables, D. P., Levin, N. W., and Suzman, M. M.:
     The prophylactic value of propranolol in angina pectoris. Am. J.
     Cardiol. 18:370, 1966.

115. Diamond, S., and Medina, J. L.: Double blind study of pro-
     pranolol for migraine prophylaxis. Headache 16:24, 1976.

116. Caviness, V. S., Jr., and O'Brien, P.: Headache. N. Engl. J.
     Med. 302:446, 1980.

117. Borgesen, S. E.: Propranolol for migraine. Compr. Ther. 3(4):
     53, 1977.

118. Forssman, B., Henriksson, K.-G., Johannsson, V., et al.:
     Propranolol for migraine prophylaxis. Headache 16:238, 1976.

119. Behan, P. O., and Reid, M.: Propranolol in the treatment of
     migraine. Practitioner 224:201, 1980.

120. Phillips, C. I., Howitt, G., and Rowlands, D. J.: Propranolol
     as ocular hypotensive agent. Br. J. Ophthalmol. 51:222, 1967.

121. Bucci, M. G., Missiroli, A., Giraldi, J. P., and Virno, M.:
     Local administration of propranolol in the treatment of glaucoma.
     Boll. Oculist. 47:51, 1968.

122. Heel, R. C., Brogden, R. N., Speight, T. M., and Avery, G. S.:
     Timolol: A review of its therapeutic efficacy in the topical treat-
     ment of glaucoma. Drugs 17:38, 1979.

123. Friedman, L. M.: How do the various beta-blockers compare in
     type, frequency, and severity of their adverse effects? Circula-
     tion (Suppl. I) 67:89, 1983.

124. Frishman, W., Silverman, R., Strom, J., et al.: Clinical Pharma-
     cology of the new beta-adrenoceptor blocking drugs. Part 4.
     Adverse effects. Choosing a β-adrenoceptor blocker. Am. Heart
     J. 98:256, 1979.

125. Vaughan Williams, E. M., Baywell, E. E., and Singh, B. N.:
     Cardiospecificity of beta-receptor blockade. A comparison of
     the relative potencies on cardiac and peripheral vascular beta-
     adrenoceptors of propranolol, of practolol and its ortho-substituted
     isomer and of oxprenolol and its para-substituted isomer. Cardio-
     vasc. Res. 7:226, 1973.

126. Frishman, W. H., and Kostis, J.: The significance of intrinsic
     sympathomimetic activity in beta-adrenoceptor blocking drugs.
     Cardiovasc. Rev. Rep. 4:503, 1982.

127. Frishman, W., Kostis, J., Strom, J., et al.: Clinical pharmacology
     of the new beta-adrenergic blocking drugs. Part 6. A comparison

of pindolol and propranolol in treatment of patients with angina pectoris. The role of intrinsic sympathomimetic activity. Am. Heart J. 98:526, 1979.

128. Giudicelli, J. F., and Lhoste, F.: β-Adrenoceptor blockade and atrioventricular conduction in dogs. Role of intrinsic sympathomimetic activity. Br. J. Clin. Pharmacol. (Suppl. 2) 13:167, 1982.

129. Alderman, E. L., Coltart, D. J., Wettach, G. E., and Harrison, D. C.: Coronary artery syndromes after sudden propranolol withdrawal. Ann. Intern. Med. 81:925, 1974.

130. Miller, R. R., Olson, H. G., Amsterdam, E. A., and Mason, D. T.: Propranolol withdrawal rebound phenomenon: Exacerbation of coronary events after abrupt cessation of anti-anginal therapy. N. Engl. J. Med. 293:416, 1975.

131. Frishman, W. H., Christodoulou, J., Weksler, B., et al.: Abrupt propranolol withdrawal in angina pectoris: Effects on platelet aggregation and exercise tolerance. Am. Heart J. 95:169, 1978.

132. Frishman, W. H., Klein, N., Strom, J., et al.: Comparative effects of abrupt propranolol and verapamil withdrawal in angina pectoris. Am. J. Cardiol. 50:1191, 1982.

133. Kristensen, B. O., Steiness, E., and Weeke, J.: Propranolol withdrawal and thyroid hormones in patients with essential hypertension. Clin. Pharmacol. Ther. 23:624, 1978.

134. Rangno, R., and Nattel, S.: Prevention of propranolol withdrawal phenomena by gradual dose reduction. Clin. Res. 28:214A, 1980.

135. Dunlop, D., and Shanks, R. G.: Selective blockade of adrenoceptive antagonist with partial agonist activity. N. Engl. J. Med. 308:940, 1983.

136. George, R. B., Manocha, K., Burford, J., et al.: Effects of labetalol in hypertensive patients with chronic obstructive pulmonary disease. Chest 83:457, 1983.

137. Benson, M. K., Berrill, W. T., Cruickshank, J. M., and Sterling, G. S.: A comparison of four adrenoceptor antagonists in patients with asthma. Br. J. Clin. Pharmacol. 5:415, 1978.

138. Frohlich, E. D., Tarazi, R. C., and Dustan, H. P.: Peripheral arterial insufficiency: A complication of beta-adrenergic blocking therapy. J.A.M.A. 208:2471, 1969.

139. Lundvall, J., and Jarhult, J.: Beta-adrenergic dilator component of the sympathetic vascular response in skeletal muscle. Acta Physiol. Scand. 96:180, 1976.

140. Simpson, F. O.: β-Adrenergic receptor blocking drugs in hypertension. Drugs 7:85, 1974.

141. Rodger, J., Sheldon, C. D., Lerski, R. A., and Livingston, W. R.: Intermittent claudication complicating beta-blockade. Br. Med. J. 1:1125, 1976.

142. Reveno, W. S., and Rosenbaum, H.: Propranolol hypoglycemia. Lancet 1:920, 1968.

143. Allison, S. P., Chamberlain, M. I., and Miller, J. E.: Effects of propranolol on blood sugar, insulin, and free fatty acids. Diabetologia 5:339, 1969.

144. Deacon, S. P., and Barnett, D.: Comparison of atenolol and propranolol during insulin-induced hypoglycaemia. Br. Med. J. 2:7, 1976.

145. Lloyd-Mostyn, R. H., and Oram, S.: Modification by propranolol of cardiovascular effects of induced hypoglycaemia. Lancet 2:1213, 1975.

146. Frishman, W. H., Razin, A., Swencionis, C., and Sonnenblick, E. H.: Beta-adrenoceptor blockade in anxiety states: A new approach to therapy? Cardiovasc. Rev. Rep. 2:447, 1981.

147. Jacob, H., Brandt, L. J., Farkas, P., and Frishman, W.: Beta-adrenergic blockade and the gastrointestinal system. Am. J. Med. 74:1042, 1983.

148. Stephen, S. A.: Unwanted effects of propranolol. Am. J. Cardiol. 18:463, 1966.

149. Nawabi, I. U., and Ritz, N. D.: Agranulocytosis due to propranolol. J.A.M.A. 223:1376, 1973.

150. Bailey, R., and Neale, T. J.: Rapid clonidine withdrawal with blood pressure overshoot exaggerated by beta-blockade. Br. Med. J. 1:942, 1976.

151. Agabiti-Rosei, E., Brown, J. J., Lever, A. F., et al.: Treatment of phaeochromocytoma and clonidine withdrawal hypertension with labetalol. Br. J. Clin. Pharmacol. (Suppl. 3) 3:809, 1976.

152. Wright, P.: Untoward effect associated with practolol administration. Oculomucocutaneous syndrome. Br. Med. J. 1:595, 1975.

153. Missri, J. C.: How do beta-blockers interact with other commonly used drugs? Cardiovasc. Med. 8:668, 1983.

154. Hansten, P.: Drug interactions, 4th ed. Philadelphia, Lea & Febiger, 1979, pp. 13-24.

4

Adrenergic Receptors as Pharmacological Targets: The Alpha-Adrenergic Blocking Drugs

WILLIAM H. FRISHMAN   *The Albert Einstein College of Medicine, Bronx, New York*

SHLOMO CHARLAP   *Long Island College Hospital, and Downstate Medical Center, Brooklyn, New York*

I. INTRODUCTION

Catecholamines mediate a variety of physiological and metabolic responses in human beings as a result of their interactions with receptors located on the plasma membrane. The differences in the ability of the various catecholamines to stimulate a number of physiological processes are the criteria used to separate these receptors into two distinct types, termed alpha- and beta-adrenergic (1) (see Chap. 1). Alpha-receptor sites respond preferentially to epinephrine and norepinephrine and much less to isoproterenol, whereas beta-receptor sites respond preferentially to isoproterenol and less so to norepinephrine and epinephrine.
 Investigators have documented that beta-adrenergic receptors exist as two discrete subtypes called $beta_1$ and $beta_2$ (2). More recently, it has also been appreciated that there are two subtypes of alpha-receptors, designated $alpha_1$ and $alpha_2$ (3,4). Specific drugs are available that will inhibit or block these receptors (5). In this chapter, we will (1) examine the alpha-adrenergic receptor and the drugs that can inhibit its function, and (2) discuss the rationale for the use and the clinical experience with alpha-adrenergic blocking drugs in the treatment of various cardiovascular and non-cardiovascular disorders.

II. ALPHA-ADRENERGIC RECEPTOR AND ITS INHIBITORS

When a nerve is stimulated, norepinephrine is released from its storage granules in the adrenergic neuron, enters the synaptic cleft, and binds to alpha-receptors on the effector cell, i.e., postsynaptic receptors, to evoke a response; the magnitude of the response depends chiefly on the concentration of norepinephrine in the snaptic cleft (6). A feedback loop exists by which the amount of neurotransmitter released can be regulated: accumulation of norepinephrine in the synaptic cleft leads to stimulation of alpha receptors in the neuronal surface, i.e., presynaptic receptors, and thereby, inhibition of further norepinephrine release (6). Norepinephrine and epinephrine from the circulation can also enter the synaptic cleft and bind to presynaptic or postsynaptic receptors.

In general, alpha-receptors located postsynaptically on the vascular smooth muscle are alpha$_1$ subtypes and help mediate smooth muscle contraction, whereas alpha-receptors located presynaptically on the sympathetic nerve endings are alpha$_2$ subtypes (5,6). Activation of the presynaptic alpha$_2$ receptors inhibits stimulus-induced endogenous norepinephrine release and helps to maintain the local level of sympathetic activity and, hence, vascular tone. It appears that at least in some blood vessels, the postsynaptic receptors are also of the alpha$_2$ subtype, and like the alpha$_1$ receptors, help mediate smooth muscle contraction (5,6). The physiological and pharmacological significance of the observation is not totally clear (6).

Drugs having alpha-adrenergic blocking properties are of several types (Fig. 1):

1. Nonselective alpha-blockers, having prominent effects on both the alpha$_1$ and alpha$_2$ receptors, e.g., the older drugs such as phenoxybenzamine, a noncompetitive antagonist, and phentolamine, a competitive antagonist (7). Phenoxybenzomine is actually 30-100 times more potent at alpha$_1$-receptors (8), but this does not appear to be of clinical significance. While virtually all of the clinical effects of phenoxybenzamine are explicable in terms of alpha blockade (7), this is not the case with phentolamine, which also possesses several other properties, including a direct vasodilator action and sympathomimetic and parasympathomimetic effects (7-9).

2. Selective alpha$_1$-blockers having little affinity for the alpha$_2$-receptor, e.g., prazosin. Originally introduced as a direct acting vasodilator, it now appears clear that prazosin exerts its major effect by reversible blockade of postsynaptic alpha$_1$ receptors (10,11). In ligand binding studies, the drug has been demonstrated to have up to a 5000-fold greater affinity for the alpha$_1$ than the alpha$_2$ receptor (12). High concentrations of the drugs could potentially also inhibit presynaptic alpha$_2$ receptors (11).

3. Selective alpha$_2$-blockers, e.g., yohimbine (5). Its primary use has been as a tool in experimental pharmacology. At present, there are no selective alpha$_2$-blockers in clinical use in the United States.

4. Blockers that inhibit both alpha- and beta-adrenergic receptors, e.g., labetalol. Labetalol, like prazosin, is a selective $alpha_1$-blocker (13,14). As this agent is more potent as a beta-blocker than an alpha-blocker (14,15), it is discussed in greater detail in the chapter on beta-blockers (see Chap. 4).

5. Agents having alpha-adrenergic blocking properties but whose major clinical use is felt unrelated to these properties, e.g., chloropromazine and haloperidol (16). The alpha-adrenergic blockade produced by these agents can lead to orthostatic hypotension, primarily on first use of the drugs.

All the alpha-blockers in clinical use inhibit the postsynaptic $alpha_1$ receptor and result in relaxation of vascular smooth muscle and vasodilation. However, the nonselective alpha-blockers also can antagonize presynaptic $alpha_2$ receptors (5,6,17). This leads to

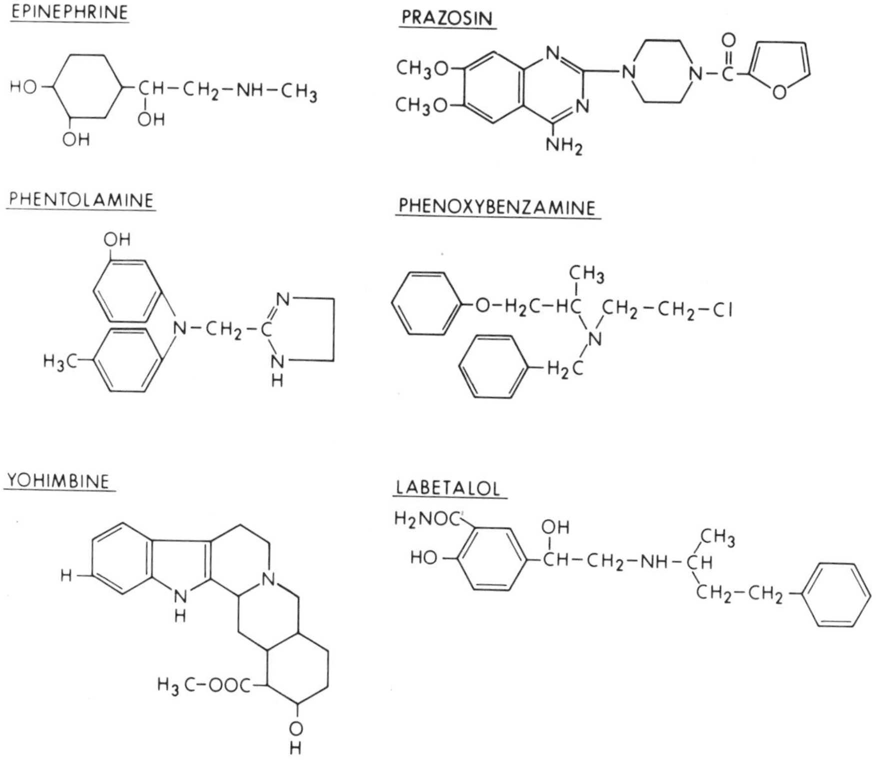

EPINEPHRINE

PRAZOSIN

PHENTOLAMINE

PHENOXYBENZAMINE

YOHIMBINE

LABETALOL

FIGURE 1 Molecular structure of the alpha-adrenergic agonist epinephrine and some alpha-adrenergic blocking drugs.

increased release of neuronal norepinephrine, resulting in (1) attenuation of the desired postsynaptic blockade, and (2) spillover stimulation of the beta receptors, and, consequently, troublesome side effects such as tachycardia and tremulousness and increased renin release. As such, animal studies have demonstrated that phentolamine infusion in doses not producing noteworthy hypotension (which would in itself elicit reflex sympathetic stimulation) results in increases in heart rate, cardiac output, and myocardial contractility, effects consistent with beta-receptor stimulation (18,19). These cardiac effects are not seen in animals treated with reserpine (which prevents norepinephrine release) or in those receiving beta-blockers.

The alpha$_1$-selective agents which preserve the alpha$_2$-mediated presynaptic feedback loop prevent excessive norepinephrine release and thereby avoid the adverse cardiac and systemic effects that would result. Consistent with this postulate, phenoxybenzamine, a less selective alpha-blocker, produces a greater increase in plasma norepinephrine concentration than equivalent vasodepressor doses of prazosin, an alpha$_1$-selective agent (20). Also, while use of prazosin in hypertension produces no change or a decrease in plasma renin activity, use of phenoxybenzamine or phentolamine is uniformly associated with increased renin activity (21).

Because of potent peripheral vasodilatory properties, one would anticipate that selective alpha$_1$-blockers would induce reflex stimulation of the sympathetic and renin-angiotensin system similar to that seen with other vasodilators such as hydralazine and minoxidil. The explanation for the relative lack of tachycardia and renin release observed after prazosin may in part be due to the drug's combined action of reducing vascular tone in both resistance (arteries) and capacitance (veins) beds (18). Such a dual action may prevent the marked increases in venous return and cardiac output observed with agents that act more selectively to reduce vascular tone only in the resistance vessels. In this respect, prazosin resembles nitroprusside: the ratio of arteriolar to venous relaxation induced by oral prazosin is comparable to that induced by intravenous nitroprusside and both drugs have been demonstrated to produce similar hemodynamic changes (22).

The lack of tachycardia with prazosin use has also been attributed by some investigators to a significant negative chronotropic action of the drug independent of its peripheral vascular effect (23,24).

III. CLINICAL APPLICATION OF THE ALPHA-ADRENERGIC BLOCKING DRUGS (Table 1)

A. Hypertension

Increased peripheral vascular resistance is the primary hemodynamic derangement in the majority of patients with long-standing hypertension. Since dilation of constricted arterioles should result in lowering of

TABLE 1 Potential Clinical Applications of Alpha-Adrenergic Blockers

Cardiac disorders	Noncardiac disorders
Hypertension	Pheochromocytoma
Congestive heart failure	Bronchospasm
Angina	Pulmonary hypertension
Cardiac arrhythmias	Shock
	Raynaud's phenomenon
	Benign prostatic hypertrophy

elevated blood pressure, use of alpha-adrenergic blockers in the medical treatment of systemic hypertension has been proposed. Attempts to use the older nonselective alpha-blockers, phenoxybenzamine and phentolamine, in the treatment of hypertension were disappointing. While these drugs were effective in acutely lowering blood pressure, primarily catecholamine-dependent hypertension, their effects were offset by accompanying reflex stimulation of the sympathetic and renin-angiotensin system, resulting in frequent side effects and limited long-term antihypertensive efficacy (23,25). Interest in the use of alpha-antagonists in hypertension was revived with the discovery of prazosin, a selective alpha-blocker whose use elicited little, if any, reflex stimulation.

Therapeutic trials have shown prazosin to be effective in lowering blood pressure in all grades of hypertension: In mild and sometimes moderate hypertension, when used alone; and in moderate and severe hypertension, when used in combination with other agents (26-28). Because its antihypertensive effect is accompanied by little or no increase in heart rate, plasma renin activity, or circulating catecholamines, prazosin has been found useful as a step 1 agent in hypertension. However, although less pronounced than with other vasodilators, prazosin monotherapy does promote sodium and water retention in some patients (29,30). The concomitant use of a diuretic prevents fluid retention, and in many cases, also markedly enhances the antihypertensive effect of the drug (11).

In clinical practice, prazosin has had its widest application as an adjunct to one or more established antihypertensive drugs in treating moderate to severe hypertension (10,11). Prazosin's effects are additive to those of diuretics, beta-blockers, α-methyldopa, and the direct-acting vasodilators (11). The drug causes little change in glomerular filtration rate or renal plasma flow, and can be used safely in patients with severe renal hypertension (29,31). There is no evidence of attenuation of prazosin's antihypertensive effect during chronic therapy (29).

Terazosin, trimazosin and indoramin are new selective $alpha_1$-blockers that have also been found to be effective in the treatment of systemic hypertension. Indoramin produces many unwanted effects, e.g., lethargy and impotence, which may limit its clinical value (32,33). These drugs are not yet available for clinical use in the United States; however, terazosin, a long-acting drug which can be used once daily, will soon be released for marketing.

B. Congestive Heart Failure

The objectives of vasodilator therapy in congestive heart failure are to improve cardiac output by reducing left ventricular impedance and to decrease pulmonary congestion by lowering increased preload. Alpha-adrenergic blocking drugs appear particularly attractive for use in the treatment of heart failure because they hold the possibility of producing balanced reductions in resistance (afterload) and venous tone (preload).

In fact, phentolamine was the first vasodilator to be shown to be effective in the treatment of heart failure (34,35). The drug was infused into normotensive patients with persistent left ventricular dysfunction after a myocardial infarction, and found to induce a significant fall in systemic vascular resistance accompanied by considerable elevation of cardiac output and a reduction in pulmonary artery pressures (35). Phentolamine appears to be less of a venodilator than an arteriolar dilator because it exerts a greater effect on impedance to ventricular emptying than on preload (36). Its effectiveness in the treatment of heart failure has been attributed not only to its alpha-receptor blocking properties but also to at least two other actions of the drug: (1) a direct relaxant effect on vascular smooth muscle, and (2) a mild inotropic effect secondary to release of cardiac norepinephrine stores (37). Because of its high cost and the frequent side effects that it produces, especially tachycardia, phentolamine is no longer used in the treatment of heart failure. Oral phenoxybenzamine has also been of use as vasodilator therapy in heart failure (38). Like phentolamine, phenoxybenzamine has been replaced by newer vasodilator agents.

Studies evaluating the acute hemodynamic effects of prazosin in patients with congestive heart failure consistently find significant reductions in systemic and pulmonary vascular resistances and left ventricular filling pressures associated with increases in stroke volume (11,24,39,40). In most studies, there is no change or a decrease in heart rate (11). The response pattern seen with prazosin is similar to that observed with nitroprusside with the exception that heart rate tends to be higher with the use of nitroprusside and, therefore, the observed increases in cardiac output are also higher (24).

Controversy still exists as to whether the initial clinical and hemodynamic improvements seen with prazosin are sustained during long-term therapy (11,41). Whereas some studies have demonstrated continued efficacy of prazosin therapy after chronic use (42,43), others have found little hemodynamic difference between prazosin and placebo-treated patients (44,45). Some investigators believe that whatever tolerance to the drug does develop is most likely secondary to activation of counterposing neurohumoral forces (11,46,47): If the dose is raised and the tendency toward sodium and water retention is countered by appropriate increases in diuretic dose, prazosin is likely to remain effective. Others argue that sustained increases in plasma renin activity or plasma catecholamines are not seen during long-term therapy and that tolerance is not prevented or reversed by a diuretic (41,45,48). Further study may not resolve this controversy. What does appear clear is the need to individually evaluate patients as to the continued efficacy of their prazosin therapy. Whether there are subgroups of patients with heart failure, e.g., those with highly activated sympathetic nervous systems, who are more likely to respond to prazosin or other alpha-blockers remains to be determined (11,49).

Trimazosin, another selective alpha$_1$-blocker, produces hemodynamic effects in patients with heart failure similar to those of prazosin (50,51). The overall experience with the drug in heart failure is limited. Indoramin, also a selective alpha$_1$-blocker, is now being evaluated in the United States for clinical use in congestive heart failure.

C. Angina

1. Variant Angina

Alpha-adrenergic receptors help mediate coronary vasoconstriction (52). It has been suggested that a pathological alteration of the alpha-adrenergic system may be the mechanism of coronary spasm in some patients with variant angina (53). In uncontrolled studies, the administration of alpha-adrenergic blockers, both acutely and chronically, has been shown to be effective in reversing and preventing coronary spasm (53-55). The addition of prazosin to six patients with refractory variant angina abolished all attacks in four patients, markedly reduced the frequency of attacks in one, and had to be discontinued because of hypotension in a sixth (55). However, a recent long-term, randomized, double-blind trial found prazosin to exert no obvious beneficial effect in patients with variant angina (56). The demonstration of an important role for the postsynaptic alpha$_2$ receptors in determining coronary vascular tone may help explain prazosin's lack of efficacy (57). Further study in this area is anticipated.

2. Chronic Stable Angina

Investigators have also suggested the use of alpha-blockers in the treatment of angina of effort (58). It would appear that in patients with chronic angina, the alpha-blockers have a greater potential of worsening myocardial ischemia, both by their hypotensive properties and the cardiac stimulation they elicit (59).

D. Cardiac Arrhythmias

Investigators have postulated that enhanced alpha-adrenergic responsiveness occurs during myocardial ischemia and that it is a primary mediator of the electrophysiological derangements and resulting malignant dysrhythmias induced by catecholamines during myocardial ischemia and reperfusion (60). Supporting this hypothesis is the demonstration of an increase in alpha-adrenergic receptors in ischemic myocardium 30 min after coronary occlusion (61). The use of the alpha-blockers phentolamine and prazosin has been shown in cats to abolish ventricular fibrillation induced by either coronary occlusion or subsequent reperfusion (60). In man, there are favorable reports with the use of an alpha-blocker in the treatment of supraventricular and ventricular ectopy (62,63). In a double-blind study of 39 patients with uncomplicated myocardial infarction, the use of phentolamine afforded a highly significant protection from ventricular and supraventricular premature beats (63). The exact role of alpha-adrenergic blockers in the treatment of cardiac arrhythmias will be determined through further clinical study.

E. Pheochromocytoma

Alpha receptor blockers have been used in the treatment of pheochromocytoma to control the peripheral effects of the excess catecholamines, e.g., hypertension, hemoconcentration, and decreased vascular space (64,65). In fact, intravenous phentolamine has been used as a test for this disorder, but the test is now rarely done because of reported cases of cardiovascular collapse and death in patients who exhibited exaggerated sensitivity to the drug (7,66). Sudden expansion of the vascular space as the vasoconstriction is counteracted is felt to be the primary cause of the profound hypotension and hemodynamic deterioration seen with the alpha-blockers; appropriate volume infusions before initiating therapy can usually prevent the hypotensive response. Intravenous phentolamine is still rarely used in cases of pheochromocytoma-related hypertensive crisis. For long-term therapy, oral phenoxybenzamine is the preferred agent.

Beta-blocking agents may also be needed in pheochromocytoma for the control of tachycardia and arrhythmias. All beta-blockers, but primarily the nonselective agents, should not be initiated prior to adequate alpha blockade, since severe hypertension may occur as a result of the unopposed alpha-stimulating activity of the circulating catecholamines (65). Although there are reports of the use of labetalol, an alpha- and beta-blocker, as monotherapy in the treatment of pheochromocytoma, it would appear that like other beta-blockers, the drug should be avoided unless alpha-adrenergic blockers have already been given (14).

F. Shock

A common denominator in shock is hyperactivity of the sympathetic nervous system as a compensatory reflex response to reduced blood pressure. With persistent reduction of cardiac output and a decrease in circulatory blood volume, the adrenergic response continues inappropriately to a point of severe ischemia in the viscera (67). Effective exchange of metabolites with tissue cells is prevented because of constriction of the microcirculatory vessels.

Use of alpha-blockers in shock has been advocated as a means of lowering peripheral vascular resistance and increasing vascular capacitance while not antagonizing the cardiotonic effects of the sympathomimetic amines, e.g., norepinephrine and dopamine, which are frequently administered to patients in shock (67-69). A lowered vascular resistance will result in redistribution of blood to the viscera and a decreased external pressure workload on the heart. Lower capillary hydrostatic pressures and opening of precapillary sphincters will allow perfusion of nutritional capillaries. Although investigated for many years for the treatment of shock, alpha-blockers still have not been approved for this purpose. Clinical trials evaluating these drugs in the treatment of shock have met with limited success (67,70, 71). A prime concern of the use of alpha-blockers in shock is that the rapid drug-induced increase in vascular capacitance may lead to inadequate cardiac filling and profound hypotension, especially in the hypovolemic patient (67). Adequate amounts of fluid replacement prior to the use of an alpha-blocker can minimize this concern.

G. Lung Disease

1. Bronchospasm

Bronchoconstriction may be mediated in part through catecholamine stimulation of alpha receptors in the lung (5,72). In the normal bronchial tree, the balance in the alpha-adrenergic and $beta_2$-adrenergic system appears to favor $beta_2$-activity (73). However, in patients with allergic asthma, it has been postulated that a deficient beta-adrenergic system or an enhanced alpha-adrenergic responsiveness could result in alpha-adrenergic activity being the main mechanism of bronchoconstriction (73,74). Several studies have shown bronchodilation or inhibition of histamine, allergen, or exercise-induced bronchospasm with a variety of alpha-blockers (75-77). In a double-

blind crossover study, prazosin used by inhalation in 10 asthmatic
children significantly reduced the severity of postexercise bronchoconstriction (77). Additional studies are needed to define more fully
the role of alpha-blockers for use as bronchodilators.

2. Pulmonary Hypertension

The part played by endogenous circulating catecholamines in the
maintenance of pulmonary vascular tone appears to be minimal. Studies
evaluating the effects of norepinephrine administration on pulmonary
vascular resistance have found the drug to have little or no effect (78).
However, phentolamine has been demonstrated to produce significant
reductions in pulmonary artery pressure and pulmonary vascular
resistance (78,79). In a patient with primary pulmonary hypertension,
intravenous phentolamine markedly attenuated the pulmonary hypertensive response to exercise; chronic oral therapy with phentolamine
produced sustained improvements in symptoms and exercise tolerance
(80). Whatever beneficial effects phentolamine and other alpha-blockers,
e.g., tolazoline (81), have on the pulmonary circulation is most likely
primarily owing to their direct vasodilatory actions rather than to alpha
blockade. Like other vasodilators, in patients with pulmonary hypertension due to fixed anatomical vascular changes, alpha-blockers can
produce hemodynamic deterioration secondary to their systemic vasodilatory properties (82).

H. Arterioconstriction

Oral alpha-adrenergic blockers can produce subjective and clinical
improvement in patients experiencing episodic arterioconstriction (Raynaud's phenomenon) (83,84), but the improvement may only be partial
and patients may develop early tolerance to the drugs' effects (84).
Symptomatic arteriospasm with intense pallor or cyanosis that fails to
clear within several hours may respond to intravenous phentolamine.
Alpha-blockers may also be of value in the treatment of severe peripheral ischemia caused by an alpha agonist, e.g., norepinephrine or
ergotomine overdose (85). In cases of inadvertent infiltration of a
norepinephrine infusion, phentolamine can be given intradermally to
avoid tissue sloughing (17).

I. Benign Prostatic Obstruction

Alpha-adrenergic receptors have been identified in the bladder neck
and prostatic capsule of male patients (86). It has been suggested
that urinary retention in benign prostatic hyperplasia may be caused
partly by stimulation of prostatic alpha-adrenoceptors. Supporting
this view, favorable results have been reported during treatment with
alpha-blockers (87-89). In a double-blind crossover study in 20
patients with benign prostatic obstruction, use of prazosin resulted

in increased urinary flow rates and reductions in residual volume and obstructive symptoms (89). Alpha-blockers appear to have an important role to play in the medical treatment of patients with benign prostatic obstruction.

IV. CLINICAL USE

Oral phenoxybenzamine has a rapid onset of action with the maximal effect from a single dose seen in 1-2 hr (90). The gastrointestinal absorption is incomplete and only 20-30% of an oral dose reaches the systemic circulation in active form. The half-life of the drug is 24 hr, with the usual dose varying between 20 and 200 mg/day in one or two doses (91). Intravenous phentolamine is initially administered at 0.1 mg/min and is then increased at increments of 0.1 mg/min every 5-10 min until the desired hemodynamic effect is reached. The drug has a short duration of action of 3-10 min. Little is known about the pharmacokinetics of long-term oral use of phentolamine (91). The main side effects of the drugs include postural hypotension, tachycardia, gastrointestinal disturbances, and sexual dysfunction (90,91). Intravenous infusion of norepinephrine can be used to combat severe hypotensive reactions. Oral phenoxybenzamine is approved for use in pheochromocytoma.

Prazosin is almost completely absorbed following oral administration, with peak plasma levels being achieved at 2-3 hr (92). The drug is 90% protein bound. Prazosin is extensively metabolized in the liver. The usual half-life of the drug is 2.5-4.0 hr (25), in patients with heart failure, the half-life increases to the range of 5-7 hr (93).

The major side effect of prazosin is the first-dose phenomenon: severe postural hypotension occasionally associated with syncope seen after the initial dose or after a rapid dose increment (10,11). The reason for this phenomenon has not been clearly established, but it may involve the rapid induction of venous and arteriolar dilation by a drug that elicits little reflex sympathetic stimulation (10). It is seen more widely when the drug is administered as a tablet rather than a capsule, possibly related to the variable bioavailability or rates of absorption of the two formulations (10). (In the United States, the drug is available in capsule form.) The postural hypotension can be minimized if the initial dose of prazosin is not higher than 1 mg and if it is given at bed time. In treating hypertension, a dose of 2-3 mg/day should be maintained for 1-2 weeks, followed by a gradual increase in dosage titrated to achieve the desired reductions in pressures, usually up to 20-30 mg/day, given in two or three doses. In treating heart failure, larger doses (2-7 mg) may be used to initiate therapy in recumbent patients, but the maintenance dose is also usually not more than 30 mg (10). Higher doses do not seem to produce additional clinical improvement (49).

Other side effects of prazosin include dizziness, headache, and drowsiness (25). The drug produces no deleterious effects on the clinical course of diabetes mellitus, chronic obstructive pulmonary disease, renal failure, or gout. It does not adversely affect the lipid profile (94). Prazosin is presently approved for use in hypertension.

REFERENCES

1. Ahlquist, R. P.: Study of the adrenotropic receptors. Am. J. Physiol. 153:586, 1948.

2. Lands, A. M., Luduena, F. P., and Buzzo, H. J.: Differentiation of receptor systems responsive to isoproterenol. Life Sci. 6:2241, 1967.

3. Berthelsen, S., and Pettinger, W. A.: A functional basis for classification of alpha-adrenergic receptors. Life Sci. 21:596, 1977.

4. Langer, S. Z.: Presynaptic regulation of the release of catechol-amines. Pharmacol. Rev. 32:337, 1981.

5. Andersson, K. E.: Drugs blocking adrenoceptors. Acta Med. Scand. (Suppl.) 665:9, 1982.

6. Langer, S. Z., and Armstrong, J. M.: Prejunctional receptors and the cardiovascular system: Pharmacological and therapeutic relevance. In: Cardiovascular Pharmacology, 2nd ed., M. Antonaccio (ed.). New York, Raven, 1984, p. 197.

7. Oates, J. A., Robertson, D., Wood, A. J. J., Woosley, R. L.: Alpha and beta-adrenergic agonists and antagonists. In: Cardiac Therapy. M. R. Rosen and B. F. Hoffman (eds.). Boston, Martinus Nijhoff, 1983, p. 145.

8. Cubeddu, L. X., Barnes, E. M., Langer, S. Z., and Weiner, N.: Release of norepinephrine and dopamine-B-hydroxylase by nerve stimulation. I. Role of neuronal and extraneuronal uptake and of alpha presynaptic receptors. J. Pharmacol. Exp. Ther. 190: 431, 1975.

9. Gould, L., and Reddy, C. V. R.: Phentolamine. Am. Heart J. 92:392, 1976.

10. Graham, R. M., and Pettinger, W. A.: Prazosin. N. Engl. J. Med. 300:232, 1979.

11. Colucci, W. S.: Alpha-adrenergic receptor blockade with prazosin. Consideration of hypertension, heart failure and potential new applications. Ann. Intern. Med. 97:67, 1982.

12. U'Prichard, D. C., Charness, M. E., Robertson, D., and Snyder, S. H.: Prazosin: Differential affinities for two populations of α-noradrenergic receptor binding sites. Eur. J. Pharmacol. 50:87, 1978.

13. Levy, G. P., and Richards, D. A.: Labetalol. In: *Pharmacology of Antihypertensive Agents.* A. Scrabine (ed.). New York, Raven, 1980, p. 325.

14. Frishman, W. H., MacCarthy, P. E., Kimmel, B., Lazar, E., Michelson, E. L., and Bloomfield, S. S.: In: *Clinical Pharmacology of the β-Adrenoceptor Blocking Drugs,* 2nd ed. Norwalk, Connecticut, Appleton-Century-Crofts, 1984, p. 205.

15. Brittian, R. T., and Levy, G. P.: A review of the animal pharmacology of labetalol, a combined α- and β-adrenoceptor blocking drug. Br. J. Pharmacol. (Suppl. 3) 3:681, 1976.

16. Nickerson, M.: The pharmacology of adrenergic blockade. Pharmacol. Rev. 1:27, 1949.

17. Hoffman, B. B., and Lefkowitz, R. J.: Alpha-adrenergic receptor subtypes. N. Engl. J. Med. 302:1390, 1981.

18. Singh, J. B., Hood, W. B. Jr., and Abelman, W. H.: Beta-adrenergic mediated inotropic and chronotropic actions of phentolamine. Am. J. Cardiol. 26:660, 1970 (abstract).

19. Das, P. K., and Pratt, J. R.: Myocardial and hemodynamic effects of phentolamine. Br. J. Pharmacol. 41:437, 1971.

20. Graham, R. M., Kennedy, P., Stephenson, W., et al.: In vivo evidence for the presynaptic α-adrenergic receptor controlling norepinephrine release. Fed. Proc. 37:308, 1978.

21. Keeton, T. K., and Campell, W. B.: Control of renin release and its alteration by drugs. In: *Cardiovascular Pharmacology,* 2nd ed. M. Antonaccio (ed.). New York, Raven, 1984, p. 65.

22. Awan, N. A., Miller, R. R., and Mason, D. T.: Comparison of effects of nitroprusside and prazosin on left ventricular function and the peripheral circulation in chronic refractory heart failure. Circulation 57:152, 1978.

23. Ribner, H. S., Bresnahan, D., Hsieh, A. M., et al.: Acute hemodynamic responses to vasodilation therapy in congestive heart failure. Prog. Cardiovasc. Dis. 25:1, 1982.

24. Packer, M., Meller, J., Gorlin, R., and Herman, M. V.: Differences in the hemodynamic effects of nitroprusside and prazosin in severe chronic congestive heart failure: Evidence for a direct negative chronotropic effect of prazosin. Am. J. Cardiol. 44:310, 1979.

25. Rand, M. J., McCulloch, W. W., and Story, D. F.: Prejunctional modulation of noradrenergic transmission by noradrenaline, dopamine and acetylcholmine. In: *Central Action of Drugs in Blood Pressure Regulation.* Proceedings of an International Symposium on Central Action of Drugs in the Regulation of Blood Pressure, London, Pitman, 1975, p. 94.

26. Brogden, R. N., Heel, R. C., Speight, T. M., and Avery, G. S.: Prazosin: A review of its pharmacological properties and therapeutic efficacy in hypertension. Drugs 14:163, 1977.

27. Colucci, W. S.: New developments in alpha-adrenergic receptor pharmacology: Implications for the initial treatment of hypertension. Am. J. Cardiol. 51:639, 1983.

28. Okun, R.: Effectiveness of prazosin as initial antihypertensive therapy. Am. J. Cardiol. 51:646, 1983.

29. Bauer, J. H., Jones, L. B., and Gaddy, P.: Effects of prazosin therapy on blood pressure, renal function and body fluid composition. Arch. Intern. Med. 144:1196, 1981.

30. Koshy, M. C., Mickley, D., Bourgoignie, J., et al.: Physiologic evaluation of a new antihypertensive agent: Prazosin HCl. Circulation 55:533, 1977.

31. Gunnels, J. C., Jr.: Treating the patient with mild hypertension and renal insufficiency. Am. J. Cardiol. 51:651, 1983.

32. Lewis, P. J., George, C. F., and Dollery, C. T.: Clinical evaluation of indoramin, a new antihypertensive agent. Eur. J. Clin. Pharmacol. 6:211, 1973.

33. Gould, B. A., Mann, S., Davies, A. B., Altman, D. G., and Raftery, E. B.: α-Adrenoreceptor blockade with indoramin in hypertension. J. Cardiovasc. Pharmacol. 5:343, 1983.

34. Gould, L., Zahir, M., and Ettingers, S.: Phentolamine and cardiovascular performance. Br. Heart J. 31:154, 1969.

35. Majid, P. A., Sharma, B., and Taylor, S. H.: Phentolamine for vasocilator treatment of severe heart failure. Lancet 2:719, 1971.

36. Williams, D. O., Hilliard, G. K., Cantor, S. A., Miller, R., and Mason, D. T.: Comparative mechanisms of ventricular unloading by systemic vasodilator agents in therapy of cardiac failure: Nitro-prusside versus phentolamine. Am. J. Cardiol. 35:177, 1975.

37. Taylor, S. H., Sutherland, G. R., MacKenzie, G. J., Stannton, H. P., and Donald, K. W.: The circulatory effects of intravenous phentolamine in man. Circulation 31:741, 1965.

38. Kovick, R. B., Tillisch, J. H., Berens, S. C., Bramowitz, A. D., and Shine, K. I.: Vasodilator therapy for chronic left ventricular failure. Circulation 53:322, 1976.

39. Miller, R. R., Awan, N. A., Maxwell, K. S., and Mason, D. T.: Sustained reduction of cardiac impedance and preload in congestive heart failure with the antihypertensive vasodilator prazosin. N. Engl. J. Med. 297:303, 1977.

40. Feldman, R. C., Ball, R. M., Winchester, M. A., Jaillon, P., Kates, R. E., and Harrison, D. C.: Beneficial hemodynamic response to chronic prazosin therapy in congestive heart failure. Am. Heart J. 101:534, 1981.

41. Packer, M.: Vasodilator and inotropic therapy for severe chronic heart failure: Passion and skepticism. J. Am. Coll. Cardiol. 2: 841, 1983.

42. Arnow, W., Lurie, M., Turbow, M., Whittaker, K., van Camp, S., and Hughes, D.: Effect of prazosin versus placebo on chronic left ventricular heart failure. Circulation 59:344, 1979.

43. Colucci, W. S., Wynne, J., Holman, B. L., and Braunwald, E.: Long-term therapy of heart failure with prazosin: A randomized double-blind trial. Am. J. Cardiol. 45:337, 1980.

44. Harper, R. W., Claxton, H., Anderson, S., and Pitt, A.: The acute and chronic hemodynamic effects of prazosin in severe congestive heart failure. Med. J. Aust. (Suppl.) 2:36, 1980.

45. Markham, R. V., Corbett, J. R., Gilmore, A., Pettinger, W. A., and Firth, B. G.: Efficacy of prazosin in the management of chronic heart failure: A 6 month randomized, double-blind, placebo-controlled study. Am. J. Cardiol. 51:1346, 1983.

46. Awan, N. A., Needham, K. E., Evenson, M. K., Amsterdam, E. A., and Mason, D. T.: Therapeutic application of prazosin in chronic refractory heart failure tolerance and "tachyphylaxis" in perspective. Am. J. Med. 71:153, 1981.

47. Stein, L., Henry, D. P., and Weinberger, M. H.: Increase in plasma norepinephrine during prazosin therapy for chronic congestive heart failure. Am. J. Med. 70:825, 1981.

48. Packer, M., Meller, J., Medina, N., Yushak, M., and Gorlin, R.: Serial hemodynamic studies indicate that early tolerance to prazosin in heart failure is not reversible. Circulation (Suppl. II.) 66: II-210, 1982 (abstract).

49. Packer, M., and LeJemtel, T. H.: Physiologic and pharmacologic determinants of vasodilator response: A conceptual framework for national drug therapy for chronic heart failure. Prog. Cardiovasc. Dis. 24:275, 1982.

50. Awan, N. A., Hermanovich, J., Whitcomb, C., Skinner, P., and
    Mason, D. T.: Cardiocirculatory effects of afterload reduction with
    oral trimazosin in severe chronic congestive heart failure. Am. J.
    Cardiol. 44:126, 1979.

51. Weber, K. T., Kinasewitz, G. T., West, J. S., Janicki, J. S.,
    Rerchek, N., and Fishman, A. P.: Long-term vasodilator therapy
    with trimazosin in chronic cardiac failure. N. Engl. J. Med. 303:
    242, 1980.

52. Orlick, A. E., Ricci, D. R., Cipriano, P. R., Guthaner, D. F.,
    Alderman, E. L., and Harrison, D. C.: The contribution of
    alpha-adrenergic tone to resting coronary vascular resistance in
    man. J. Clin. Invest. 62:459, 1978.

53. Ricci, D. R., Orlick, A. E., Cipriano, P. R., Guthaner, D. F.,
    and Harrison, D. C.: Altered adrenergic activity in coronary
    arterial spasm: Insight into mechanism based on study of coronary
    hemodynamics and the electrocardiogram. Am. J. Cardiol. 43:
    1073, 1979.

54. Levine, D. L., Freeman, M. R.: Alpha-adrenoceptor mediated
    coronary artery spasm. J.A.M.A. 236:1018, 1976.

55. Tzivoni, D., Keren, A., Benhorin, J., Gottlieb, S., Atlas, D.,
    and Stern, S.: Prazosin therapy for refractory variant angina.
    Am. Heart J. 105:262, 1983.

56. Winniford, M. D., Flipchuk, N., and Hillis, D. L.: Alpha-
    adrenergic blockade for variant angina: A long-term double-blind
    randomized trial. Circulation 67:1185, 1983.

57. Bassenage, E., Holtz, J., Sommer, O., and Saeed, M.: Vascular
    α_2-adrenoceptors mediate coronary constrictions induced by
    norepinephrine and by sympathetic nerve stimulation. Circulation
    (Suppl. II) 66:II-153, 1982 (abstract).

58. Gould, L., Reddy, C. V. R., and Gomprecht, R. F.: Oral phentol-
    amine in angina pectoris. Jpn. Heart J. 14:393, 1973.

59. Charness, M. E., Fishman, J. A., and Robertson, D.: Exacerba-
    tion of angina pectoris by prazosin. South. Med. J. 72:1213,
    1979.

60. Sheridan, D. J., Penkoske, P. A., Sobel, B. E., and Corr, P. B.:
    Alpha-adrenergic contributions to dysrhythmia during myocardial
    ischemia and reperfusion in cats. J. Clin. Invest. 65:161, 1980.

61. Corr, P. B., Shayman, J. A., Kramer, J. B., and Kipnis, R. J.:
    Increased α-adrenergic receptors in ischemic cat myocardium: A
    potential mediator of electrophysiologic derangements. J. Clin.
    Invest. 67:1232, 1981.

62. Gould, L., Gomprecht, R. F., and Zahir, M.: Oral phentolamine for treatment of ventricular premature contractions. Br. Heart J. 33:101, 1971.

63. Gould, L., Reddy, C. V. R., Weinstein, T., and Gomprecht, R. F.: Antiarrhythmic prophylaxis with phentolamine in acute myocardial infarction. J. Clin. Pharmacol. 15:191, 1975.

64. Sjoerdsma, A., Engelman, K., Waldmann, T. A., Cooperman, L. H., and Hammond, W. G.: Pheochromocytoma: Current concepts of diagnosis and treatment. Ann. Intern. Med. 65:1302, 1966.

65. Manger, W. M., and Gifford, R. W.: *Pheochromocytoma*. New York, Springer-Verlag, 1977, p. 304.

66. Roland, C. R.: Pheochromocytoma in pregnancy: Report of a fatal reaction to phentolamine (Regitine) methanesulfonate. J.A.M.A. 171:1806, 1959.

67. Bloch, J. H., Pierce, L. H., and Lillehe, R. C.: Adrenergic blocking agents in the treatment of shock. Ann. Rev. Med. 17: 483, 1966.

68. Wilson, R. F.: Combined use of norepinephrine and dibenzyline in clinical shock. Surg. Forum 15:30, 164.

69. Goldberg, L. I., Talley, R. C., and McNay, J. C.: The potential role of dopamine in the treatment of shock. Prog. Cardiovasc. Dis. 12:40, 1969.

70. DaLuz, P. L., Shubin, H., and Weil, M. H.: Effectiveness of phentolamine for reversal of circulatory failure. Crit. Care Med. 1:135, 1973.

71. Honston, M. C., Thompson, W. L., and Robertson, D.: Shock. Arch. Intern. Med. 144:1433, 1984.

72. Mathe, A. A., Astrom, A., and Persson, N. A.: Some broncho-constricting and bronchodilating responses of the isolated human bronchi: Evidence for the existence of alpha-adrenoceptors. J. Pharmacol. 23:905, 1971.

73. Shiner, R. J., and Molho, M. I.: Comparison between an α-adrenergic antagonist and a β_2-adrenergic agonist in bronchial asthma. Chest 83:602, 1983.

74. Henderson, W. R., Shelhamer, J. H., Reingold, D. B., Smith, L. J., Evan, R., and Kaliner, M.: Alpha-adrenergic hyper-responsiveness in asthma. N. Engl. J. Med. 300:642, 1979.

75. Gaddie, L., Legge, J. S., Petrie, G., and Palmer, K. N. V.: The effect of an alpha-adrenergic receptor blocking drug on histamine sensitivity in bronchial asthma. Br. J. Dis. Chest 66: 141, 1972.

76. Marcelle, R.: Traitment de l'etat de mal asthmatique par la phentolamine. Acta Allereg. 24:432, 1969.

77. Barnes, P. J., Wilson, N. M., and Vickers, H.: Prazosin, an alpha$_1$-adrenoceptor antagonist, partially inhibits exercise-induced asthma. J. Allergy Clin. Immunol. 68:411, 1981.

78. Taylor, S. H., MacKenzie, G. J., George, M., and McDonald, A.: Effects of adrenergic blockade on the pulmonary circulation in man. Br. Heart J. 27:627, 1965.

79. Gould, L., DeMartino, A., Gomprecht, R. F., Umali, T., and Michael, A.: Hemodynamic effects of phentolamine in cor pulmonale. J. Clin. Pharmacol. 12:153, 1972.

80. Ruskin, J. N., and Hutter, A. M.: Primary pulmonary hypertension treated with oral phentolamine. Ann. Intern. Med. 90:772, 1979.

81. Grossman, W., Alpert, J. S., and Braunwald, E.: Pulmonary hypertension. In: *Heart Disease*. E. Braunwald (ed.). Philadelphia, Saunders, 1984, p. 823.

82. Cohen, M. L., and Kronzon, I.: Adverse hemodynamic effects of phentolamine in primary and pulmonary hypertension. Ann. Intern. Med. 95:591, 1981.

83. Harper, F. E., and Leroy, E. C.: Raynaud's phenomenon: An update on treatment. J. Cardiovasc. Med. 7:282, 1982.

84. Nielsen, S. L., Vitting, K., and Rasmussen, K.: Prazosin treatment of primary Raynaud's phenomenon. Eur. J. Clin. Pharmacol. 24:421, 1983.

85. Cobaugh, D. S.: Prazosin treatment of ergotamine-induced peripheral ischemia. J.A.M.A. 244:1360, 1980.

86. Caine, M., Raz, S., and Ziegler, M.: Adrenergic and cholinergic receptors in the human prostate, prostatic capsule and bladder neck. Br. J. Urol. 47:193, 1975.

87. Caine, M., Perlberg, S., and Meretyk, S.: A placebo-controlled double-blind study of the effect of phenoxybenzamine in benign prostatic obstruction. Br. J. Urol. 50:551, 1978.

88. Abrams, P. H., Shah, P. J. R., and Choa, R. G.: Bladder outflow obstruction treated with phenoxybenzamine. Br. J. Urol. 54:527, 1982.

89. Hedlund, H., Andersson, K. E., and Ek, A.: Effects of prazosin in patients with benign prostatic obstruction. J. Urol. 130:275, 1983.

90. Westfall, D. P.: Adrenoceptor Antagonists. In: *Modern Pharmacology*. C. R. Craig and R. E. Zitzel (eds.). Boston, Little, Brown, 1982, p. 141.

91. Nickerson, M., and Collier, B.: Drugs inhibiting adrenergic nerves and structures innervated by them. In: *The Pharmacological Basis of Therapeutics*, 5th ed., L. S. Goodman, and A. Gilman (eds.). New York, Macmillan, 1975, p. 533.

92. Bateman, D. N., Hobbs, D. C., Twomey, T. M., Stevens, E. A., and Rawlins, R. M.: Prazosin, pharmacokinetics and concentration effect. Eur. J. Clin. Pharmacol. 16:177, 1979.

93. Wood, A. J., Bolli, P., and Simpson, F. O.: Prazosin in normal subjects: Plasma levels, blood pressure and heart rate. Br. J. Clin. Pharmacol. 3:199, 1976.

94. Culter, R.: Effect of antihypertensive agents on lipid metabolism. Am. J. Cardiol. 51:628, 1983.

5

In Vitro Methods for Studying Human Adrenergic Receptors: Methods and Applications

HARVEY J. MOTULSKY and PAUL A. INSEL   *University of California, San Diego, La Jolla, California*

I. INTRODUCTION

The catecholamines epinephrine and norepinephrine exert their effects on target cells only after binding to adrenergic receptors on the cell surface. These receptors are thus key loci of regulation and pathology in the adrenergic system. In this chapter, we will discuss the methods used to study adrenergic receptors in vitro. Several methods are available (Table 1). Because indirect methods have been available for many years, and previous authors have discussed techniques and problems with such methods, we will devote most of this chapter to the use of radioligand binding, the one direct method currently available for investigating adrenergic receptors in vitro. Other detailed descriptions of methodology for radioligand binding are available (1-3), and we will therefore explain the methods only briefly and will devote more space to discussing some of the problems inherent in these techniques. Finally, we will speculate about new methods that may become more useful in the future.

II. HOW TO PROVE THAT A RADIOLIGAND BINDS TO ADRENERGIC RECEPTORS

The underlying principle of radioligand binding is straightforward. A radioactively labeled compound (radioligand) selective for the receptor of interest is incubated with the tissue, unbound radioligand is removed by filtration or centrifugation, and the radioactivity asso-

TABLE 1 Methods for Studying Human Adrenergic Receptors in Vitro

Direct
  Radioligand binding

Indirect
  Second-messenger generation
  Tissue-specific responses, such as muscle contraction (alpha, beta),
  platelet aggregation/secretion (alpha$_2$), neutrophil and lymphocyte
  function (beta$_2$), lipolysis (alpha$_2$, beta$_1$)

___

ciated with the tissue is counted as a measure of radioligand bound
to receptors. Before such an assay can be useful, it is necessary to
establish optimal experimental conditions and to prove that tissue-
associated radioactivity represents radioligand bound to receptors,
and not radioligand (or derivatives) bound to other types of receptors,
transport proteins, metabolic enzymes, or nonspecific sites. The
necessary experiments are designed to answer the following questions:

1. *How is nonspecific binding determined?* Radioligands invariably
bind not only to the "specific" receptors but also to other "nonspecific"
sites. The amount of nonspecific binding is usually quantitated by
incubating the radioligand with the tissue in the presence of an excess
of an unlabeled compound that binds to the specific receptor sites,
usually 1 µM (-)-propranolol or 0.1-1.0 mM (-)-isoproterenol for
studies of beta-adrenergic receptors, and 10 µM phentolamine or 100
µm (-)-epinephrine or (-)-norepinephrine for studies of alpha-
adrenergic receptors. Under these conditions, the unlabeled compound
rather than the radioligand occupies the receptors, and bound radioligand is therefore considered to be on nonspecific sites. When one
subtracts this nonspecific binding from radioligand bound in the
absence of competing compounds (total binding), one obtains an
estimate of the specific radioligand binding to the receptors. Clearly,
the choices of which compound and which concentration of that compound to use to define nonspecific binding are crucial. One wants to
use enough of the unlabeled compound to occupy virtually all the
receptors, but not enough to cause more widespread perturbations
to the membranes. When inappropriate choices are made, the resulting
specific binding may overestimate or underestimate the radioligand
binding to receptors.

Inevitably, the definition of nonspecific binding can be somewhat
circular. One can only be confident of one's choice for the initial
definition of specific binding when later experiments yield consistent
and appropriate results. Ideally, identical results will be obtained
when nonspecific binding is defined using appropriate concentrations

of several different unlabeled compounds. In molecular terms, non-specific binding in target cell membranes is low-affinity binding of the radioligand to nonreceptor sites. This low-affinity binding does not saturate over the typical concentration range of radioligand tested, and is usually proportional to tissue and radioligand concentrations. As operationally defined above, however, nonspecific binding also includes background radioactivity and radioligand binding to receptors not recognized by the unlabeled compound. It may also include binding to the filters (used in filtration assays) or test tubes (used in centrifugation assays).

2. *Does the radioligand bind to the adrenergic receptors reversibly and with appropriate kinetics?* Many properties of adrenergic receptors are adequately described by a simple model, the law of mass action:

$$R + L \rightleftarrows RL$$

Here the unbound adrenergic receptors (R) reversibly bind the ligand (L) to form the receptor-ligand complex (RL). This model is the basis of most analyses of radioligand-binding data. Equilibrium is reached when the rate at which the receptor-ligand complex forms is equal to the rate at which it dissociates. Kinetic experiments are used to measure these rates and to demonstrate that the radioligand binding follows the model.

One measures the rate of association (the "on-rate") of radioligand and receptor by incubating a radioligand with an adrenergic receptor-containing tissue and measuring the amount of specific binding at various times thereafter. The binding gradually reaches a plateau as the formation of RL from radioligand and receptors achieves equilibrium (or steady state). The time required for equilibrium to be established depends in part on the concentrations of radioligand and receptor used, higher concentrations leading to more rapid achievement of equilibrium. Typically, adrenergic radioligands have association rate constants of 10^7 - 10^9 M^{-1} min^{-1}, and under the usual conditions assays are conducted (typically at 25-37°), equilibrium is reached in 30-90 min.

Visual inspection of data in an on-rate experiment allows one to estimate how long it takes for the binding reaction to reach apparent equilibrium. This information is then useful in designing protocols for other experiments that must proceed to equilibrium. Two caveats: First, the time course of binding is influenced by the concentration of radioligand. Therefore, if one assesses binding at only a single concentration of radioligand, the on-rate experiment is best performed with a low concentration of radioligand. This will help prevent the underestimate of equilibration time for subsequent experiments.

Second, competitive binding experiments take longer to reach equilibrium than would be predicted from a simple on-rate experiment (4-5). Thus, adjustment may be required to assure achievement of equilibrium in competitive binding experiments.

One measures the rate of dissociation as follows. First the radioligand is allowed to bind to receptors. Then the binding ("forward") reaction is blocked either by adding an excess of an unlabeled compound so that it (rather than the radioligand) will bind to the free receptors, or by greatly ($\geq 1:100$) diluting the reaction mixture (to reduce the concentration of radioligand). With either approach, appreciable amounts of radioligand will no longer bind to the receptors, but radioligand that is already bound to the receptors will continue to dissociate. By measuring the amount of specific radioligand binding at various times thereafter, one can test the reversibility of the binding and can determine the rate of dissociation.

The dissociation of radioligand binding typically follows a monoexponential decay curve (being dependent on disruption of a single molecular interaction), and the "half-life" of radioligand-receptor complexes can be easily estimated. One can readily determine how much dissociation occurs during the few seconds required for separation of bound and free ligand. If the dissociation is substantial during this time, then the results of other experiments will be distorted. This problem can be obviated either by speeding up the separation process (perhaps using centrifugation rather than filtration) or by retarding the dissociation (perhaps by using ice-cold buffer when washing filters).

More quantitative analyses of kinetic experiments are also possible. Because dissociation curves decay as simple exponentials, plotting the natural logarithm of the specific binding as a function of time yields a line and the dissociation rate constant (in units of time^{-1}) is the slope of this line. Typically the dissociation rate constants for adrenergic radioligands are $0.01\text{-}1.0 \text{ min}^{-1}$. Dissociation rate constants are temperature dependent, with radioligands dissociating from receptors more rapidly at warmer temperatures.

From the law of mass action, the equilibrium dissociation constant (K_d, expressed in units of concentration) is defined as equal to the dissociation rate constant divided by the association rate constant. The terminology is potentially confusing: the "equilibrium dissociation constant" and the "dissociation rate constant" have distinct meanings and are measured in different units. Another term is the "equilibrium association constant" (K_a, in units of $\text{concentration}^{-1}$), which is the inverse of the K_d and is different than the "association rate constant."

The association and dissociation of radioligands from adrenergic receptors usually follow the characteristic patterns shown in Figure 1.

3. *Is specific binding saturable with increasing radioligand concentration?* Binding experiments are conducted with a fixed concentra-

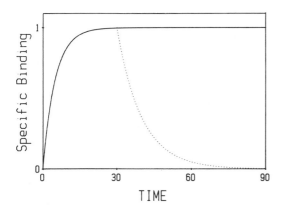

FIGURE 1 Idealized association and dissociation kinetics for the binding of ligands to adrenergic receptors. The ligand and cells are mixed at time 0 and the specific binding is plotted thereafter. At 30 min, an unlabeled antagonist is added in excess to block the "forward" association reaction. The dissociation is shown in dotted lines.

tion of tissue and varying concentrations of radioligand under steady-state conditions. This type of experiment is performed to demonstrate that specific binding saturates as concentrations of radioligand increase. Typically, one assesses both total and nonspecific binding at many concentrations of radioligand and then calculates specific binding as the difference. Two parameters can be estimated from the saturation binding curve: the maximal binding (B_{max}), which is a measure of the total number of adrenergic receptors present in the tissue, and the equilibrium dissociation constant (K_d) of the receptors for the radioligand. K_d, which represents the concentration of radioligand required to bind half the receptors, should be the same in saturation binding experiments and in the kinetic studies described above. The K_d is inversely related to receptor affinity, the "strength" of binding: a low K_d means high-affinity binding; a high K_d means low-affinity binding.

The relationship of radioligand concentration to specific receptor binding is expressed by the following equation when the binding follows the law of mass action:

$$[RL] = B_{max}[L]/(K_d + [L])$$

This equation has the expected properties: When the radioligand is present at a concentration equal to its K_d, the binding equals half the B_{max}; at higher radioligand concentrations the binding approaches the B_{max}. The form of this equation is similar to that derived by Langmuir to describe the absorption of a gas to a surface at constant temperatures; thus, the binding curves are often called "binding isotherms." Mathematically this equation is described as a rectangular hyperbola. As shown in Figure 2, when the radioligand is present at five times its K_d, it binds to 83% of the receptors. When the radioligand concentration equals 10 times its K_d, then 90% of the receptors will be bound. The concentration of radioligand must exceed 100 times its K_d to occupy over 99% of the receptors. Often, investigators attempt to measure changes in adrenergic receptor number by measuring specific binding of a single "saturating" concentration of radioligand. Since a very high concentration of radioligand is required to "saturate" the receptors, such an experiment may force one to use a large amount of radioactivity, thereby increasing nonspecific binding as well as the cost of the experiment. In addition, changes in binding observed at a single concentration of radioligand do not distinguish alterations in K_d from those in B_{max}.

Saturation binding experiments are often analyzed by the method of Scatchard, a method more accurately but less commonly attributed to Rosenthal (7). When the radioligand binds to a single class of binding sites (and that binding follows the law of mass action, and thus matches the binding equation described above), the Scatchard transformation will linearize the data points as shown in Figure 2. The intercept of the line with the abscissa will be equal to B_{max}, and the slope of the line will be equal to the negative inverse of the K_d. Typically, radioligand binding to adrenergic receptors on human tissues displays a straight line in Scatchard analyses of saturation isotherms. A better means to analyze such experiments—computer modeling to the law of mass action—is described below.

4. *Do unlabeled drugs compete for specific radioligand binding with appropriate potencies appropriate for adrenergic receptors?* The several types and subtypes of alpha- and beta-adrenergic receptors are defined by the relative potencies of various agonists and antagonists. To demonstrate that specific radioligand binding is to one of those receptor types, it is necessary to demonstrate that the binding is competed for appropriately by a series of unlabeled compounds. One typically incubates the tissue with a single concentration of radioligand and many concentrations of the competing drug until equilibrium is reached. Radioligand binding is then determined at each concentration of competitor. As the concentration of competitor increases, more of it will bind to the receptor and, thus, less of the radioligand can bind. A schematic of such an experiment is shown in Figure 3.

The distinction between alpha- and beta-adrenergic receptors is defined by the relative potencies of several catecholamines (alpha-adrenergic receptors: epinephrine > norepinephrine >> isoproterenol;

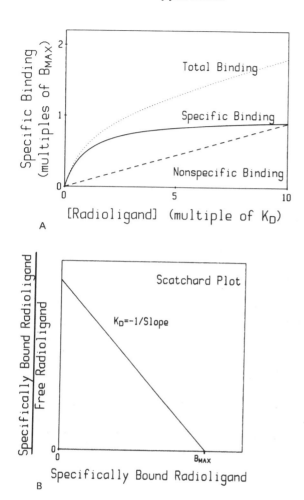

FIGURE 2 Idealized saturation binding isotherm. A. Various amounts of radioligand are added to the tissue and the total binding at each concentration is plotted in dotted lines. In parallel, the nonspecific binding is determined by including an excess of an unlabeled compound that will block all the receptors and this nonspecific binding is shown as dashed lines. The difference between these two curves, therefore, is specific binding to the receptors. The dissociation constant (K_d) is the concentration of the ligand required to bind half the receptors. The B_{max} is the total number of receptors. B. A Scatchard plot of the specific binding is shown. The B_{max} is the intercept with the X axis; the K_d is inversely proportional to the negative of the slope.

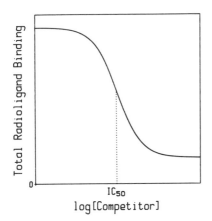

FIGURE 3 An idealized competitive binding curve. Tissue is incubated with a constant concentration of radioligand, and variable amounts of a competing drug as plotted on a log scale on the X axis. The total binding of the radioligand is plotted on the Y axis. The competitor competes for specific, but not nonspecific, binding sites and thus reaches a plateau above 0.

beta-adrenergic receptors: isoproterenol > epinephrine ≥ norepine-phrine). Competitive binding experiments are also used to determine the subtype of receptors by demonstrating that adrenergic agonists and antagonists compete for radioligand binding with the appropriate potencies. Furthermore, catecholamines and certain other adrenergic agents (e.g., propranolol) are asymmetric molecules and the potency of each levo (or -) enantiomer is much greater than that of the dextro (or +) enantiomer. This stereoselectivity can also be demonstrated in competitive binding experiments.

From inspection of a competitive binding curve, one can readily determine the IC_{50}, the concentration of competitor that competes for half the radioligand binding. This concentration is related to the dissociation constant of the receptors for the competitor, the K_i (i for "inhibitor"). This relationship is not fixed; it depends on the concentration of radioligand used, and the K_d of the receptors for the radioligand. A simple equation, attributed to Cheng and Prussoff, (8), relates these variables:

$$K_i = IC_{50}/(1 + [L]/K_d).$$

Thus, the K_i will be very close to the IC_{50} when a low concentration of the radioligand (L) is used (compared to its K_d), and will be less than the IC_{50} when higher radioligand concentrations are used.

The steepness of a competition curve is quantitated by a "Hill slope"; also called "slope factor" or "pseudo-Hill slope" (because Hill's equation was not really designed for competition curves). When both radioligand and competition bind to a single class of receptors, the Hill slope is 1.0 and the 10-90% portion of the competition occurs over a hundredfold concentration range of competitor. In some situations, competition curves are more shallow and Hill slopes are less than 1.

5. *Do independent methods of deriving the equilibrium dissociation constant agree?* From the law of mass action, the equilibrium dissociation constant (K_d, expressed in units of concentration) is defined to be equal to the dissociation rate constant divided by the association rate constant. Both of these rate constants can be determined from kinetic experiments as described above, so the K_d can be determined solely from kinetic experiments. The K_d can also be determined from saturation binding experiments as described above. Demonstrating that these two methods yield similar values for the K_d confirms the consistency of the data, and the applicability of the law of mass action.

6. *Is specific binding proportional to tissue concentration?* Clearly, if one uses twice as much tissue, one ought to find twice as many adrenergic receptors. Demonstrating a linear relationship between tissue concentration and receptor number rules out several types of experimental artifacts. This relationship holds when only a small (< 10%) fraction of the radioligand binds to receptors; that is, the free (unbound) concentration of radioligand (which is in equilibrium with bound radioligand) is essentially equal to the amount added. Under such conditions, one does not need to worry about "depletion" of the radioligand as it binds to receptors.

7. *Does the radioligand remain unaltered during the experiment?* It is necessary to prove that the radioactivity bound to the tissue represents the radioligand and not breakdown products. Thin-layer chromatography of radioligand extracted from the tissue is the technique most frequently used for this purpose. High-pressure liquid chromatography or rebinding of extracted ligand has also been used.

III. DATA ANALYSIS

There are three approaches to analyzing the results of radioligand binding experiments: first, one can usually ascertain the major findings of an experiment by visually inspecting the data. This is an important skill to develop, as it allows a check of subsequent calculations and computer programs, and it facilitates critical evaluation of published data. Second, one can mathematically transform data so that they are linear and thus easier to analyze graphically (as in the Scatchard plot). Third, one can analyze the untransformed data using nonlinear regression to fit the data points to equations describing a particular molecular model of receptor binding. We will discuss all three methods.

Most analyses of radioligand binding data are based on the law of mass action: $R + L \rightleftarrows RL$. These reversible binding reactions are similar to the reactions describing enzyme kinetics and the ionization of acids and bases. Despite its simplicity, the law of mass action appears to adequately describe many types of receptor-ligand interactions. However, this model is not applicable in certain situations— such as if the ligand binds irreversibly to the receptor; if the receptor-ligand complex undergoes further changes such as internalization; if the receptors interact with one another so that the binding of the ligand displays positive or negative cooperativity; or if the radioligand or receptors are degraded.

Experimental data points are commonly transformed mathematically so that they form a straight line. Examples of such linearizing transformations include the Scatchard plot, the Hill plot, and the Eadie-Hofstee plot. An advantage of using these transformations is that they are easy to perform and allow data analysis with a small computer or calculator (or even by hand). Another advantage of these methods is that they make it easy to visually interpret the data. Analyzing the transformed data by linear regression is convenient, but these analyses may be only roughly accurate. One problem is that the transformations distort the data in nonobvious ways: Small experimental errors are transformed into larger deviations. Another problem is that the relationship between independent and dependent variables are transformed along with the data. Thus, the assumptions inherent in a rigorous interpretation of linear regression are not met when Scatchard, Hill, or Eadie-Hofstee plots are analyzed, and the resulting estimates of receptor number and affinities are not statistically optimal. In many situations, these simple methods are satisfactory, but in other circumstances more complicated methods of data analysis are required. This is especially true when several classes of receptor binding sites are present.

Nonlinear regression computer programs allow one to fit experimental data directly to mathematical equations describing the assumed model, usually the law of mass action. The data need not be transformed, and the experimental errors are therefore not distorted. Thus, these nonlinear methods are the most powerful means of analyzing radioligand binding data. The principles required to use and understand these programs are straightforward: (1) The programs need a starting place. They must be given initial values (first guesses) of all the parameters to be fit (e.g., the K_d and B_{max}). These may be obtained from previous experience, from visual inspection of the data, or from a cruder analytic method (such as a Scatchard plot). (2) The programs are iterative. Starting with the initial values, the variables are systematically varied. With each combination of values, the equation describing the model is evaluated to determine how well it describes the data, using the sum of the squares of the distances of the data points from the values predicted by the model. When this sum of squares is minimal, the program "converges" and defines a set of variables that "fit" the data to the models. If the initial values of the parameters are not reasonable, or if the model does not come close to describing the data, then the program will halt without determining "best values" for the variables. (3) The programs must explicitly be given an equation describing the model to which the data are being fit. The equations describing the law of mass action are built into several available programs. General-purpose curve fitting programs do not find the "best curve" that describes the data. They merely attempt to find values for the variables that make a specified model fit the data most closely. (4) Some programs can compare one model with another. This is done by using analysis of variance to compare the mean square of deviations of model estimated values versus observed data for one model with that of another. The complexity of a model is accounted for in the degrees of freedom, so that a more complicated model may sometimes not fit the data as well (as measured by "mean square") as does a simple model. A nonlinear regression program cannot reject a more complicated model; instead, given the quality of the experimental data, it determines that the more complicated model does not fit the data significantly better than the simpler model. The number of data points is important. Given a minimal number of data points, nonlinear programs will virtually always make the simplest model appear best.

IV. USE OF RADIOLIGAND BINDING IN STUDIES OF ADRENERGIC RECEPTORS IN MAN

A. Counting Receptors

The most straightforward use of radioligand binding assays is to quantitate the number of adrenergic receptors present in a tissue in order to count the receptors. This is done by using saturation binding experiments as discussed above. Receptor number is usually expressed as femtomoles or picomoles of receptors per milligram of membrane protein. Typical values are ~100-1000 fmol adrenergic receptor/mg protein. When intact cells are studied, the appropriate unit is number of receptors per cell. For example, human platelets each have 200-400 $alpha_2$-adrenergic receptors. The number of receptors varies about fivefold between individuals; much of this variation is due to genetic factors (27).

When quantitating receptors on tissue from humans, one needs to beware of the possibility that catecholamines or drugs are retained with the tissue. Retained ligands that reversibly bind to the receptors will reduce the apparent affinity of the receptors for the radioligand (thus increasing the measured K_d) without altering the B_{max}. This is a particular problem with platelets which actively store catecholamines. Thus alpha$_2$-adrenergic receptors on platelets that have been exposed to elevated concentrations of catecholamines demonstrate an apparently reduced affinity for radioligands (9). Similar findings have been reported for beta receptors on human lymphocytes (10). Note that one can be misled about the number of receptors if the radioligand binding is performed with only a single concentration of radioligand.

Most often these types of experiments are performed on membrane preparations. The assumption (often unstated) is that all the adrenergic receptors present in the original tissue will be present in the membrane preparation. Usually such preparations are made by homogenizing the cells, centrifuging at 20,000-30,000 g, and saving the pellet. Not all adrenergic receptors from some tissues, however, are present in those pellets. Instead, some receptors are located in small membrane vesicles that can only be pelleted with centrifugation at >100,000 × g. Thus, in conventional membrane preparations, adrenergic receptors may be lost.

Many studies have demonstrated that the number of adrenergic receptors can dynamically vary. Often the number of detectable receptors (B_{max}) decreases after the tissue is exposed to a receptor agonist. This phenomena is termed "down-regulation." In addition, the number of receptors sometimes increases when the tissue is exposed to decreased amounts of agonist (or if the receptors are occupied by an antagonist); this phenomena is termed "up-regulation." However, down- and upregulation of receptors does not occur with all types of adrenergic receptors in all tissues (9).

B. Measuring the Affinity of Receptors
   for Agonists

Most of the available adrenergic radioligands are antagonists. That is, they bind to receptors but do not initiate receptor-mediated responses. Agonists are compounds that initiate such responses, but for a variety of reasons, radiolabeled adrenergic agonists are difficult to use in radioligand binding studies. To assess the interaction of agonists with receptors, therefore, competitive binding studies are often performed using a single concentration of radiolabeled antagonist and several concentrations of an unlabeled agonist.

Beta-adrenergic receptors (both subtypes) are linked to a stimulation of adenylate cyclase and alpha$_2$-receptors are linked to

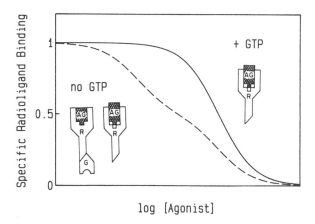

log [Agonist]

FIGURE 4 Effects of GTP on agonist binding. This shows an idealized competitive binding experiment with an adrenergic agonist competing for binding to alpha- or beta-adrenergic receptors on washed membranes. In the presence of GTP (often an analog is used) the curve is steep. In the absence of GTP (dashed lines) the curve is biphasic, demonstrating that the interaction of the agonist with the receptors is more complex. As shown, about half the receptors interact with the agonist with high affinity; these are thought to represent receptors attached to the G protein. The other half interact with lower affinity. In the presence of GTP the interaction of receptors with G is transient, so virtually all of the competition is of low affinity.

an inhibition of adenylate cyclase. The receptors are linked to adenylate cyclase through coupling proteins that bind guanine nucleotides. The coupling protein that stimulates adenylate cyclase is known as G_s, the one that inhibits adenylate cyclase as G_i (G, guanine nucleotide binding; s, stimulatory; i, inhibitory). The terms N_s and N_i are also used. Both these coupling proteins have been purified and biochemically characterized (11,12). The effects of these coupling proteins have been observed in radioligand binding studies employing membrane preparations when the competition of an agonist with antagonist radioligand binding is assessed. As shown in Figure 4, the competitive interaction of an agonist with alpha$_2$- or beta-adrenergic receptors on washed membranes is complex, and the competition curve is shallow. When GTP, or a nonhydrolyzable analog like Gpp(NH)p, is added to the incubation, the resulting competition curve is shifted to the right, indicating a reduced affinity of the receptors for epinephrine. When guanine nucleotides are present, the competitive binding curve is also of normal steepness, indicating that the receptors are homogeneous.

These data have been explained by a "ternary complex model." In the absence of guanine nucleotides the binding of agonist to the receptor induces or stabilizes the interaction of receptors with the appropriate coupling protein, resulting in the formation of a ternary complex of agonist, receptor, and coupling protein. When the receptors and coupling proteins are attached, the receptors have a high affinity for the agonist. In the presence of guanine nucleotides, however, the receptors dissociate from the G proteins and the resulting binding curve reflects only low-affinity interaction of receptors and agonist. Further details are discussed in Chapter 6.

Additional complications are introduced when intact cells rather than isolated membranes are used in competitive binding assays with agonists. When these experiments are performed at physiological temperatures, the agonists exert effects on the cells, and the number, location, or properties of the receptors may change during the incubation. Thus, one cannot view the results as reflecting only competition between agonist and radioligand for binding to the receptor (12a). Another feature of studies of intact cells is that the receptors may be exposed to organelles and constituents in the intracellular environment. It is known, for example, that intracellular guanine nucleotides and ions can alter receptor binding.

In many cases, the results of radioligand binding experiments with adrenergic agonists appear to conflict with results of pharmacological studies: The apparent K_d of the adrenergic receptors for agonists determined in radioligand binding experiments with intact cells is commonly several orders of magnitude greater than the concentration of agonist that initiates a half-maximal biological or biochemical response (K_{act}). Several explanations have been proposed for this "K_d/K_{act} discrepancy." One proposed explanation is the existence of "spare receptors"—receptors whose occupation by agonist is not required for a maximal receptor-mediated response to occur. Thus, the maximal response can occur when the agonist concentration is less than its K_d and only a small proportion of the receptors are occupied. A second proposed explanation for the K_d/K_{act} discrepancy is an agonist-promoted receptor "state change" from a state with a high affinity for agonists to a state with a low affinity for agonists. Thus, during the first seconds to minutes (when functional studies are generally performed) the agonist may bind to the receptors with high affinity. But after a longer interval (when binding studies are terminated) the receptors may have a lower affinity for the agonist. Recent studies of beta receptors in several cell types (including human lymphocytes) have demonstrated high-affinity agonist binding at early time points, and thus have supported the notion of an agonist-promoted receptor state change (13,14). The mechanism for this change is unknown. One possibility is that the receptors are altered (perhaps phosphorylated) so that their affinity for agonists is reduced. Another possibility is that the physical location of the receptors is altered so that agonists have reduced access to the receptors.

C. Determining Receptor Subtypes

The concept of adrenergic receptor subtypes emerged from studies of the relative potencies of various adrenergic agonists and antagonists in physiological and biochemical assays. It is now clear that there are two types of beta receptors (beta$_1$ and beta$_2$), and two types of alpha-adrenergic receptors (alpha$_1$ and alpha$_2$). It is possible that other subtypes exist as well (15).

Highly selective radioligands are available for alpha$_1$ and alpha$_2$ receptors. Examples include [^{3}H]prazosin and [^{125}I]BE2254 for alpha$_1$ receptors and [^{3}H]yohimbine, [^{3}H]rauwolscine, and [^{3}H]idazoxan for alpha$_2$ receptors. Under usual experimental conditions, each ligand binds exclusively to one subtype of alpha-adrenergic receptor; the data analysis is straightforward.

Subtype-selective radioligands are not yet available for the study of beta-adrenergic receptors, although such ligands are under development. All currently available beta-adrenergic radioligands bind equally to beta$_1$ and beta$_2$ receptors. Selective unlabeled compounds are available, but these compounds are only moderately selective: They bind to both beta$_1$ and beta$_2$ receptors but with different affinities. Competitive binding studies between a nonselective radioligand and one of these selective unlabeled compounds can be used to determine receptor subtypes (Fig. 5). In a competition curve in which the radioligand and competitor both bind to a single class of sites, the bulk of the competition (from 10 to 90%) occurs as the competitor concentration increases 100-fold. A larger span of concentrations is necessary when the unlabeled compound competes for two classes of binding sites with different affinities, and thus the competitive binding curve is more shallow. As discussed above, dissecting out the two components of the curve is best performed by a computer program that does nonlinear regression (16,17). An alternative way of displaying and analyzing the data is an Eadie-Hofstee plot.

When a tissue contains exclusively one subtype of beta-adrenergic receptor, then the competition curve by a subtype-selective drug will be monophasic and of normal steepness. From the relative potencies of drugs known to be selective for beta$_1$ or beta$_2$ receptors, one can determine which subtype is present. If both beta$_1$ and beta$_2$ receptors are present, then a competition curve with a subtype-selective compound will be shallow (Fig. 5). As noted above, computer analysis can be used to determine the relative numbers of the two subtypes. In this manner, it has been shown that human lymphocytes contain only beta$_2$ receptors and atria from human heart about 75% beta$_1$ and 25% beta$_2$ receptors (18,19).

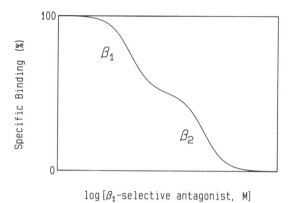

FIGURE 5 Discriminating between beta-receptor subtypes. Tissue
is incubated with a single concentration of radioligand and various
concentrations of radioligand and various concentrations of competing
drug. The radioligand (e.g., [^{125}I]ICYP) binds equally to both
receptor subtypes, but the competitor binds to one subtype with
higher affinity than to the other.

D. Assessing Mechanisms of Drug Action

Competitive radioligand binding experiments can be used to determine
whether an unlabeled drug interacts with the receptor. Such radioligand binding experiments are often easier and more direct than
classic pharmacological experiments examining the effect of the unlabeled drug in a physiological or biochemical assay. Further experiments are required to test whether this interaction represents direct
competition for binding to the receptors. Such experiments include
studies examining whether competition is reversible and whether the
interaction of drug alters K_d and not B_{max} of the radioligand (19a).

E. Measuring the Effects of Modulators on Agonist Binding

The binding of catecholamines to the adrenergic receptors is usually
assessed through competitive binding experiments. The pattern of
competition in such an experiment often depends on the composition
of the incubation buffer. Guanine nucleotides and monovalent and
divalent cations are the best-studied modulators of agonist binding
(Fig. 4). These modulators do not generally alter competitive binding
curves of antagonists, but they can markedly alter the competitive
binding curves of agonists. This type of experiment allows one to
approach the fundamental question of how receptors recognize agonists
and antagonists differently.

F. Receptor Turnover and Redistribution

Investigators studying receptors on cultured cells have begun to elucidate the "life cycle" of adrenergic receptors (20, 21). Such studies have explored the rate at which receptors appear in the membranes, the rate at which they are removed from the membranes, and their transit through various compartments in the cell. It is becoming clear that receptors may follow complex itineraries that can be regulated at several loci. Several approaches have been used. One approach is to block protein synthesis and then follow changes in receptor number. The assumption is that this will block creation of new receptors without altering disposition of the existing ones, but it may be difficult to prove unequivocally this assumption. A second approach is to inactivate binding sites on receptors with irreversible antagonists, and then follow the appearance of new binding sites over time. A third approach is to fractionate the cells and examine the subcellular location of the receptors after various perturbations. A fourth approach is to use radioligands with restrictive access to intracellular compartments. [^{3}H]CGP-12177 is a beta-adrenergic antagonist that binds to all beta receptors in control cells but to many fewer receptors in cells that have been incubated with agonists ("desensitized cells") (23). The explanation for this restricted binding has yet to be unequivocally defined, but several investigators have suggested that desensitized receptors are internalized to an intracellular location to which [^{3}H]CGP-12177 cannot penetrate because it is hydrophilic. In addition to binding [^{3}H]CGP-12177, one can also use [^{125}I]ICYP sites and quantitate sites with which unlabeled CGP-12177 can compete. This approach takes advantage of the higher specific activity of ^{125}I. These studies have not been extensively performed with adrenergic receptors from humans, and our knowledge of the "life cycle" of these receptors in man is thus rather limited.

G. Radioreceptor Assay

A competitive radioligand binding experiment can be used to assay drug concentrations if the affinity of the drug for the receptors identified by the radioligand is known. The underlying principle is similar to that of a radioimmunoassay. The drug is extracted from blood or serum, and the ability of several dilutions of this extract to compete for radioligand binding is compared with known standards. This method has not been widely used, but its potential for assaying plasma propranolol concentrations has been demonstrated in several studies (27). A feature of such assays is that they measure all compounds in the extract that compete for receptor binding. Thus, active metabolites are usually measured as well as the parent compound.

V. OTHER DIRECT METHODS FOR STUDYING ADRENERGIC RECEPTORS

Although radioligand binding has been the most useful methodology for direct in vitro studies of adrenergic receptors, other methods have also been developed. These other methods have not yet been applied to human tissue, so they will be only briefly summarized here.

A. Autoradiography

With conventional radioligand binding one examines the total amount of radioligand binding to a suspension of cells or cell fragments. With autoradiography one binds radioligand to a tissue slice or to fixed cells and then exposes the material to a photographic emulsion or film. By examining the location of silver grains that were exposed to radioactivity, the location of the receptors within the tissue can be determined. This method has not yet been used with human tissue, but it has been used with several different tissues from experimental animals. Potentially, it should be possible to perform these types of experiments with biopsy specimens of adrenergic receptor-bearing tissue.

B. Photoaffinity Labeling

Most radioligands bind to receptors reversibly. For some purposes, however, it is more useful to use a ligand that binds irreversibly. Photoaffinity probes combine both approaches. When used in the dark they bind reversibly. Exposure to light activates a functional group (usually a nitrene) which binds to adjacent molecules. Several probes are now available for beta-adrenergic receptors—these bind selectively but the efficiency of photolabeling with these probes is $\leq 20\%$. Thus, it is important to prove that the probes are not merely detecting properties of a distinct subpopulation of receptors. To the extent the probe binds specifically to a given receptor, receptor proteins will be irreversibly labeled. Because the receptors are covalently labeled, they can be readily examined using denaturing conditions. This allows one to determine the molecular size of receptor binding proteins by chromatographic or gel electrophoretic techniques. Although application of photoaffinity labeling to human adrenergic receptors has not yet been reported, the straightforward nature of the experiments suggests such studies will soon be undertaken.

C. Solubilization, Purification, and Reconstitution

Most radioligand-binding studies are performed with cells or cell fragments. It is also possible to solubilize the membranes with detergents and still maintain the ability of the receptors to bind radioligands. The advantage of this approach is that the receptor is removed from its membrane environment, and the properties of the native receptor can be studied. This is also an essential first step toward receptor purification. Both $beta_1$ and $beta_2$ receptors from several tissues have been purified; both are single polypeptide chains with a molecular weight of about 60,000 daltons. During the purification schemes, the receptor is tracked either by following a radiolabeled photoaffinity probe or by its ability to bind radioligand. Solubilized receptors have also been reconstituted into membranes lacking beta-adrenergic receptors, and their ability to stimulate adenylate cyclase has been demonstrated (24).

Less work has been done with alpha receptors than with beta receptors, but partial purification of $alpha_2$ receptors from human platelets has been reported (25).

D. Radionuclide Scans of Adrenergic Receptors

Ideally, one would like to examine the number and distribution of adrenergic receptors in vivo. One way of doing this is to measure receptors on blood cells, but these are usually not the receptors of interest. Another approach is to use nuclear medicine techniques to develop "receptor scans." The idea is to bind radioligands to receptors on internal tissues and to monitor the distribution and amount of binding by external detection devices. A small amount of work has been done with animals; the potential is enormous.

E. Antireceptor Antibodies

Understanding of many types of receptors has been aided by the availability of antibodies directed against receptors. There are many advantages to such an approach. Techniques are available for coupling antibodies to fluorescent or electron-dense moieties so that the distribution of the antibody binding can be monitored by fluorescent or electron microscopy. This would be particularly useful for tracking the intracellular location of receptors. Antibodies may be directed at the ligand binding site on the receptor, but perhaps even more useful are antibodies directed at other sites. Such antibodies may enable one to locate newly synthesized receptors, perhaps at a stage before they are able to bind radioligand, and also to identify receptors—perhaps after modification by agonists and/or other treatments—that have lost their ability to bind radioligands.

Progress at obtaining antibodies directed at adrenergic receptors has been disappointing. Several groups have reported obtaining such antibodies, but the affinity has been poor. Therefore, these antibodies have not yet been particularly useful in extending our understanding of adrenergic receptors. Considerable future efforts are likely to be directed at generating and studying adrenergic receptors with antibodies.

REFERENCES

1. Yamamura, H., Enna, S., and Kuhar, M. J.: *Neurotransmitter Receptor Binding*. New York, Raven, 1978.

2. Williams, L. T., and Lefkowitz, R. J.: *Receptor Binding Studies in Adrenergic Pharmacology*. New York, Raven, 1978.

3. Motulsky, H. J., and Insel, P. A.: The study of cell surface receptors with radioligand binding: Methodology, data analysis and experimental problems. In: *Receptor Science in Cardiology*. J. I. Haft and J. S. Karliner (eds.). Mount Kisco, New York, Futura, 1984.

4. Motulsky, H. J., and Mahan, L. C.: The kinetics of competitive radioligand binding predicted by the law of mass action. Mol. Pharmacol. 25:1-9, 1984.

5. Ehlert, F. J., Roeske, W. R., and Yamamura, H. I.: Mathematical analysis of the kinetics of competitive inhibition in neurotransmitter receptor binding assays. Mol. Pharmacol. 19:367-371, 1981.

6. Scatchard, G.: The attraction of proteins for small molecules and ions. Ann. N.Y. Acad. Sci. 51:660-672, 1949.

7. Rosenthal, H. E.: A graphic method for the determination and presentation of binding parameters in complex systems. Anal. Biochem. 20:525-6532, 1967.

8. Cheng, Y., and Prusoff, W. H.: Relationship between the inhibition constant (K_i) and the concentration of an inhibitor that causes a 50% inhibition of an enzymatic reaction. Biochem. Pharmacol. 22: 3009-3108, 1973.

9. Karliner, J. S., Motulsky, H. J., and Insel, P. A.: Apparent "down-regulation" of human platelet $alpha_2$-adrenergic receptors is due to retained agonist. Mol. Pharmacol. 21:36-43, 1982.

10. Aarons, R. D., Nes, A. S., Gerber, J. G., and Molinoff, P. B.: Decreased beta-adrenergic receptor density on human lymphocytes after chronic treatment with agonists. J. Pharmacol. Exp. Ther. 224:1-6, 1983.

11. Gilman, A. G.: G proteins and dual control of acetylcholine. Cell 36:577-579, 1984.

12. Hildebrandt, J. D., Codima, J., Rosenthal, W., Sunyer, T., Iyengar, R., and Birnbauer, L.: Properties of human erythrocyte N_S and N_i, the regulatory components of adenylate cyclase, as purified without regulatory ligand. Adv. Cyclic Nucleotide Res. 19:87-102, 1985.

12a. Motulsky, H. J., Mahan, L. C., and Insel, P. A.: Radioligands, agonists, and membrane receptors on intact cells: Data analysis in a bind. Trends Pharmacol. Sci. 6:317-319, 1985.

13. Insel, P. A., Mahan, L. C., Motulsky, H. J., Stoolman, L. M., and Koachman, A. M.: Time-dependent decreases in binding affinity of agonists for beta-adrenergic receptors of intact S49 cells. J. Biol. Chem. 258:13597-13605, 1984.

14. Toews, M. L., Harden, T. K., and Perkins, J. P.: High affinity binding of agonists to β-adrenergic receptors on intact cells. Proc. Natl. Acad. Sci. U.S.A. 80:3553-3557, 1983.

15. Agnus, J. A.: Sympathetic vasoconstriction—No role for alpha-adrenoceptors. Trends Pharmacol. Sci. 3:464-465, 1982.

16. Munson, P. J., and Rodbard, D.: LIGAND: A versatile computerized approach for characterization of ligand binding systems. Anal. Biochem. 107:220-239, 1980.

17. Delean, A., Hancock, A. A., and Lefkowitz, R. J.: Validation and statistical analysis of a computer modeling method for quantitative analysis of radioligand binding data for mixtures of pharmacological recepter subtypes. Mol. Pharmacol. 21:5-16, 1982.

18. Brodde, O.-E., Engel, G., Hoyer, D., Bock, K. D., and Weber, F.: The beta-adrenergic receptor in human lymphocytes: Subclassification by the use of a new radioligand, [^{125}I]iodocyano-pindolol. Life Sci. 29:2189-2198, 1981.

19. Robberecht, P., Delhaye, M., Tator, G., et al.: The human heart beta-adrenergic receptors: Heterogeneity of the binding site: Presence of 50% beta$_1$- and 50% beta$_2$-adrenergic receptor. Mol. Pharmacol. 24:169, 1983.

19a. Motulsky, H. J., Maisel, A. S., Snavely, M.D., and Insel, P. A.: Quinidine is a competitive antagonist at α_1- and β_2-adrenergic receptors. Circ. Res. 55:376-81, 1984.

20. Insel, P. A., Hughes, R. J., Meier, K. E., Mahan, L. C., and Snavely, M. D.: Cellular metabolism of adrenergic receptors. In: *Adrenergic receptors: Molecular Characterization and Therapeutic Implication*. R. J. Lefkowitz (ed.). F. K. Schattauer Verlag, 1985.

21. Hertel, C., and Perkins, J. P., Receptor-specific mechanisms of desensitization of -adrenergic receptors. Mol. Cell Endocrinol. 37:245-256, 1984.

22. Rochester, C. L., Gammon, D. E., Shae, E., and Bilezikian, J. P. A radioreceptor assay for propranolol and 4-hydroxypropranolol. Clin. Pharmacol. Ther. 28:32-39, 1980.

23. Hertel, C., Staehelin, M., and Perkins, J. P.: Evidence for intravesicular beta-adrenergic receptors in membrane fractions from desensitized cells: Binding of the hydrophilic ligand CGP-12177 only in the presence of alamethicin. J. Cyclic Nucleotide Res. 9:119-128, 1983.

24. Cerione, R. A., Strulovici, B., Benovic, J. L., Lefkowitz, R. J., and Caron, M. G.: Pure beta-adrenergic receptor: The single polypeptide confers catecholamine responsiveness to adenylate cyclase. Nature 306:562-566, 1983.

25. Welstein, A., Palm, D., Weimer, G., Shafer-Karting, M., and Mutscler, E. Simple and reliable radioreceptor assay for beta-adrenoreceptor antagonists and active metabolites in native human plasma. Eur. J. Clin. Pharmacol. 27:545-553, 1984.

26. Propping, P., and Friedt, W.: Genetic control of adrenergic receptors on human platelets. A twin study. Human Genet. 64: 105-109, 1983.

27. Regan, J. W., Nakata, H., Damarinis, R. M., Caron, M. G., and Lefkowitz, R. J. Purification and characterization of the human platelet alpha$_2$-adrenergic receptor. J. Biol. Chem. 261: 3894-3900, 1986.

6

Biochemical Characterization of Human Adrenergic Receptors

ROSS FELDMAN   *University of Iowa, Iowa City, Iowa*

LEE E. LIMBIRD   *Vanderbilt University School of Medicine, Nashville, Tennessee*

I. INTRODUCTION

Over the past 10 years, great advances have been made in the biochemical characterization of adrenergic receptors in the target issues where these receptors had earlier been identified using classic pharmacological techniques. Additionally, we have a much greater understanding of the mechanisms by which binding of catecholamines to their specific receptors initiates adrenergic-mediated physiological effects.

This chapter will focus on what is known about the biochemical properties of adrenergic receptors in human tissues and the molecular consequences of receptor occupancy by agonists. As indicated elsewhere in this volume, adrenergic receptors have not only been classified into alpha- and beta-adrenergic receptors based on the order of potency of agonists in eliciting particular responses and the selectivity of antagonists in blocking these responses, but they also have been further subdivided into receptor subpopulations; i.e., $beta_1$- and $beta_2$-adrenergic receptors and $alpha_1$- and $alpha_2$-adrenergic receptors. Both $beta_1$- and $beta_2$-adrenergic receptors are linked to the activation of the membrane-bound enzyme adenylate cyclase, which catalyzes the conversion of adenosine triphosphate (ATP) to cyclic AMP (cAMP), resulting in the elevation of intracellular cAMP, which is felt to mediate, at least in part, the physiological effects of beta-adrenergic stimuli. In contrast, $alpha_1$- and $alpha_2$-adrenergic receptors appear, at the present time, to be linked to two independent effector mechanisms. $Alpha_1$ receptors have been demonstrated to mobilize cellular Ca^{2+}, which either precedes or is a consequence of transient changes in the turnover of a small fraction of membrane lipids, the phosphatidylinositols. Alpha$_2$ receptors on the other hand are linked to inhibition of adenylate cyclase activity, and this biochemical effect may explain how alpha$_2$-adrenergic agents attenuate those effects mediated by elevations in cAMP.

Since this volume is a compendium of information concerning adrenergic receptors in human tissues, we will try to emphasize findings obtained in human preparations. However, since the framework for the understanding of how beta receptors stimulate adenylate cyclase has derived from studies in avian and amphibian erythrocytes and cell culture model systems, it will be essential to provide an overview of the findings from these systems because a great many of the conclusions made from studies in human tissues rely heavily on analogies to the better-characterized model systems. In contrast, much of the basic information on the properties of alpha$_2$ receptors and how alpha$_2$-adrenergic receptor activation results in the inhibition of adenylate cyclase activity has come from the study of the human platelet system. Thus, in the discussion of human alpha$_2$-receptor systems, further detail concerning the methods and conceptual approaches used to characterize alpha$_2$-adrenergic receptor-linked inhibition of adenylate cyclase will be provided. At present, alpha$_1$ receptors have been studied to only a limited extent in human tissues, and thus only a cursory overview of alpha$_1$-receptor mechanisms will be provided at the end of the chapter.

II. BETA-ADRENERGIC RECEPTORS IN HUMAN TISSUES

A. Molecular Basis for the Effects of Beta–Adrenergic Agents in Activation of Adenylate Cyclase

As shown in Figure 1, the beta receptor-adenylate cyclase complex consists of at least three distinct protein components: (1) the receptor component, R, which binds adrenergic agents at a recognition site on the external face of the cell membrane; (2) the catalytic subunit, C, which converts ATP to cyclic AMP; and (3) the multisubunit guanine nucleotide regulatory protein, G, which couples receptor occupancy to the stimulation of catalytic activity. Guanine nucleotides are absolutely required for receptor-mediated activation of adenylate cyclase, and appear to elicit their crucial effects via binding to a site(s) on the GTP-binding protein (see Refs. 1 and 2 for a review).

Figure 1 provides a schematic diagram of the molecular interactions which are felt to accompany beta-adrenergic stimulation of adenylate cyclase. Agonist occupancy of the beta receptor stabilizes interactions between the receptor and the GTP-binding protein (Ag · R · G com-

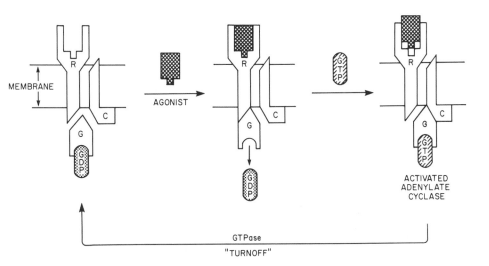

FIGURE 1 Molecular interactions that appear to accompany hormone-stimulated and guanine nucleotide-stimulated adenylate cyclase activity. R, receptor for hormones and drugs; G, multisubunit GTP-binding protein that binds guanine nucleotides and confers sensitivity to these agents; C, catalytic component. (From Ref. 2, used with permission.)

plex) (3). Presumably, through formation of this Ag $\cdot$ R $\cdot$ G complex, agonists facilitate the exchange of GDP for GTP at its binding site on G ($G_{GDP} \to G_{GTP}$) (4). This G_{GTP} species then dissociates from the Ag $\cdot$ R $\cdot$ G complex (5) and interacts in a more stable manner with C, forming $G_{GTP} \cdot C$ (6). It is this $G_{GTP} \cdot C$ species that is capable of synthesizing cAMP from ATP using the physiological substrate ATP-Mg^{2+}. It is important to understand that the naked catalytic subunit, C, can only synthesize cAMP in in vitro situations when presented with the substrate ATP-Mn^{2+} (1). Thus, an important role of hormones and agonists in activating adenylate cyclase is to continually replenish GTP at G so that the $G_{GTP} \cdot C$ complex, which represents catalytic activity, can be formed. Synthesis of cAMP continues until GTP is hydrolyzed to GDP by a GTPase activity within the adenylate cyclase system (7).

The above molecular interactions have interesting consequences for the interaction of agonists with the beta receptor. Thus, a variety of biochemical and radioligand binding data suggest that the R $\cdot$ G complex possesses a higher affinity for agonist than the isolated R species (8). This higher affinity interaction of agonists can be detected in agonist competition binding studies for radiolabeled antagonist binding to well-washed membrane preparations devoid (or possessing a reduced concentration) of endogenous guanine nucleotides. In this technique, membrane preparations (the source of receptors) are incubated with a single concentration of a radiolabeled beta-adrenergic antagonist and varying concentrations of an agonist competitor. With increasing concentrations of the competitor, radioligand binding is inhibited. This "indirect" technique allows an estimate of the affinity of the receptor for the agonist competitor to be made (cf. Chap. 00). In the absence of added guanine nucleotides, agonist competition curves for radiolabeled antagonist binding at beta receptors are of higher potency (shifted to the left) and "shallow" in shape (proceed from 90 to 10% competition over a greater than 81-fold range of competitor; pseudo-Hill coefficient 1). Computer curve-fitting analysis of the shallow competition curves allows these curves to be "dissected" into two interconvertible affinity states of the receptor, and a determination made for the K_d for the high-affinity state (K_H) and the low-affinity state (K_L) and an estimate of the fraction of the receptors binding agonist with the high-affinity state (%R_H) (8,9). Correlative biochemical studies suggest that receptor-agonist interactions in the high-affinity state probably represent a ternary complex of agonist · R · G. Competition curves in the presence of GTP or the hydrolysis-resistant analog of GTP, Gpp(NH)p, are of "normal" steepness (pseudo-Hill coefficient = 1). These observations are interpreted to result from the ability of guanine nucleotides to destabilize R · G complexes (cf. Fig. 1), and thus promote the formation of a homogeneous population of R, all functionally dissociated from G, which possesses a lower affinity for agonists.

It is important to keep in mind that the detection of these high- and low-affinity states of the receptor is only observed for agonist agents. Antagonist binding is characterized by competition curves of "normal steepness" (90-10% competition proceeds over an 81-fold range of competitor), indicating that the antagonist binds to the beta receptor with a single affinity which is not modulated by exogenous guanine nucleotides. Partial agonists exhibit a situation intermediate between that observed for agonists and antagonists. Thus, partial agonists demonstrate a lesser ability to form or perceive the "high-affinity state" of the receptor than full agonists and are less affected by the addition of exogenous guanine nucleotides (8,9).

B. Identification and Classification of Human Beta-Adrenergic Receptors

The vast majority of studies characterizing beta-adrenergic receptors in human tissues by radioligand binding techniques have employed radiolabeled beta-receptor antagonists because of their greater potency than available radiolabeled agonists and lower binding to nonreceptor binding sites in target membranes. Table 1 provides a summary of the human tissues in which beta-adrenergic receptors have been identified using both radiolabeled antagonists as well as agonists. It is important to emphasize that radiolabeled agonists identify only the high-affinity state of the beta receptor, since this affinity state has a K_d (equilibrium dissociation constant) in the nanomolar concentration. In contrast, the lower affinity state of the receptor for agonists has a K_d in the micromolar range. Using available techniques for separating receptor-bound from free radioligand to terminate the binding incubation, only binding to the high-affinity state can be measured because ligand dissociates so rapidly from the low-affinity state that it is lost during the separation procedure. Thus, agonist ligands tend to demonstrate lower values for receptor density than antagonist radioligands, which bind to the whole population of receptors with a single K_d in the picomolar to nanomolar range.

Table 1 also indicates the pharmacological subtype of beta-adrenergic receptor identified in particular human tissues. To evaluate the existence of either a single or mixed population of beta-adrenergic receptor subtypes in a given target membrane the biological preparation is incubated with a radiolabeled antagonist which possesses equal potency at beta$_1$- and beta$_2$-receptors and competition for this binding by subtype-selective agents is determined. When receptors of a single subtype are present, the competition curve is of "normal steepness," indicating a homogeneous receptor population exists. In contrast, in tissues possessing both beta$_1$ and beta$_2$ receptors, the competition curves of subtype-selective agonists or antagonists will be shallow, indicating that heterogeneity of receptor binding is occurring. Using computer curve fitting techniques completely analogous to those for the discrimination of high- and low-affinity states of the beta receptors for the binding of agonists, one can estimate the proportion of beta$_1$ and beta$_2$ receptors present in the tissue studied. For example, Stiles and co-workers (16) characterized the proportion of beta$_1$ and beta$_2$ receptors in human cardiac tissue for various chambers of the heart. They detected a significantly higher proportion of beta$_2$ receptors in the right atrium ($26 \pm 6\%$) than in the left ventricle ($14 \pm 1\%$). Heitz and co-workers using similar techniques have reported a β_1 / β_2 ratio of 65:35 in human myocardium, with no differences in beta-receptor subtype proportions between left atrial and left ventricular samples (17). Although these data are consistent with a role for beta$_2$-adrenergic receptors in cardiac function (e.g., in beta-adrenergic-mediated chronotropic response), their interpretation is limited, since the exact histological location of the beta$_2$ receptors cannot of course be characterized from myocardial homogenates. Thus, it is possible that these beta$_2$ receptors represent populations in coronary vascular

TABLE 1 Characterization of Beta-Adrenergic Receptors in
Human Tissues

Tissue	Radioligand	Subtype	References
Brain			
cerebellum	[³H]DHA[a]	N.C.[g]	10
cerebral microvessels	[¹²⁵I]HYP[b]	N.C.	11
Lung	[³H]DHA	beta$_2$	12
	[³I]CYP[c]	beta$_1$ and beta$_2$	13
Myocardium	[³H]bupranolol	N.C.	14
	[¹²⁵I]pABC[d]	beta	15
	[¹²⁵I]CYP	beta (pre-dominantly)	16
		beta$_1$, beta$_2$	17
Myometrium	[³H]DHA	beta$_2$	18
Placenta	[³H]DHA	beta$_2$	19
	[³H]DHA	beta$_1$	20
Adipocytes	[³H]DHA	beta$_1$	21
	[³H]ISO[e]	beta$_2$.	22
	[³H]DHA	N.C.	23
Penile corpus cavernosum	[³H]DHA	N.C.	24
Adrenocortical adenoma	[³H]DHA	N.C.	25
Circulating cells			
granulocytes	[³H]DHA	beta$_2$	26
	[¹²⁵I]HYP	N.C.	27
	[¹²⁵I]CYP	N.C.	45
lymphocytes	[³H]DHA	beta$_2$	28
	[¹²⁵I]HYP	beta$_2$	29
	[¹²⁵I]CYP	N.C.	30
	[³H]HBI[f]	N.C.	31
erythrocytes	[³H]DHA	beta$_2$	32, 33

Antagonist radioligands:
[a] [³H]Dihydroalpranolol.
[b] [¹²⁵I]Iodohydroxybenzylpindolol.
[c] [¹²⁵I]iodocyanopindolol.
[d] [¹²⁵I]p-azido benzyl carazolol (photolabel).

Agonist radioligands:
[e] [³H]isoproterenol.
[f] [³H]Hydroxybenzylisoproterenol.

Not a radioligand:
[g] Not characterized.

smooth muscle or fibrous tissue and may not be directly involved with the regulation of cardiac myocyte function.

Further characterization of the human cardiac beta receptor has been achieved using recently developed covalent labeling techniques which employ the photoaffinity ligand $[^{125}I]$p-azido-iodobenzylcarazolol ($[^{125}I]$pABC). In the absence of light, pABC acts as a reversible competitive antagonist at the beta receptor. However, when photo-activated, a nitrene is generated from the arylazido group on pABC, resulting in covalent binding of the ligand to the beta-receptor binding site, at least for some of the ligand-receptor complexes. This irreversi-ble covalent labeling technique allows the identification and character-ization of beta receptors under denaturing conditions such as SDS-polyacrylamide gel electrophoresis. Using this procedure, Stiles and co-workers (15) demonstrated that a 62,000 mol. wt. peptide incorporated $[^{125}I]$pABC and that propranolol prevented incorporation into this band. Bands of similar molecular size were also found in studies with canine, porcine, rabbit, and rat ventricular tissue, suggesting a common molecular weight of the $beta_1$ receptors in these species. An additional component of 55,000 mol. wt. appeared to be a proteolytic degradation product of the 62,000 mol. wt. band, since its appearance in SDS-polyacrylamide gels could be attenuated by protease inhibitors. Studies with frog ventricular tissue, which possesses beta receptors predomi-nantly of the $beta_2$ subtype, also demonstrated photoincorporation of $[^{125}I]$pABC into 62,000 and 55,000 mol. wt. bands. These data have been interpreted to suggest that there may be a structure common to $beta_1$- and $beta_2$-adrenergic receptor recognition sites. It should be mentioned, however, that similar molecular weight values for $beta_1$- and $beta_2$-receptor populations had not been obtained in earlier studies, and Stiles et al. suggest that the differences in the molecular weight reported previously for beta-receptor subtypes might be, at least in part, due to a variable degree of proteolysis in the different tissue preparations as well as subtle variations in methodology in preparing and analyzing biochemical material in different laboratories (15,34,35).

C. Beta-Adrenergic Receptors on Circulating Cells: Models for Regulation of Human Beta Receptors Both Physiologically and Pathologically?

One limitation of studies in human tissues using biopsy/autopsy speci-mens is that although they are useful for the identification and partial characterization of adrenergic receptors in a particular tissue, these studies can offer only limited insights into the regulation of the target organ physiologically or pathologically. This is because the specimens derived from either biopsies or autopsy examinations offer little oppor-tunity to control for the many variables which may regulate adrenergic receptor systems. Furthermore, because of the inevitable "lag time" involved in examination of these tissues, particularly for autopsy material, artifactual alterations in the system by proteolysis, for example, may supervene. In the case of biopsy materials, information derived from these specimens may be further limited because they are often from the area around a "pathological" site and "normal tissue" may not be available for comparison.

Beta receptors have also been identified on circulating blood cells, including lymphocytes and granulocytes (see Table 1). At the time of their earliest identification, these receptor populations were suggested as a potential model system for evaluating changes in beta-receptor function that might accompany particular pathological states. In contrast, for the platelet, which has been used extensively as a model for the alpha$_2$-adrenergic receptor, the evidence for a physiological beta-receptor population is less clear. Earlier studies of platelet aggregation had suggested the existence of beta receptors (36,37). However, radioligand binding studies to identify physiological platelet beta receptors have been inconclusive. Using [^{3}H]DHA and [^{3}H]acetobutalol, Kerry and Scrutton failed to identify physiological beta-adrenergic binding sites (38). In contrast, Steer and Atlas have reported the characterization of a platelet beta-receptor population (39).

Catecholamine stimulation of leukocytes has been reported to alter hemolytic plaque formation, mitogenesis of lymphocytes, and release of hydrolytic enzymes by granulocytes (40). These functions are felt to be subserved by the beta receptor. Preliminary reports have suggested the existence of an alpha$_2$-receptor population of lymphocytes (41,42); however, it is now felt that these observations were due to the unavoidable contamination of the mononuclear cell preparation by platelet fragments containing alpha$_2$ receptors (43). In contrast, one report has demonstrated the characterization of alpha$_2$ receptors on granulocyte membranes (44).

The first report of the characterization of beta receptors on leukocytes using radioligand binding techniques was by Williams and coworkers (28). Since this first report numerous investigators have confirmed the existence of a relatively homogeneous population of beta$_2$ receptors on both lymphocytes and granulocytes which is linked to stimulation of adenylate cyclase (26-31,45,46). Identification of leukocyte beta receptors has relied on several ligands, most commonly [^{3}H]dihydroalprenolol (DHA) or [^{125}I]iodohydroxybenzlpindol (HYP). Leukocyte beta receptors have been analyzed in both intact cell and broken cell preparations. (The reader is referred to Table 3 in Ref. 47 for a more complete listing of lymphocyte beta-receptor binding studies.)

As postulated in early studies, human leukocytes have been useful in studying the regulation of the beta-adrenergic receptor-adenylate cyclase system in pathological situations, particularly in patients with cardiovascular disease. In these studies, the implicit assumption is made that alterations of leukocyte beta-receptor properties reflect alterations at sites more critical in cardiovascular/respiratory regulation but less accessible for direct study in humans. The basis of this assumption rests on two lines of evidence. First, several authors have demonstrated that alterations in lymphocyte beta receptors parallel changes in beta-receptor-mediated cardiovascular function. Thus, Fraser and co-workers demonstrated that chronic elevations in circulating catecholamines were accompanied by parallel reductions in lymphocyte beta-receptor density and in beta-adrenergic-mediated cardiac chronotropic response (48). Similarly, Colucci et al. demonstrated that chronic administration of an adrenergic agonist resulted in tachyphylaxis of the inotropic effect in parallel with reduced leukocyte beta-receptor density and beta-adrenergic-stimulated adenylate cyclase activity (49). The above studies suggest that regulation of the cardiac beta-adrenergic system may be reflected by alterations in properties of lymphocyte beta receptors; however, the relatively scant evidence to date cannot be viewed as definitive at this time.

Another line of evidence suggesting that the properties of the beta-adrenergic receptor-adenylate cyclase system of circulating leukocytes might reflect changes in the properties of less accessible target organs comes from animal studies in which alterations in lymphocyte beta receptors under different conditions have been compared to changes in cardiac and/or lung tissue. Aarons and Molinoff reported that chronic administration of propranolol to rats resulted in increases in beta-receptor density in lymphocytes, heart, and lungs, suggesting a parallel regulation in these three tissues under these conditions (50). This parallel regulation, however, has not been universally observed. For example, Scarpace and Abrass, in their study of rat beta-receptor alterations in hyperthyroidism, found that T_3-induced increases in myocardial beta-receptor density were not accompanied by changes in beta-receptor density in lymphocytes or lung, suggesting organ-specific effects on beta-receptor regulation under these circumstances (51). Furthermore, it appears that the time course for regulation of the beta-adrenergic system may be species and/or organ specific. In a study of the acute effects of cortisone acetate administration in vivo on leukocyte beta-receptor density, Davies and Lefkowitz reported an increase in granulocyte beta-receptor density but reduced lymphocyte beta-receptor density at 4 hr posttreatment. However, at 24 hr posttreatment, receptor density was increased in both lymphocytes and granulocytes (52). The implication is that although the effect of steroids on beta-receptor density may be qualitatively the same in the two systems, the time course of regulation may differ. It should be noted that in the former studies in which parallel alterations in

target organ and circulating cell beta receptors were noted, changes
in beta-receptor density were provoked by a beta-adrenergic agent
(homologous effect), whereas in the latter examples, changes in beta-
receptor density detected in some tissues but not others was provoked
by a hormone acting via distinct receptors, e.g., T_3 or cortisone
(heterologous effect). Thus, it is possible that circulating leukocytes
may be a confident reflection of effects at other target organs only in
situations in which homologous regulation of the beta-adrenergic system
is suspected.

Although leukocytes have been used extensively as a model system
for evaluating the operation of the beta receptor-adenylate cyclase
complex in various clinical states, certain limitations to this system
should be appreciated. First, both mononuclear and polymorphonuclear
cell preparations are heterogeneous. Lymphocyte preparations consist
of a variable proportion of T cells, B cells, and null cells. The differential distribution of beta receptors on T versus B cells is argued.
Sheppard et al. showed that T cells have fewer beta receptors than
B cells (53). However, Bishopric et al., in a larger study, did not
find any differences in either the density of [^{3}H]DHA binding sites
or beta-adrenergic-mediated cyclic AMP accumulation between T and
B cells (54). More recently, Krawietz and co-workers in a study of
lymphocytes from three normal subjects reported a higher density of
[^{125}I]ICYP binding sites in B-cell-enriched fractions compared to
T-cell-enriched fractions (55). These kind of studies are however
difficult to interpret because the methods used to separate T and B
cells, i.e., selective absorption or separation of one cell type because
of its unique surface properties, could themselves affect membrane-
bound receptor systems, and differ between studies. Granulocytes
are usually assumed to be a more homogeneous cell population. However, recent studies have suggested that these cells may also exist
as subpopulations, at least as defined by expression of specific F_c
receptors. Furthermore, the shorter half-life of granulocytes may
make the granulocyte a less attractive model for the study of longer
term regulation of the beta-receptor system.

In both granulocyte and lymphocyte systems, studies are often
limited by the concentration of receptors available for assay. The
relatively low concentration of beta receptors (~ 50 fmol/mg protein)
has required large blood samples to be drawn (up to 700 ml of blood
in some studies). However, this limitation has been overcome with
the use of higher specific activity iodinated radioligands, e.g.,
[^{125}I]iodocyanopindolol and [^{125}I]iodohydroxybenzylopindolol. These
newer probes have enabled investigators to perform experiments with
as little as 10 ml of blood.

D. Beta Receptors in Other Human Tissues.
   Potential Use As Monitors of
   Pathophysiological States

Two other human tissues have also been used as model systems and
may be more extensively used in the future to monitor changes in the
beta-adrenergic receptor-adenylate cyclase system in pathological
states: the adipocyte beta receptor and erythrocyte beta receptor.
    Adipocytes have been demonstrated to possess beta and perhaps
beta receptors as well as $alpha_2$ receptors (21, 22, 56). The effect of
fasting on the human adipocyte alpha receptor was studied by Burns
et al. using sequential biopsies in seven normal volunteers (57).
The workers reported elevations in catecholamine-stimulated cAMP
accumulation with fasting consistent with a change in the balance
between $alpha_2$- and beta-receptor-mediated effects on the adipocyte.
This model system may thus be useful in further studies of adrenergic-
receptor regulation.
    $Beta_2$ receptors have been identified on human erythrocytes
(32-33). The authors also demonstrated that under standard binding
conditions using ultrafiltration techniques that up to 92% of the erythrocyte membrane protein in each incubation tube may be filtrated through
the glass fibers (perhaps related to the deformability of the erythrocyte
membrane). This may account for the previous difficulties of other
investigators in characterizing this system and may lead to the more
widespread use of this model system for human study of beta-receptor
regulation.

E. Characterization of R · G Coupling in Human
   Beta-Adrenergic Receptor Systems

As outlined earlier, the use of antagonist radioligands has permitted
the confident localization of beta receptors in many human tissues.
However, the technical approaches summarized to this point primarily
allow characterization of the recognition unit of the receptor. As
noted in Figure 1, important mechanisms in the regulation of beta-
adrenergic systems occur distal to the recognition unit of the receptor.
    Over the past several years, it has been appreciated that altera-
tions in the interaction between the beta receptor and the GTP-binding
protein result in changes in the functional "coupling" of the system,
i.e., the efficiency with which the signal of agonist binding to the
receptor is translated into adenylate cyclase activation. Furthermore,
alterations in coupling may be an important regulator of the system.
Alterations in coupling between the components of the beta-adrenergic
receptor systems have been studied most extensively in the context

of the phenomenon of desensitization. Two basic approaches have
been taken to infer whether or not changes in coupling are occurring.
An earlier technique employed for determining possible changes
in receptor-cyclase coupling in parallel with desensitization to catecholamines was to compare changes in catecholamine-stimulated adenylate
cyclase activity with changes in beta-receptor density. When decreases
in catalytic activity were quantitatively greater than could be accounted
for by decreases in receptors available for binding of catecholamines,
then an "uncoupling" of the system distal to receptor binding was said
to occur. This could be due to decreases in the density of other components in the adenylate cyclase system, i.e., G or C, or to changes
in functional communication of R with G or G with C or any combination
of the above. The disadvantage of this indirect comparison of receptor
binding with extent of cyclase activation, however, is that it cannot
localize the "lesion" in the beta receptor-adenylate cyclase system
further than distal to the recognition site. Nonetheless, this technique
was used by Krall and his colleagues to characterize alterations in
human lymphocyte beta receptors which occur following exposure to
beta-adrenergic agonists (58). They measured both beta-receptor
density as well as cAMP accumulation in intact lymphocytes following
exposure to 0.1 nM isoproterenol. They found a rapid reduction in
cAMP accumulation without any changes in beta-receptor density, and
thus concluded that uncoupling was an important regulator of the
response to persistent agonist exposure in the lymphocyte beta-
adrenergic receptor system.

A second approach for monitoring changes in coupling in beta-
adrenergic receptor-adenylate cyclase systems focuses on receptor-
GTP-binding protein "communication." As indicated schmatically in
Figure 1, the beta receptor binds agonists with one of two affinity
states: the high-affinity state corresponds to the binding by the
R · G complex and the lower affinity state corresponds to binding by
the isolated R species. As indicated above, the extent of formation
of the high-affinity state ($\%R_H$) can be determined by computer analysis
of heterogeneous agonist competition curves for radiolabeled antagonist
binding. The ability of the receptor to form the high-affinity state
correlates with its ability to activate adenylate cyclase. Thus, agonists
induce or perceive more R · G complexes than partial agonists, and
antagonists do not induce or stabilize this interaction between R and
G at all. Furthermore, desensitization of beta-adrenergic-stimulated
adenylate cyclase activity provoked by prolonged exposure to catecholamines has been linked to a reduction in beta-receptor affinity
for agonists in several in vitro model systems (9,59). This reduction
in affinity has been demonstrated to result from an impairment in the
ability to form the high-affinity state of the receptor for agonist.

Recent studies have demonstrated that high- and low-affinity
states of the beta receptor for agonists are also detected in the human

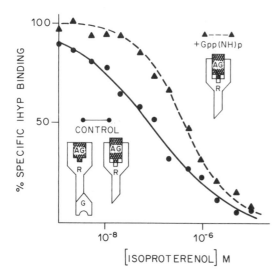

FIGURE 2 Competition of the agonist isoproterenol for the binding of the radiolabeled antagonist, [^{125}I]HYP, to human leukocyte membranes. In the absence of guanine nucleotides, the competition curve is shallow (pseudo-Hill coefficient < 1). This is consistent with receptor binding agonists (Ag) with both a high affinity (Ag-R-G complex) and a low affinity (Ag-R complex). With the addition of the hydrolysis-resistant GTP analog, Gpp(NH)p, the competition curve is "shifted" rightward and steepened, consistent with receptors binding agonist with one homogeneous "low" affinity under these conditions (all Ag-R complexes). (Data are from Ref. 46, used with permission.)

leukocyte. It has been demonstrated by Davies et al. (41) in the polymorphonuclear leukocyte and by Feldman and co-workers in the lymphocyte (42) that agonist competition for radiolabeled antagonist binding to beta receptors resulted in "shallow" curves (pseudo-Hill Coefficient < 1) which were shifted to the right and steepened by guanine nucleotides. Figure 2 provides an example of the type of competition binding data obtained for radiolabeled antagonist binding to human leukocytes. When analyzed using computer curve fitting programs, the data in human leukocytes are consistent with beta receptors binding agonist with two affinities and with guanine nucleotides mediating the transition from the high- to the low-affinity state of the receptor agonist. Feldman et al. (46) further demonstrated that the transition to the low-affinity complex was guanine nucleotide specific and was not mimicked by the hydrolysis-resistant adenosine triphosphate (ATP) analog App(NH)p. Furthermore the "shift" in competition curves with guanine nucleotides was agonist specific;

i.e., competition for radioligand binding by the antagonist propranol resulted in curves of normal steepness which were not altered by the further addition of guanine nucleotides. These data thus demonstrate that an agonist-stabilized high-affinity state of the receptor can be demonstrated in the circulating leukocyte. More recent comparisons of the percent of beta receptors that can perceive or induce the high-affinity state ($\%R_H$) of the receptor suggest that the $\%R_H$ detected on circulating leukocytes correlates inversely with the concentration of circulating catecholamines. Davies et al. (41) observed a similar reduction in the formation of the high-affinity state for agonists following exposure of polymorphonuclear cells to catecholamines in vitro. Thus, it appears possible that following the exposure of beta receptors to catecholamines and receptor-mediated activation of adenylate cyclase, the receptor might functionally "uncouple" from the GTP-binding protein. Since activation of cyclase proceeds from $R \rightarrow G \rightarrow C$, this functional uncoupling of R from G could provide a mechanism for tachyphylaxis or desensitization of the beta receptor-adenylate cyclase system, as receptor communication with C would be interrupted. In fact, the reciprocal decrease in $\%R_H$ that was detected in leukocytes exposed to increasing concentrations of catecholamines in the physiological range was paralleled by a decrease in the extent of stimulation of adenylate cyclase activity by the beta-adrenergic agonist isoproterenol (46). All of these changes, however, occurred without any decrease in the density of beta receptors detected using radiolabeled antagonist binding. Interestingly, a number of studies in in vitro model systems of the beta-adrenergic receptor-adenylate cyclase complex have demonstrated that desensitization to catecholamines is paralleled by a functional uncoupling of R from G that precedes later decreases in beta-adrenergic receptor density. Therefore, a more sensitive and functionally significant parameter to evaluate in situations in which sensitivity to catecholamines has declined may be the efficacy of beta-receptor-GTP-binding protein "communication" using agonist competition for the binding of radiolabeled antagonists to circulating leukocytes as an assay for this parameter rather than simply monitoring for possible changes in receptor density.

F. Identification of the GTP-Binding Protein That Mediates Activation of Adenylate Cyclase in Human Tissues: Changes in Density in Pathological Situations

The GTP-binding protein that mediates activation of adenylate cyclase is composed of at least three subunits, although it is depicted as a single molecular species in Figure 1. The 45,000 mol. wt. subunit possesses the binding site for GTP, and GTP occupancy of this site results in dissociation of the 45,000 mol. wt. subunit from its associated 35,000 and 8,000 mol. wt. subunits (60). It has been demonstrated that the 45,000 mol. wt. GTP-binding subunit can be identified using cholera toxin as a probe. Cholera toxin is an enzyme that cleaves nicotinamide adenine dinucleotide (NAD) and transfers the adenine diphosphate (ADP)-ribose moiety to the 45,000 mol. wt. subunit (61-63). This covalent modification results in the inhibition of the GTPase function of the adenylate cyclase system, thus at least in part accounting for the ability of cholera toxin to persistently activate adenylate cyclase (57). When $[^{32}P]$NAD is used as a substrate for cholera toxin, the 45,000 mol. wt. GTP-binding subunit responsible for activation of adenylate cyclase is not only covalently but also radioactively labeled as a result of the ADP-ribosylation reaction.

Two laboratories (64,65) have described a decrease in the density of the GTP-binding protein based on its diminished recognition and $[^{32}P]$ADP-ribosylation by cholera toxin in erythrocytes from patients with pseudohypoparathyroidism, type IA. This disease is characterized by a reduced responsiveness to several hormones linked to activation of adenylate cyclase activity, especially parathyroid hormone. Thus, if the decrease in the concentration of GTP-binding proteins detectable in human erythrocytes is a genuine reflection of target organs affected by pseudohypoparathyroidism, then a reasonable postulate to explain this disease is that receptor-cyclase coupling has declined secondary to a reduction in the density of the crucial intermediary protein, G.

The functional efficacy of the GTP-binding protein can be evaluated in reconstitution studies. In these experiments, the GTP-binding protein of a target tissue is removed from the membrane using a non-denaturing biological detergent, such as lubrol or cholate, and any contaminating activity of adenylate cyclase is eliminated by mild thermal inactivation. The detergent extract is then incubated with a "recipient" preparation, either membrane bound or similarly detergent solubilized, which possesses catalytic activity of adenylate cyclase that is not responsive to guanine nucleotides. The guanine nucleotide-induced activation of adenylate cyclase in the subsequent incubation of the mixed "G" and "C" preparations is felt to represent a monitor of the quantity and functional efficacy of the GTP-binding protein derived from the target tissue of interest (60-62). A commonly exploited recipient membrane for these studies is that of the cyc⁻ variant of S49 lymphoma cells, since this variant possesses beta receptors and the catalytic subunit, C, of adenylate cyclase, but lacks a functional GTP-binding protein (66-68).

Using the reconstitution approach described above, Abrass and Scarpace assayed the function of GTP-binding proteins in lymphocytes obtained from young and elderly subjects (69). These investigators

observed no alteration in the function of the GTP-binding protein with
aging. Thus, either the changes in beta-adrenergic sensitivity that
occur with aging do not result from changes in function at the level
of the GTP-binding protein or these functional changes in the GTP-
binding protein are no longer measurable following the solubilization
and reconstitution protocol employed in their studies. However, these
authors did demonstrate that the technique of reconstitution used to
characterize the function of the GTP-binding protein in nonhuman
model systems can be applied to the study of the beta-adrenergic
receptor-adenylate cyclase system in human leukocytes.

III. ALPHA-ADRENERGIC RECEPTORS
IN HUMAN TISSUES

As indicated at the outset of the chapter, catecholamines can elicit
their effects via a number of adrenergic receptors. Classically, alpha-
adrenergic receptors were characterized as those receptors which
mediated physiological effects of epinephrine and norepinephrine that
could be blocked by the selective alpha-adrenergic antagonist phentolamine. More recently, it has been appreciated that alpha-adrenergic
effects can be elicited by at least two pharmacological subtypes of
receptors, called alpha$_1$-adrenergic and alpha$_2$-adrenergic receptors.
The identification of adrenergic receptor subtypes is described in
detail elsewhere. Alpha-adrenergic receptors of the alpha$_1$ subtype
are specifically blocked by the antagonist prazosin. In contrast,
yohimbine is an alpha-adrenergic antagonist which selectively blocks
alpha$_2$-adrenergic effects. Alpha$_1$ and alpha$_2$ receptors are considerably
less selective in distinguishing among agonists. (See Ref. 70 for a
review.)
    As indicated earlier, beta-adrenergic receptors can also be divided
into two pharmacological subtypes, beta$_1$ and beta$_2$. An interesting
difference between subtypes of alpha-adrenergic receptors and subtypes of beta-adrenergic receptors is that although both beta$_1$- and
beta$_2$-adrenergic receptors are coupled to stimulation of adenylate
cyclase, each of the alpha-adrenergic receptors subtypes, at least at
the present time, appears to be coupled to a different "effector system."
Thus, alpha$_2$-adrenergic receptors are linked to inhibition of adenylate
cyclase, whereas alpha$_1$-adrenergic receptors alter calcium mobilization
and stimulate phosphatidylinositol turnover. Because the two subtypes
of alpha-adrenergic receptors appear to elicit their effects via distinct
physiological mechanisms, the identification and biochemical characterization of each of these subtypes will be discussed separately. The organi-
zation of the section related to alpha$_2$ receptors will be similar to that for
the discussion of beta-adrenergic receptors. Thus, identification of the

TABLE 2 Radioligands Used for the Identification of Alpha$_2$-Adrenergic Receptors in Human Tissues

Radioligands	Tissues
Antagonists	
[^{3}H] dihydroergocryptine	Intact platelet (72); platelet membranes (71-74); adipocytes (75)
[^{3}H] dihydroergonine	Intact platelets and platelet membranes (76)
[^{3}H] phentolamine	Platelet membranes (83,87)
[^{3}H] yohimbine	Intact platelets (77,78); platelet membranes (79-82,87); adipocytes (84-86)
[^{3}H] rauwolscine	Cerebral cortex (88); platelet membranes (121)
Agonists	
[^{3}H] epinephrine	Platelet membranes (81,82)
[^{3}H] norepinephrine	Platelet membranes (87)
Partial Agonists	
[^{3}H] clonidine	Platelet membranes (79,89); adipocytes (50,56,85); brain (90)
[^{3}H] para-aminoclondine	Platelet membranes (91)

components of the alpha$_2$ receptor-adenylate cyclase system, i.e., the receptor and GTP-binding protein, will be followed by a discussion of the experimental methods for determining the interaction between these two components.

A. Identification of Alpha$_2$-Adrenergic Receptors in Human Tissues

Alpha$_2$-adrenergic receptors have been identified in a number of human tissues and using a number of radioligands, including antagonists, agonists, and partial agonists at alpha$_2$-adrenergic receptors. Table 2 outlines the radioligands used and the tissues on which alpha$_2$-adrenergic receptors have been identified. Properties of the different radioligands will be discussed only briefly. Suffice it to say that in all findings cited, the binding sites identified appeared to represent the physiologically relevant receptor based on a number of characteristics of radioligand binding, including the saturability of binding, the specificity of a number of agents in competing for radioligand binding, and kinetics consistent with binding to the physiologically relevant alpha$_2$-adrenergic receptor.

[^{3}H]Dihydroergocryptine (DHE) was the first radiolabeled antagonist used to identify alpha$_2$-adrenergic receptors in human tissues (71). It was shown that human platelets possessed [^{3}H]DHE binding sites which demonstrated the specificity characteristic of alpha$_2$-adrenergic receptors (71-74). Later, [^{3}H]yohimbine and [^{3}H]rauwolscine, both antagonists selective for receptors of the alpha$_2$ subtype, were used to identify alpha$_2$-adrenergic receptors in both human platelet and human fat preparations (77-86). Since both [^{3}H]DHE and [^{3}H]phentolamine demonstrate a substantial amount of nonspecific binding when present at higher concentrations (77,83), it may be preferable to use [^{3}H]yohimbine or [H]rauwolscine to identify alpha$_2$-adrenergic receptors, at least in human platelets where the potency of these agents is anomalously 10-fold higher than in other tissues possessing alpha$_2$ receptors.

The radiolabeled agonists [^{3}H]epinephrine (81,82) and [^{3}H] norepinephrine (87) and the partial agonists [^{3}H]clonidine (56,79, 85, 89) and [^{3}H]para-aminoclinidine (91) have also been used to identify alpha$_2$ receptors in human tissues. In a manner analogous to agonists interacting at beta-adrenergic receptors, it has been demonstrated that agonists interacting with alpha$_2$-adrenergic receptors do so in both high- and low-affinity "states." The high-affinity state binds agonists with K_d in the nanomolar range, whereas the low-affinity state for agonists manifests a K_d in the micromolar range. As mentioned earlier for radiolabeled agonist binding to beta-adrenergic receptors, the method routinely used for separating receptor-bound from free radioligand, i.e., vacuum filtration, does not allow the low-affinity state of receptor-agonist interactions to be "trapped" when radiolabeled agonists are used to identify alpha$_2$ receptors because of the rapid rate of radioligand dissociation from these low-affinity complexes. Thus, what is identified with radiolabeled agonist and partial agonist binding is the so-called high-affinity state of the alpha$_2$-adrenergic receptor for agonist. A more thorough discussion of the affinity states of the receptor for agonist and the molecular counterparts of these affinity states is provided below.

B. Identification of the GTP-Binding Protein Mediating Inhibition of Adenylate Cyclase

Alpha$_2$-adrenergic receptors are one of a number of receptor populations which, subsequent to agonist occupancy, are capable of inhibiting adenylate cyclase activity in broken cell preparations. In a manner analogous to stimulatory adenylate cyclase systems, guanine nucleotides are required for hormonal inhibition of adenylate cyclase and also modulate receptor affinity for agonists. Despite the important role of GTP in both activation and inhibition of adenylate cyclase, a

TABLE 3 Properties of the GTP-Binding Regulatory Proteins Linked to Stimulation (G_s) and Inhibition (G_i) of Adenylate Cyclase

Property	G_s	G_i
GTP-binding subunit	45,000 mol. wt.	41,000 mol. wt.
Cholera toxin substrate	45,000 mol. wt.	—
IAP substrate	—	41,000 mol. wt.
GTP decreases receptor affinity for agonist	Yes	Yes
ED_{50} for GTP in changing cyclase activity	~0.1 μM	≤10 μM
Sensitivity to Mn^{2+} to uncouple G → C	Less sensitive	More sensitive
Sensitivity to NEM to uncouple R → G	Less sensitive	More sensitive
Sensitivity to proteases to uncouple R → G → C	Less sensitive	More sensitive

substantial amount of evidence has accumulated which suggests that distinct GTP-binding proteins are involved in activation and inhibition of cyclase. (See Ref. 92 for a review.) Thus, 10-fold higher concentrations of GTP are typically required for inhibition of adenylate cyclase than for activation of the enzyme in the same tissue. In addition, the greater sensitivity of hormone-mediated inhibition of cyclase to perturbation by divalent cations, proteases, radiation inactivation, and sulfhydryl-directed reagents than noted for GTP-dependent hormone activation of cyclase has provided further evidence that a GTP-binding protein distinct from the GTP-binding protein that mediates activation of adenylate cyclase is responsible for inhibitory effects of agonists on adenylate cyclase activity. For brevity, we will refer to the "stimulatory GTP-binding protein" as G_s and the "inhibitory GTP-binding protein" as G_i. Table 3 provides a comparison of some of the known biochemical and functional properties of G_s and G_i.

A method for the selective identification of G_i has evolved only recently. One of the toxins of *Bordetella pertussis*, termed islet-activating protein (IAP), has been shown to cause a time- and concentration-dependent inhibition of hormonal attenuation of adenylate cyclase in a number of target systems (93-96). The first system characterized was the pancreas, where it was noted that IAP eliminated

alpha$_2$-adrenergic attenuation of cAMP accumulation (93). Since this
alpha$_2$-adrenergic response is responsible for a tonic repression of
insulin secretion, the physiological effect of IAP is to enhance insulin
release—hence the appropriateness of the toxin name, islet-activating
protein. Islet-activating protein has been demonstrated to be an
enzyme possessing ADP-ribosyl transferase activity, and has been
shown to ADP-ribosylate a 41,000 mol. wt. GTP-binding protein distinct from the 42,000-45,000 mol. wt. GTP-binding protein which
stimulates adenylate cyclase and is a substrate for cholera toxin-
catalyzed ADP-ribosylation (97). It is felt that this IAP-catalyzed
covalent modification causes a change in the function of the 41,000
mol wt. GTP-binding protein which results in IAP elimination of alpha$_2$-
adrenergic attenuation of cAMP accumulation (96). Thus, the inhibitory
GTP-binding protein, G$_i$, probably contains this 41,000 mol. wt. subunit, and IAP thus represents a useful tool for specific identification
of the GTP-binding protein-mediating inhibition of adenylate cyclase.
    The IAP substrate has been purified from rat liver, and appears
to contain at least three subunits: the 41,000 mol. wt. GTP-binding
subunit ADP-ribosylated by IAP and associated 35,000 and 8,000 mol.
wt. subunits that dissociate from the 41,000 mol. wt. subunit upon binding of guanine nucleotides (98). This 35,000 mol. wt. species appears
to be identical to that subunit which is associated with the 45,000
mol. wt. GTP-binding protein which is recognized by cholera toxin
and responsible for activation of adenylate cyclase (99). Since the
concentration of the 35,000 mol. wt. species has been demonstrated
to regulate the extent of activation of cyclase by the 45,000 mol. wt.
subunit in reconstitution experiments, it has been postulated that the
mechanism whereby agents inhibit adenylate cyclase may be due to
regulation of the concentration of "free" 35,000 mol. wt. subunit,
thereby altering by mass action law the concentration of "free" 45,000
mol. wt. subunit available for activation of the catalytic moiety of
adenylate cyclase (100). This mechanism, however, is still speculative.
Thus, although it is reasonably certain that distinct GTP-binding
proteins, G$_s$ and G$_i$, mediate stimulation and inhibition of adenylate
cyclase, respectively, it is still not clear whether G$_i$-mediated inhibition
of adenylate cyclase is via direct inhibition of the catalytic subunit
or indirectly via influencing the efficacy of G$_s$-C interactions or a
combination of both mechanisms (Fig. 3).
    Despite the involvement of different GTP-binding proteins in
inhibition and activation of adenylate cyclase, the similar requirement
for guanine nucleotides to observe both processes and the ability of
guanine nucleotides to modulate receptor affinity for agonist agents
at receptors that mediate either activation or inhibition of adenylate
cyclase has suggested that the same molecular sequelae might accompany

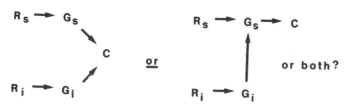

FIGURE 3 Molecular interactions between alpha$_2$ receptors and other components of the adenylate cyclase system.

hormone-induced inhibition of adenylate cyclase as have been shown to accompany activation of adenylate cyclase. The molecular events involved in activation of adenylate cyclase summarized in Figure 1 and outlined earlier were determined primarily from studies in avian and amphibian erythrocytes and cell culture systems. In contrast, some of the most thorough mechanistic and biochemical data concerning alpha$_2$-adrenergic receptors and receptor-induced function have been obtained in a human tissue, the human platelet. The platelet has provided an excellent model system for studies of the molecular basis of alpha$_2$-adrenergic receptor-adenylate cyclase coupling because of its accessibility, its availability as a single cell type, and the fortuitous situation that the human platelet possesses alpha-adrenergic receptors of only the alpha$_2$-adrenergic subtype. Because a great deal of bio-chemical evidence concerning alpha$_2$-receptor function has been obtained in human tissue, the experimental approaches will be considered in greater detail here than was discussed earlier for beta-receptor func-tion.

Studies utilizing a number of experimental approaches have demonstrated that, in fact, the molecular interactions involved in inhibition of cyclase strikingly resemble those involved in activation of adenylate cyclase. For example, receptor occupancy by agonist agents results in an interaction between the receptor and the GTP-binding protein, which can be monitored in a number of ways, each of which will be described below.

One manifestation of agonist-promoted alpha$_2$ receptor-GTP-binding protein interaction is that guanine nucleotides, via binding to the GTP-binding protein, modulate receptor affinity for agonist agents. Tsai and Lefkowitz demonstrated that GTP decreases alpha$_2$-receptor affinity for agonists by 10-20-fold (101). In studies of competition for [^{3}H]DHE binding to human platelet membranes by agonists, partial agonists, and antagonists, it was noted that the ability of guanine nucleotides to modulate receptor-agonist interactions directly correlated with the intrinsic activity of the agonists in mediating inhibition of adenylate cyclase. Thus, greater shifts in competition curves promoted by GTP, or its hydrolysis-resistant analog Gpp(NH)p, were observed for full agonists than for partial agonists and no Gpp(NH)p-mediated decreases in receptor affinity were noted for receptor-antagonist interactions (101). Agonist competition curves were characterized by a "shallow" shape. In a manner analogous to studies on beta-adrenergic receptors, computer modeling of these shallow competition curves has dissected out a high affinity and low affinity component of the binding curve, and has allowed the determination of the K_d of the high-affinity state (K_H) and of low-affinity state (K_L) as well as the fraction of the total receptor population existing in either affinity state ($\%R_H$, $\%R_L$) (76,82,102). The addition of guanine nucleotides to the competition binding incubation simultaneously decreases receptor affinity for agonist and causes the agonist competition curve to become one of normal steepness, i.e., that expected of a single, homogeneous population of receptors interacting with an agonist with a single affinity. Computer modeling of the competition curves obtained in the presence of guanine nucleotides has demonstrated that the addition of Gpp(NH)p converts all receptors to the low affinity state by the addition of guanine nucleotides (82,102). Figure 4 demonstrates the effects of guanine nucleotide on agonist competition for [³H]yohimbine binding to alpha₂-adrenergic receptors in human platelet membranes.

One consequence of the interaction between alpha₂ receptors and the GTP-binding protein is that alpha-adrenergic agonists promote the exchange of GDP for GTP and other guanosine triphosphates at the GTP-binding protein modulating alpha₂-receptor affinity for agonists, in a manner entirely analogous to effects of agonists noted in previously reported studies on beta receptors and shown schematically in Figure 1. Thus, alpha₂-agonists, but not antagonists, provoke the release of prebound [³H]Gpp(NH)p from human platelet membranes in a time- and concentration-dependent manner, and partial agonists facilitate [³H]Gpp(NH)p release to a lesser extent (103,104). PGE₁, an agent that activates adenylate cyclase in the human platelet, also promotes the release of [³H]Gpp(NH)p from human platelet membranes, but the quantity of [³H]Gpp(NH)p released in response to PGE₁ is additive to that released by epinephrine (103). These findings are consistent with findings discussed above indicating that the adenylate cyclase-coupled "stimulatory" PGE₁ receptor and "inhibitory" alpha₂-adrenergic receptor, although both regulated by guanine nucleotides, are probably associated with distinct pools of GTP-binding proteins.

The ability of agonists to promote or stabilize an interaction between the alpha₂-receptor and the GTP-binding protein modulating receptor affinity for agonists can also be demonstrated by more molecu-

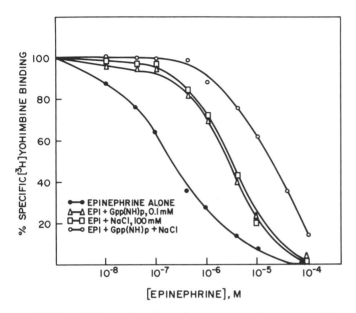

FIGURE 4 Effect of sodium ion and guanine nucleotides on human platelet alpha$_2$-receptor-agonist interactions. Human platelet membranes were incubated with 7.5 mM [^{3}H]yohimbine as described in Ref. 113.

lar techniques than those described above. Thus, the alpha$_2$ receptor can be solubilized from human platelet membranes using the gentle biological detergent digitonin. Incubation of human platelet membranes with 1% digitonin-containing buffers for 30 min at 4-15° results in the solubilization of the major fraction of the alpha$_2$ receptors from the human platelet membrane (81). The detergent-solubilized receptors can be recovered in a 100,000 × g supernatant fraction, and have been demonstrated to be free of the biological membrane by a number of functional and morphological criteria (81). The digitonin-solubilized receptor possesses an affinity for antagonists identical to that characteristic of the intact membrane. In contrast, solubilization of the human platelet alpha$_2$ receptor results in a selective decrease in alpha$_2$-receptor affinity for agonist agents. Thus, agonist competition curves in solubilized preparations are shifted to the right compared to human platelet membranes, and to a position characteristic of agonist competition curves obtained in membranes incubated in the presence of guanine nucleotides. Furthermore, guanine nucleotides no longer modulate receptor-agonist interactions following solubilization (81). One interpretation of these findings is that solubilization of the human platelet

membrane has resulted in a loss of the receptor-GTP binding protein
interactions responsible for both higher affinity receptor-agonist
interactions as well as the effects of the guanine nucleotides on these
interactions in the native platelet membrane. Occupancy of the human
platelet alpha$_2$-adrenergic receptor with the agonist [^{3}H]epinephrine
prior to solubilization results in a retention of higher affinity receptor-
agonist interactions as well as their sensitivity to guanine nucleotides.
The sensitivity of prelabeled receptor-agonist complexes to guanine
nucleotides can be demonstrated by the ability of Gpp(NH)p to facilitate
[^{3}H]epinephrine dissociation from these complexes. The ability of
agonist occupancy of the human platelet alpha$_2$-adrenergic receptor
prior to detergent solubilization to maintain receptor sensitivity to
guanine nucleotides is paralleled by an agonist-promoted (or -stabilized)
increase in apparent receptor size (75,96). Thus, agonist receptor
complexes sediment more rapidly in sucrose gradients than either
antagonist-receptor complexes (81,105) or receptors solubilized from
the human platelet membrane in an unoccupied state (81). These
findings provide molecular evidence that when the alpha$_2$-adrenergic
receptor is occupied by an agonist it has a more stable interaction
with the GTP-binding protein that modulates receptor affinity than
when it is without agonist occupancy. In earlier studies of the beta-
adrenergic receptor, it had been demonstrated that the agonist-
stabilized beta-receptor complex of larger molecular size contained
the GTP-binding protein which mediates activation of cyclase and
which was recognized by cholera toxin as a substrate for ADP-
ribosylation (3). In contrast, it has been demonstrated that the
alpha$_2$ receptor of the human platelet does not appear to associate
with the cholera toxin substrate upon occupancy with an agonist (106).
Thus, these findings lend additional support to the hypothesis that
distinct GTP-binding proteins mediate activation and inhibition of
adenylate cyclase. As mentioned earlier, a possible candidate for
the GTP-binding protein that mediates inhibition of adenylate cyclase,
and thus may associate with the agonist-occupied alpha$_2$-adrenergic
receptor, is the 41,000 mol. wt. GTP-binding protein that is recognized
as a substrate for ADP-ribosylation by IAP.
    Another possible monitor of agonist-promoted receptor-GTP-
binding protein interactions is agonist-activated GTP hydrolysis.
As shown in Figure 1, GTP hydrolysis is the mechanism by which
activation of adenylate cyclase activity is felt to terminate. Hormones
that activate adenylate cyclase activity also activate GTPase activity.
This may seem anomalous, since one might not expect that a stimulatory
hormone would simultaneously activate the adenylate cyclase system
and the GTPase "turn-off" mechanism. However, since one consequence
of the agonist-facilitated release of guanine nucleotides from the GTP-
binding protein is an acceleration of the rate at which nucleoside tri-

phosphates can rebind to the GTP-binding site, it is likely that activation of GTPase by hormones that stimulate adenylate cyclase is a manifestation of agonist-promoted substrate replenishment at the GTPase substrate binding site, rather than an increase in the velocity of the hydrolysis reaction per se.

Recently, hormones that are capable of inhibiting adenylate cyclase activity, like alpha$_2$-adrenergic agents in human platelets (98), have been demonstrated to activate GTP hydrolysis, and this effect on GTPase can be selectively inhibited by IAP (108) but not cholera toxin (107). This latter finding suggests that the GTPase being stimulated by alpha$_2$-adrenergic agents is at G_i and not G_s. However, the data are still inconclusive as to whether or not alpha$_2$-adrenergic-stimulated GTP hydrolysis results from agonist-promoted substrate replenishment at the GTP-binding site or agonist-activated hydrolysis per se. As noted above, alpha$_2$ agents accelerate [^{3}H]Gpp(NH)p release from human platelet membranes, indicating that these agents can facilitate GDP/GTP exchange. Furthermore, the observation that IAP inhibits alpha$_2$-adrenergic GTPase activity may result from the ability of IAP to destabilize receptor-GTP-binding protein interactions, an effect of IAP which has been observed using agonist competition binding studies in cultured cell (95) and human platelet (109) preparations. Thus, although it is a reasonable postulate that hormones and agonist drugs which inhibit adenylate cyclase activity may do so by accelerating the turn-off mechanism of the overall adenylate cyclase system, it is entirely possible that the ability of alpha$_2$ agonists to activate GTPase activity in in vitro incubations may be yet another monitor of effective receptor-GTP binding protein interactions.

C. The Role of Sodium in Alpha$_2$-Adrenergic-Mediated Inhibition of Adenylate Cyclase

Guanine nucleotides are not the only amplifiers of hormone-mediated inhibition of adenylate cyclase. Sodium ion has been demonstrated to be required for or to at least enhance hormonal inhibition of adenylate cyclase in a number of systems (110). Sodium ion also decreases receptor affinity for agonists at receptors coupled to inhibition of cyclase (cf. Fig. 4). These effects of sodium on receptor-agonist interactions demonstrate a specificity of $Na^+ > Li^+ > K^+ \gg$ choline and an EC_{50} for Na^+ of ≤ 40 mM. In Rabbit platelet membranes, it has been demonstrated that although sodium decreases receptor affinity for agonist agents, the competition curve detected in the presence of Na^+ is still not of normal steepness like that observed in the presence of GTP (111). The findings have been interpreted to suggest that the heterogeneity of receptor-agonist interactions may not be entirely eliminated by the addition of sodium ion and that the mechanism by which Na^+

modulates receptor-agonist interactions differs from the mechanism by which GTP decreases receptor affinity for agonists. In both human platelet and rabbit platelet membranes, it has been demonstrated that the addition of Na^+ and GTP together further reduces $alpha_2$-receptor affinity for agonists than either agent alone (112,113). These data suggest that there is either an additive or a synergistic effect of sodium and GTP on receptor-agonist interactions. Sodium, like GTP, does not decrease receptor affinity for antagonist agents, and if anything, causes a slight increase in receptor affinity for antagonists over the same concentration range that Na^+ mediates decreases in $alpha_2$-receptor affinity for agonists (113).

The phenomenological similarity of the effects of sodium and guanine nucleotides on human platelet receptor-agonist interactions and the synergistic effects of sodium and GTP in mediating inhibition of adenylate cyclase has suggested to some investigators that the effects of sodium may be mediated via a monovalent cation binding site that exists on the GTP-binding protein which "couples" the $alpha_2$-adrenergic receptor to the adenylate cyclase complex (91,114). However, a number of perturbants are able to differentially resolve the effects of GTP from those of sodium on receptor-agonist interactions in the human platelet system (115). Thus, incubation of intact platelets with the sulfhydryl-directed reagent N-ethyl-maleimide (NEM) results in an uncoupling of $alpha_2$-adrenergic-mediated receptors from the inhibition of adenylate cyclase (116). In membranes subsequently prepared from NEM-treated human platelets, there is no change in the recognition of the receptor for antagonist agents but there is a decrease in receptor affinity for agonists. This decrease in receptor affinity for agonists is paralleled by a loss in the ability of guanine nucleotides to modulate receptor-agonist interactions. Taken together, these findings suggest that NEM treatment of the intact platelet destabilizes receptor interactions with the GTP-binding protein responsible both for higher affinity receptor-agonist interactions and for the effects of guanine nucleotides on these interactions. In contrast, however, the ability of sodium to decrease receptor affinity for agonists is not altered in membranes derived from NEM-treated platelets. Another perturbant which eliminates the effects of GTP on receptor-agonist interactions without affecting sodium-promoted decreases in receptor affinity for agonists is elevated temperature (115). Thus, incubation of human platelet membranes at 45° for 30 min (or 60° for 5 min) causes no alterations in [^{3}H]yohimbine binding but causes a selective decrease in receptor affinity for agonists, which again is paralleled by a loss in the ability by GTP to modulate receptor-agonist interactions. However, as with NEM, elevated temperature does not modify the ability of sodium ion to decrease receptor affinity for agonist agents. These above two findings provide indirect data consistent with the hypothesis that distinct components in the human platelet membrane are mediating the effects of GTP and sodium on receptor-agonist interactions. However, these data cannot rule out the possibility that the same molecular component mediates the effects of GTP and Na^+, but via different domains, and that these domains are differentially sensitive to alkylation or elevated temperature. The observation that sodium still affects alpha$_2$ receptor-agonist interactions in detergent-solubilized extracts of human platelets has provided more convincing evidence that distinct molecular components mediate the effects of GTP and sodium on the human platelet alpha$_2$-adrenergic receptor system, since previous studies have suggested that solubilization of the alpha$_2$ receptor from the human platelet membrane in an unoccupied state resolves the receptor from the GTP-binding protein modulating receptor affinity for agonist (81). The conclusion that Na^+ and GTP elicit their effects via distinct molecular mechanisms, and presumably distinct components, has also been drawn for muscarinic receptor systems that mediate inhibition of adenylate cyclase (117). Thus, it may be a general property of inhibitory adenylate cyclase systems that the effects of sodium are mediated by a component distinct from the GTP-binding subunit that modulates receptor-agonist interactions. It has not been demonstrated by any laboratory at present, however, whether the effects of sodium result from interaction with monovalent cation binding site existing on the receptors themselves, on an entirely unique peptide that is part of an already identified component, or on a not yet characterized component within the adenylate cyclase system architecture.

D. The Potential Usefulness of the Human Platelet Alpha$_2$-Adrenergic Receptor–Adenylate Cyclase System for the Monitoring of Pathophysiological Situations

Since the human platelet is a readily accessible circulating cell, the possibility arises that investigation of the platelet alpha$_2$-adrenergic receptor in certain pathophysiological situations may shed some light on possible receptor-associated lesions responsible for or associated with particular disease states. In fact, changes in alpha$_2$-receptor density have been monitored in several clinical situations, e.g., hypertension, and results and interpretations of these studies are discussed in further detail elsewhere. One potential difficulty with these studies, however, is that the platelet alpha$_2$ receptor is somewhat atypical in its recognition properties, at least for antagonists. Thus, yohimbine is equipotent with or slightly more potent than phentolamine in the human platelet, but is 10-fold less potent than phentolamine in other target membranes possessing alpha$_2$ receptors (70). Furthermore, it must also be pointed out that it is only appro-

priate to speculate that the platelet may be a potential monitor of
alpha-adrenergic function in less accessible target organs when the
physiological effect in the other target organ is mediated via alpha-
adrenergic receptors of the $alpha_2$ subtype, and not the $alpha_1$
subtype. This is a more critical distinction when studying alpha
receptors than when studying beta receptors, e.g., on circulating
lymphocytes, because, as stressed earlier, alpha receptors of the
$alpha_2$ subtype mediate the physiological effects of catecholamines by
an effector mechanism entirely distinct from that utilized by $alpha_1$
receptors, whereas both $beta_1$ and $beta_2$ receptors mediate the effects
of catecholamines by stimulating cAMP accumulation. Whether $alpha_2$-
adrenergic receptors, e.g., are important physiological mediators of
catecholamine regulation of cardiac function or vascular tone has not
yet been rigorously documented, and thus monitoring possible changes
in $alpha_2$-receptor function associated with cardiovascular disease by
characterizing changes in platelet $alpha_2$ receptors and receptor-
cyclase coupling may be premature. However, it is worth noting that
the study of $alpha_2$ receptors on human platelets may be useful beyond
the scope of cardiovascular disease. For example, since $alpha_2$
receptors are known to be negative feedback regulators of norepinephrine release at synapses in the central nervous system, the study
of human platelet $alpha_2$ receptors has been suggested as a potential
monitor of various psychiatric disorders (118).

E. Identification of Alpha₁-Adrenergic Receptors in Human Tissues

$Alpha_1$-adrenergic receptors were first differentiated from receptors
of the $alpha_2$ subtype based on their pharmacological specificity. The
most striking selectivity for $alpha_1$-adrenergic receptors is demonstrated
by the antagonist prazosin, which at reasonable concentrations (≤ 0.1
μM) is entirely selective for alpha receptors of the $alpha_1$ subtype.
Yohimbine has a higher potency at $alpha_2$-adrenergic receptors than
at $alpha_1$-adrenergic receptors, but can interact with $alpha_1$-adrenergic
receptors at higher concentrations, so that the most useful criterion
for differentiating the two subtypes is the order of potency of adrenergic antagonists (prazosin >>> yohimbine) in blocking the effects of
catecholamines. As indicated earlier, the extant data suggest that
$alpha_1$-adrenergic receptors are coupled to phosphatidylinositol turnover and to calcium mobilization (cf. Ref. 119 for an excellent didactic
review of these postulated mechanisms). The biochemical characterization of the sequelae which follow agonist occupancy of $alpha_1$-adrenergic
receptors has lagged somewhat behind that for $alpha_2$-adrenergic
receptors, perhaps because the effects of $alpha_1$-adrenergic agents
on changes in membrane lipid composition and Ca^{2+} distribution cannot

be studied in the broken cell and thus are not amenable to extensive characterization of receptor-provoked molecular events.

At the time of the writing of this chapter, $alpha_1$-adrenergic receptors had been identified directly with radioligand binding techniques in only two human tissues; the brain (90); using [^{3}H]WB-4101, and the uterine myometrium, using [^{3}H]prazosin (120). Since [^{3}H]prazosin will selectively identify $alpha_1$-receptors, the authors of this study hope that it will become possible to determine whether the reported influence of sex steroid hormones on the total alpha-adrenergic receptor population, identified using [^{3}H]dihydroergocryptine, applies to the $alpha_1$ subpopulation or not.

Unfortunately, although possible changes in the properties of $alpha_1$ receptors and of $alpha_1$ receptor-mediated function(s) are of extreme interest for the understanding of a number of pathological states, a suitable model system in a circulating cell has yet to be identified which would permit the facile monitoring of $alpha_1$ receptors in clinical situations.

IV. CONCLUSIONS

Beta-adrenergic agents stimulate and $alpha_2$-adrenergic agents inhibit adenylate cyclase activity. The regulation of adenylate cyclase by adrenergic agents is mediated via GTP-binding proteins. The multisubunit GTP-binding protein that conveys activation of cyclase posses possesses a 45,000 mol. wt. GTP-binding subunit that can be identified by cholera-toxin catalyzed [^{32}P]ADP-ribosylation, whereas the inhibitory GTP-binding protein possesses a distinct 41,000 mol. wt. GTP-binding subunit that can be identified by *Bordetella pertussis* (IAP)-catalyzed [^{32}P]ADP-ribosylation. Agonist occupancy of its specific receptor initiates the flow of communication from the receptor to the GTP-binding protein to the catalytic moiety ($R \rightarrow G \rightarrow C$). In pathological situations characterized by decreased sensitivity to catecholamines, changes in receptor or GTP-binding protein density or changes in communication between any of the components of the adenylate cyclase system could account for the decline in responsiveness of target tissues. Radioligand binding studies can provide both information on receptor density and R-G coupling, since the ability of guanine nucleotides binding at G to influence the potency of agonists binding to R can be assessed in competition binding studies performed in the absence and presence of guanine nucleotides. The circulating leukocyte has been exploited as a readily accessible tissue for monitoring changes in the beta-adrenergic receptor-adenylate cyclase system in clinical situations. In contrast, the validity and usefulness of the human platelet as a model system for monitoring changes in $alpha_2$-receptor functions in pathological states is only now being rigorously evaluated. Finally, the identification of alpha$_1$-adrenergic receptors in human tissues and their possible coupling to Ca^{2+} mobilization and phosphatidylinositol turnover is just beginning.

REFERENCES

1. Ross, E. M., and Gilman, A. G.: Biochemical properties of hormone-sensitive adenylate cyclase. Ann. Rev. Biochem. J. 49:533-564, 1980.

2. Limbird, L. E.: Activation and attenuation of adenylate cyclase. Biochem. J. 195:1-13, 1981.

3. Limbird, L., Gill, D. M., and Lefkowitz, R. J.: Agonist-promoted coupling of the β-adrenergic receptor with the guanine nucleotide regulatory protein of the adenylate cyclase system. Proc. Natl. Acad. Sci. U.S.A. 77:775-779, 1980.

4. Cassel, D., and Selinger, Z.: Mechanism of adenylate cyclase activation through the β-adrenergic receptor: Catecholamine-induced displacement of bound GDP by GTP. Proc. Natl. Acad. Sci. U.S.A. 75:4155-4159, 1978.

5. Limbird, L. E., Gill, D. M., Stadel, J. M., Hickey, A. R., and Lefkowitz, R. J.: Loss of β-adrenergic receptor-guanine nucleotide regulatory protein interactions accompanies decline in catecholamine responsiveness of adenylate cyclase in maturing rat erythrocytes. J. Biol. Chem. 255:1854-1861, 1980.

6. Pfeuffer, T.: Guanine nucleotide-controlled interactions between components of adenylate cyclase. F.E.B.S. Lett. 101:85-89, 1979.

7. Cassel, D., Levkovitz, H., and Selinger, Z.: The regulatory GTPase cycle of turkey erythrocyte adenylate cyclase. J. Cyclic Nucleotide Res. 3:393-406, 1977.

8. DeLean, A., Stadel, J. M., and Lefkowitz, R. J.: A ternary complex model explains the agonist-specific binding properties of the adenylate cyclase-coupled β-adrenergic receptor. J. Biol. Chem. 255:7108-7117, 1980.

9. Kent, R. S., DeLean, A., and Lefkowitz, R. J.: A quantitative analysis of beta-adrenergic receptor interactions: Resolution of high and low affinity states of the receptor by computer modeling of ligand binding data. Mol. Pharmacol. 17:14-23, 1979.

10. Maggi, A., Schmidt, M. J., Ghetti, B., and Enna, S. J.: Effect of aging on neurotransmitter receptor binding in rat and human brain. Life Sci. 24:367-374, 1979.

11. Kobayashi, H., Frattola, L., Ferrarese, C., Spano, P., and Trabucchi, M.: Characterization of β-adrenergic receptors on human cerebral microvessels. Neurology 32: 1384-1387, 1982.

12. Barnes, P. J., Karliner, J. S., and Dollery, C. T.: Human lung adrenoreceptors studied by radioligand binding. Clin. Sci. 58: 457-461, 1980.

13. Engel, G.: Identification of different subgroups of beta-receptors by means of binding studies in guinea-pig and human lung. Triangle 19: 69-76, 1980.

14. Kaumann, A. J., Lemoine, H., Morris, T. H., and Schwederski, U.: An initial characterization of human heart β-adrenoceptors and their mediation of the positive inotropic effects of catecholamines. Naunyn Schmeidebergs Arch. Pharmacol. 319: 216-221, 1982.

15. Stiles, G. L., Strasser, R. H., Lavin, T. N., Jones, L. R., Caron, M. G., and Lefkowitz, R. J.: The cardiac β-adrenergic receptor: Structural similarities of β_1- and β_2-receptor subtypes demonstrated by photoaffinity labeling. J. Biol. Chem. 258: 8443-8449, 1983.

16. Stiles, G. L., Taylor, S., and Lefkowitz, R. J.: Human cardiac beta-adrenergic receptors: Subtype heterogeneity delineated by direct radioligand binding. Life. Sci. 33: 467-473, 1983.

17. Heitz, A., Schwartz, J., Velly, J. β-Adrenoceptors of the human myocardium: Determination of β_1 and β_2 subtypes by radioligand binding. Br. J. Pharmacol. 80: 711-717, 1983.

18. Hayashida, D. N., Leung, R., Goldfien, A., and Roberts, J. M.: Human myometrial adrenergic receptors: Identification of the beta-adrenergic receptor by [^{3}H]dihydroalprenolol binding. Am. J. Obstet. Gynecol. 142: 389-393, 1982.

19. Whitsett, J. A., Johnson, C. L., Noguchi, A., Darovec-Beckerman, C., and Costello, M.: β-Adrenergic receptors and catecholamine-sensitive adenylate cyclase of the human placenta. J. Clin. Endocrinol. Metab. 50: 27-32, 1980.

20. Schocken, D. D., Caron, M. G., and Lefkowitz, R. J.: The human placenta—A rich source of β-adrenergic receptors: Characterization of the receptors in particulate and solubilized preparations. J. Clin. Endocrinol. Metab. 50: 1082-1088, 1980.

21. Engfeldt, P., Arner, P., Wahrenberg, H., and Ostman, J.: An assay for beta-adrenergic receptors in isolated human fat cells. J. Lipid Res. 23: 715-719, 1982.

22. Pfeifle, B., Pfeifle, R., Faulhaber, J. D., and Ditschuneit, H.: Characterization of β-adrenergic receptors by [^{3}H]isoproterenol

in adipocytes of humans and rats. Horm. Metab. Res. 13:150-155, 1981.

23. Boriskina, G. M., and Postnov, Y. V.: Binding of [³H]-dihydroalprenolol to beta-adrenoceptors in adipocytes of spontaneously hypertensive rats and essentially hypertensive patients. Experientia 38:263-264, 1982.

24. Levin, R. M., and Wein, A. J.: Adrenergic alpha-receptors outnumber beta-receptors in human penile corpus cavernosum. Inves. Urol. 18:225-226, 1980.

25. Hirata, Y., Uchihashi, M., Sueoka, S., Matsukura, S., and Fujita, T.: Presence of ectopic β-adrenergic receptors on human adrenocortical cortisol-producing adenomas. J. Clin. Endocrinol. Metab. 53:953-957, 1981.

26. Galant, S. P., Underwood, S., Duriseti, L., and Insel, P. A.: Characterization of high-affinity β₂-adrenergic receptor binding of (-)-[³H]-dihydroalprenolol to human polymorphonuclear cell particulates. J. Lab. Clin. Med. 92:613-618, 1978.

27. Lee, T.-P., Szefler, S., and Ellis, E. F.: Beta-adrenergic receptors of human polymorphonuclear leukocytes. Res. Commun. Chem. Pathol. Pharmacol. 31:453-462, 1981.

28. Williams, L. T., Snyderman, R., and Lefkowitz, R. J.: Identification of β-adrenergic receptors in human lymphocytes by (-)[³H]alprenolol binding. J. Clin. Invest. 57:149-155, 1976.

29. Aarons, R. D., Nies, A. S., Gal, J., Hegstrand, L. R., and Molinoff, P. B.: Elevation of β-adrenergic receptor density in human lymphocytes after propranolol administration. J. Clin. Invest. 65:949-957, 1980.

30. Landmann, R., Burgisser, E., and Buhler, F. R.: Human lymphocytes as a model for beta-adrenergic receptors in clinical investigation. J. Recept. Res. 3(1&2):71-88, 1983.

31. Watanabe, Y., Lai, R-T., and Yoshida, H.: The β-adrenoreceptor in human lymphocytes. Clin. Exp. Pharmacol. Physiol. 8:273-276, 1981.

32. Sager, G.: β₂ Adrenergic receptors on intact human erythrocytes. Biochem. Pharmacol. 32:1946-1949, 1983.

33. Sager, G., Noraas, S., Jacobsen, S., Stenerud, O., Aakesson, I.: The influence of filtrability on β-adrenergic ligand binding to membrane fragments from human erythrocytes and mononuclear leukocytes. Biochem. Pharmacol. 32:1943-1946, 1983.

34. Lavin, T. N., Nambi, P., Heald, S. L., Jeffs, P. W., Lefkowitz, R. J., and Caron, M. G.: [^{125}I] Labeled p-azidobenzylcarazdol, a photoaffinity label for the β-adrenergic receptor. J. Biol. Chem. 257:12332-12340, 1982.

35. Rashidbaigi, A., and Ruoho, A. E.: Photoaffinity labeling of β-adrenergic receptors: Identification of the β-receptor binding site(s) from turkey, pigeon and frog erythrocyte. Biochem. Biophys. Res. Comm. 106:139-148, 1982.

36. Abdulla, Y. H.: β-Adrenergic receptors in human platelets. J. Atheroscler. Res. 9:171-177, 1969.

37. Jakobs, K. H., Saur, W., and Schultz, G.: Characterization of α and β-adrenergic receptors linked to human platelet adenylate cyclase. Naunyn Schmiedebergs Arch. Pharmacol. 302:285-291, 1978.

38. Kerry, R., and Scrutton, M. C.: Binding of [^{3}H]dihydroalprenolol and [^{3}H]acetobutolol to human blood platelets is not related to occupancy of β-adrenoceptors. Thrombosis Res. 29:583-594, 1983.

39. Steer, M. L., and Atlas, D.: Demonstration of human platelet β-adrenergic receptors using ^{125}I-labeled cyanopindolol and ^{125}I-labeled hydroxybenzylpindolol. Biochim. Biophys. Acta 686:240-244, 1982.

40. Bourne, H. R., Lichtenstein, L. M., Melmon, K. L., Henney, C. S., Weinstein, Y., and Shearer, G. M.: Modulation of inflammation and immunity by cyclic AMP. Science 184:19-28, 1974.

41. Sano, Y., Krokos, K., Cheng, J. B., Bewtra, A., and Townley, R.: Comparison of alpha- and beta-adrenergic receptors in asthmatics and controls: Identification and characterization of alpha-adrenergic receptors in human lymphocytes. Clin. Res. 29:172A, 1981.

42. Szentivanyi, A., Heim, O., Schultze, P., and Szentivanyi, J.: Adrenoreceptor binding studies with [^{3}H] dihydroalprenolol and [^{3}H] dihydroergocryptine on membranes of lymphocytes from patients with atopic disease. Acta Derm. Venereol. (Suppl.) 92: 19-21, 1980.

43. Casale, T. B., Halonen, M., and Kaliner, M.: Detection of beta-adrenergic receptors on rabbit mononuclear cells isolated free of significant contamination by other cell types. Life Sci. 33:971-977, 1983.

44. Panosian, J. O., and Marinetti, G. V.: α_2-Adrenergic receptors in human polymorphonuclear leukocyte membranes. Biochem. Pharmacol. 32:2243-2247, 1983.

45. Davies, A. O., and Lefkowitz, R. J.: *In vitro* desensitization of beta-adrenergic receptors in human neutrophils. Attenuation by corticosteroids. J. Clin. Invest. 71:565-571, 1983.

46. Feldman, R. D., Limbird, L. E., Nadeau, J., FitzGerald, G. A., Robertson, D., and Wood, A. J. J.: Dynamic regulation of leukocyte beta-adrenergic receptor-agonist interactions by physiological changes in circulating catecholamines. J. Clin. Invest. 72:164-170, 1983.

47. Meurs, H., van den Bogarard, W., Kauffman, H. F., Bruynzeel, P. L. B. Characterization of (-)-[^{3}H] dihydroalprenolol binding to intact and broken cell preparations of human peripheral blood lymphocytes. Eur. J. Pharmacol. 85:185-194, 1982.

48. Fraser, J., Nadeau, J., Robertson, D., and Wood, A. J. J.: Regulation of human leukocyte beta receptors by endogenous catecholamines. Relationship of leukobyte beta receptor density to the cardiac sensitivity to isoproterenol. J. Clin. Invest. 67:1777-1784, 1981.

49. Colucci, W. S., Alexander, R. W., Williams, G. H., Rude, R. E., Holman, B. L., Konstam, M. A., Wynne, J., Mudge, G. H., Jr., and Braunwald, E.: Decreased lymphocyte beta-adrenergic receptor density in patients with heart failure and tolerance to the beta-adrenergic agonist pirbuterol. N. Engl. J. Med. 305:185-190, 1981.

50. Aarons, R. D., and Molinoff, P. B.: Changes in the density of beta adrenergic receptors in rat lymphocytes, heart and lung after chronic treatment with propranolol. J. Pharmacol. Exp. Ther. 221:439-443, 1982.

51. Scarpace, P. J., and Abrass, I. B.: Thyroid hormone regulation of rat heart, lymphocyte, and lung β-adrenergic receptors. Endocrinology 108:1007-1011, 1981.

52. Davies, A. O., and Lefkowitz, R. J.: Corticosteroid-induced differential regulation of β-adrenergic receptors in circulating human polymorphonuclear leukocytes and mononuclear leukocytes. J. Clin. Endocrinol. Metab. 51:599-605, 1980.

53. Sheppard, J. R., Gormus, R., and Moldow, C. F.: Catecholamine hormone receptors are reduced on chronic lymphocytic leukemic lymphocytes. Nature 269:693-695, 1977.

54. Bishopric, N. H., Cohen, H. J., and Lefkowitz, R. J.: Beta adrenergic receptors in lymphocyte subpopulations. J. Allergy Clin. Immunol. 65:29-33, 1980.

55. Krawietz, W., Werdan, K., Schober, M., Erdmann, E., Rindfleisch, G. E., and Hannig, K. Different numbers of β-receptors in human lymphocyte subpopulations. Biochem. Pharmcol. 31:133-136, 1982.

56. Berlan, M., and La Fontan, M.: Identification of alpha$_2$-adrenergic receptors in human fat cell membranes by [^{3}H] clonidine binding. Eur. J. Pharmacol. 67:481-484, 1980.

57. Burns, T. W., Boyer, P. A., Terry, B. E., Langley, P. E., and Robison, G. A.: The effect of fasting on the adrenergic receptor activity of human adipocytes. J. Lab. Clin. Med. 94: 387-394, 1979.

58. Krall, J. F., Connelly, M., and Tuck, M. L.: Acute regulation of beta-adrenergic catecholamine sensitivity in human lymphocytes. J. Pharmacol. Exp. Ther. 214:554-560, 1980.

59. Su, Y.-T., Harden, T. K., and Perkins, J. P.: Catecholamine-specific desensitization of adenylate cyclase. Evidence for a multistep process. J. Biol. Chem. 255:7410-7419, 1980.

60. Sternweis, P. C., Northup, J. K., Smigel, M. D., and Gilman, A. G.: The regulatory component of adenylate cyclase. Purification and properties. J. Biol. Chem. 256:11517-11526, 1981.

61. Gill, D. M., and Meren, R.: ADP-Ribosylation of membrane proteins catalyzed by cholera toxin: Basis of the activation of adenylate cyclase. Proc. Natl. Acad. Sci. U.S.A. 75:3050-3054, 1978.

62. Johnson, G. L., Kaslow, H. R., and Bourne, H. R.: Reconstitution of cholera toxin-activated adenylate cyclase. Proc. Natl. Acad. Sci. U.S.A. 75:3113-3117, 1978.

63. Cassel, D., and Pfeuffer, T.: Mechanism of cholera toxin action: Covalent modification of the guanyl nucleotide-binding protein of the adenylate cyclase system. Proc. Natl. Acad. Sci. U.S.A. 75:2669-2673, 1978.

64. Farfel, Z., Brickman, A. S., Kaslow, H. R., Brothers, V. M., and Bourne, H. R.: Defect of receptor-cyclase coupling protein in pseudohypoparathyroidism. N. Engl. J. Med. 303:237-242, 1980.

65. Downs, R. W., Levine, M. A., Drezner, M. K., Burch, W. M., Jr., and Spiegel, A. M.: Deficient adenylate cyclase regulatory protein in renal membranes from a patient with pseudohypoparathyroidism. J. Clin. Invest. 71:231-235, 1983.

66. Ross, E. M., Howlett, A. C., Ferguson, K. M., and Gilman, A. G.: Reconstitution of hormone-sensitive adenylate cyclase activity with resolved components of the enzyme. J. Biol. Chem. 253:6401-6412, 1978.

67. Howlett, A. C., Sternweis, P. C., Macik, B. A., Van Arsdale,
    P. M., and Gilman, A. G.: Reconstitution of catecholamine-
    sensitive adenylate cyclase. J. Biol. Chem. 254:2287-2295, 1979.

68. Speigel, A. M., Downs, R. W., and Aurbach, G. D.: Separation
    of a guanine nucleotide regulatory unit from the adenylate cyclase
    complex with GTP affinity chromatography. J. Cyclic Nucleotide
    Res. 5:3-11, 1979.

69. Abrass, I. B., and Scarpace, P. J.: Catalytic unit of adenylate
    cyclase: Reduced activity in aged human lymphocytes. J. Clin.
    Endocrinol. Metab. 55:1026-1028, 1982.

70. Hoffman, B. B., and Lefkowitz, R. J.: Alpha-adrenergic receptor
    subtypes. N. Engl. J. Med. 302:1390-1396, 1980.

71. Kafka, M. S., Tallman, J. F., and Smith, C. C.: Alpha-adrenergic
    receptors on human platelets. Life Sci. 21:1429-1437, 1977.

72. Newman, K. D., Williams, L. T., Bishopric, N. H., and Lefkowitz,
    R. J.: Identification of α-adrenergic receptors on human platelets
    by [³H] dihydroergocryptine binding. J. Clin. Invest. 61:395-
    402, 1978.

73. Tsai, B. S., and Lefkowitz, R. J.: Agonist-specific effects of
    guanine nucleotides on Alpha-adrenergic receptors in human
    platelets. Mol. Pharmacol. 16:61-68, 1979.

74. Alexander, R. W., Cooper, B., and Handin, R. I.: Characteriza-
    tion of the human platelet α-adrenergic receptor. Correlation of
    [³H] dihydroergocryptine binding with aggregation and adenylate
    cyclase inhibition. J. Clin. Invest. 61:1136-1144, 1978.

75. Burns, T. W., Langley, P. E., Terry, B. E., Bylund, D. B.,
    Hoffman, B. B., Tharp, M. D., Lefkowitz, R. J., Garcia-Sainz,
    A., and Fain, J. N.: Pharmacological characterizations of adrenergic
    receptors in human adipocytes. J. Clin. Invest. 67:467-475, 1981.

76. Jakobs, K. H., and Raushek, R.: [³H] Dihydroergonine binding
    to α-adrenergic receptors in human platelets. Klin. Wochenschr.
    56 (Suppl. 1):139-145, 1978.

77. Macfarlane, D. E., Wright, B. L., and Stump, D. C.: Use of
    [methyl-³H]-yohimbine as a radioligand for alpha₂-adrenoreceptors
    on intact platelets. Comparison with dihydroergocryptine.
    Thrombosis Res. 24:31-43, 1981.

78. Motulsky, H., Shattil, S. J., and Insel, P. A.: Characterization
    of α₂-adrenergic receptors on human platelets using [³H]yohimbine.
    Biochem. Biophys. Res. Commun. 97:1562-1570, 1980.

79. Garcia-Sevilla, J. A., Hollingsworth, P. J., and Smith, C. B.:
    α₂-Adrenoreceptors on human platelets: Selective labeling by

[^{3}H] clonidine and [^{3}H] yohimbine and competitive inhibition of antidepressant drugs. Eur. J. Pharmacol. 74:329-341, 1981.

80. Daijuji, M., Meltzer, H. Y., and U'Prichard, D. C.: Human platelet α_2-adrenergic receptors: Labeling with [^{3}H]yohimbine, a selective antagonist ligand. Life Sci. 28:2705-2717, 1981.

81. Smith, S. K., and Limbird, L. E.: Solubilization of human platelet α-adrenergic receptors: Evidence that agonist occupancy of the receptor stabilizes receptor-effector interactions. Proc. Natl. Acad. Sci. U.S.A. 78:4026-4030, 1981.

82. Hoffman, B. B., Michel, T., Mullikin-Kilpatrick, D., Lefkowitz, R. J., Tolbert, M. E. M., Gilman, H., and Fain, J. N.: Agonist versus antagonist binding to α-adrenergic receptors. Proc. Natl. Acad. Sci. U.S.A. 77:4569-4573, 1980.

83. Steer, M. L., Khorana, J., and Galgoci, B.: Quantitation and characterization of human platelet alpha-adrenergic receptors using [^{3}H] phentolamine. Mol. Pharmacol. 16:719-728, 1979.

84. Lafontan, M., and Berlan, M.: Characterization of physiological agonist selectivity of human fat cell α_2-adrenoceptors: Adrenaline is the major stimulant of the α_2-adrenoceptors. Eur. J. Pharmacol. 82:107-111, 1982.

85. Berlan, M., and Lafontan, M.: The α_2-adrenergic receptor of human fat cells: Comparative study of α_2-adrenergic radioligand binding and biological response. J. Physiol. (Paris) 78:279-287, 1982.

86. Tharp, M. D., Hoffman, B. B., and Lefkowitz, R. J.: α-Adrenergic receptors in human adipocyte membranes. Direct determination by [^{3}H] yohimbine binding. J. Clin. Endocrinol. Metab. 52:709-714, 1981.

87. Lynch, C. J., and Steer, M. L.: Evidence for high and low affinity α_2-receptors. Comparison of [^{3}H] norepinephrine and [^{3}H] phentolamine binding to human platelet membranes. J. Biol. Chem. 256:3298-3303, 1981.

88. Summers, R. J., Barnett, D. B., and Nahorski, S. R.: The characteristics of adrenoceptors in homogenates of human cerebral cortex labelled by [^{3}H]-rauwolscine. Life Sci. 33:1105-1112, 1983.

89. Shattil, S. J., McDonough, M., Turnbull, J., and Insel, P. A.: Characterization of alpha-adrenergic receptors in human platelets using [^{3}H]clonidine. Mol. Pharmacol. 19:179-183, 1981.

90. Weinreich, P., and Seeman, P. Binding of adrenergic ligands ([^{3}H] clonidine and [^{3}H] WB-4101) to multiple sites in human brain. Biochem. Pharmacol. 30:3115-3120, 1981.

91. Mooney, J. J., Horne, W. C., Handin, R. I., Schildkraut, J. J., and Alexander, R. W.: Sodium inhibits both adenylate cyclase and high-affinity ^{3}H-labeled p-aminoclonidine binding to alpha$_2$-adrenergic receptors in purified human platelet membranes. Mol. Pharmacol. 21:600-608, 1982.

92. Cooper, D. M. F.: Bimodal regulation of adenylate cyclase. F.E.B.S. Lett. 138:157-163, 1982.

93. Katada, T., and Ui, M.: Islet-activating protein. A modifier of receptor-mediated regulation of rat islet adenylate cyclase. J. Biol. Chem. 256:8310-8317, 1981.

94. Hazeki, O., and Ui, M.: Modification by islet-activating protein of receptor-mediated regulation of cyclic AMP accumulation in isolated rat heart cells. J. Biol. Chem. 256:2856-2862, 1981.

95. Kurose, H., Katada, T., Amano, T., and Ui, M.: Specific uncoupling by islet activating protein, pertussis toxin, of negative signal transduction via α-adrenergic, cholinergic and opiate receptors in neuroblastoma x glioma hybrid cells. J. Biol. Chem. 258:4870-4875, 1983.

96. Murayama, T., and Ui, M.: Loss of the inhibitory function of the guanine nucleotide regulatory component of adenylate cyclase due to its ADP-ribosylation by islet-activating protein, pertussis toxin, in adipocyte membranes. J. Biol. Chem. 258:3319-3326, 1983.

97. Katada, T., and Ui, M.: ADP-ribosylation of the specific membrane protein of C6 cells by islet-activating protein association with modification of adenylate cyclase activity. J. Biol. Chem. 257:7210-7216, 1982.

98. Bokoch, G. M., Katada, T., Northup, J. K., Hewlett, E. L., and Gilman, A. G.: Identification of the predominant substrate for ADP-ribosylation by islet activating protein. J. Biol. Chem. 258:2072-2075, 1983.

99. Manning, D. R., and Gilman, A. G.: The regulatory components of adenylate cyclase and transducin. A family of structurally homologous guanine nucleotide-binding proteins. J. Biol. Chem. 258:7059-7063, 1983.

100. Northup, J. K., Smigel, M. D., and Gilman, A. G.: The guanine nucleotide activating site of the regulatory component of adenylate cyclase. Identification by ligand binding. J. Biol. Chem. 257:11416-11423, 1982.

101. Tsai, B.-S., and Lefkowitz, R. J.: Agonist-specific effects of guanine nucleotides on alpha-adrenergic receptors in human platelets. Mol. Pharmacol. 16: 61-68, 1979.

102. Hoffman, B. B., Michel, T., Brenneman, T. B., and Lefkowitz, R. J.: Interactions of agonists with platelet α_2-adrenergic receptors. Endocrinology 110: 926-932, 1982.

103. Michel, T. M., and Lefkowitz, R. J.: Hormonal inhibition of adenylate cyclase. α_2-Adrenergic receptors promote release of ([^{3}H] Gpp(NH)p) from platelet membranes. J. Biol. Chem. 257: 13557-13563, 1982.

104. Motulsky, H. J., and Insel, P. A. ADP- and epinephrine-elicited release of [^{3}H] guanylylimidodiphosphate from platelet membranes. Implications for receptor-N_i stoichiometry. F.E.B.S. Lett. 164: 13-16, 1983.

105. Michel, T. M., Hoffman, B. B., Lefkowitz, R. J., and Caron, M. G.: Different sedimentation properties of agonist- and antagonist-labeled platelet alpha$_2$ adrenergic receptors. Biochem. Biophys. Res. Comm. 100:1131-1136, 1981.

106. Smith, S. K., and Limbird, L. E.: Evidence that human platelet α-adrenergic receptors coupled to inhibition of adenylate cyclase are not associated with the subunit of adenylate cyclase ADP-ribosylated by cholera toxin. J. Biol. Chem. 257:10471-10478, 1982.

107. Aktories, K., Schultz, G., and Jakobs, K-H.: Cholera toxin inhibits prostaglandin E_1 but not adrenaline-induced stimulation of GTP hydrolysis in human platelet membranes. F.E.B.S. Lett. 146: 65-68, 1982.

108. Burns, D. L., Hewlett, E. L., Moss, J., and Vaughan, M.: Pertussis toxin inhibits enkephalin stimulation of GTPase of NG108-15 cells. J. Biol. Chem. 258:1435-1438, 1983.

109. Limbird, L. E., Bokoch, G. M., and Gilman, A. G.: Unpublished observations, 1983.

110. Jakobs, K.-H.: Inhibition of adenylate cyclase by hormones and neurotransmitters. Mol. Cell. Endocrinol. 16:147-156, 1979.

111. Michel, T. M., Hoffman, B. B., and Lefkowitz, R. J.: Differential regulation of the α_2-adrenergic receptor by Na^+ and guanine nucleotides. Nature 288:709-711, 1980.

112. Tsai, B. S., and Lefkowitz, R. J.: Agonist-specific effects of monovalent and divalent cations on adenylate cyclase-coupled alpha-adrenergic receptors in rabbit platelets. Mol. Pharmacol. 14: 540-548, 1978.

113. Limbird, L. E., Speck, J. L., and Smith, S. K.: Sodium ion modulates agonist and antagonist interactions with the human platelet alpha$_2$-adrenergic receptor in membrane and solubilized preparations. Mol. Pharmacol. 41:607-619, 1982.

114. Aktories, K., Schultz, G., and Jakobs, K.-H.: The hamster adipocyte adenylate cyclase system II. Regulation of enzyme stimulation and inhibition by monovalent cations. Biochim. Biophys. Acta 676:59-67, 1981.

115. Limbird, L. E., and Speck, J. L.: N-Ethylmaleimide, elevated temperature and digitonin solubilization eliminate guanine nucleotide but not sodium effects on human platelet α_2-adrenergic receptor-agonist interactions. J. Cyclic Nucleotide Protein Phosphorylation Res. 9:191-201, 1983.

116. Jakobs, K.-H., Lasch, P., Minuth, M., Aktories, K., and Schultz, G.: Uncoupling of α-adrenoceptor-mediated inhibition of human platelet adenylate cyclase by N-ethylmaleimide. J. Biol. Chem. 257:2829-2833, 1982.

117. McMahon, K. K., and Hosey, M. M.: Potentiation of monovalent cation effects on ligand binding to cardiac muscarinic receptors in N-ethylmaleimide treated membranes. Biochem. Biophys. Res. Commun. 111:41-46, 1983.

118. Stahl, S. M.: The human platelet. Arch. Gen. Psychiatry 34: 509-516, 1977.

119. Berridge, M. J.: Phosphatidylinositol hydrolysis and calcium signaling. Adv. Cyclic Nucleotide Res. 14:289-299, 1981.

120. Bottari, S. P., Vauquelin, G., Lescrainier, J. P., Kaivez, E., and Vokaer, A.: Identification and characterization of α_1-adrenergic receptors in human myometrium by [^{3}H] prazosin binding. Biochem. Pharmacol. 32:925-928, 1983.

121. Motulsky, H. J., and Insel, P. A.: [^{3}H]Dihydroergocryptine binding to alpha-adrenergic receptors of human platelets: A reassessment using the selective radioligands [^{3}H]prazosin, [^{3}H]yohimbine, and [^{3}H]rauwolscine. Biochem. Pharmacol. 31: 2591-2597, 1982.

7

Physiologic and Pharmacologic Regulation of Adrenergic Receptors

PAUL A. INSEL and HARVEY J. MOTULSKY   *University of California, San Diego, La Jolla, California*

I. INTRODUCTION

In this chapter, we will discuss two aspects of the regulation of adrenergic receptors in man: changes that are attributable to physiologic alterations and pharmacologic interventions. As will become obvious during the chapter, this is a somewhat arbitrary distinction as common mechanisms are likely to mediate both types of changes. In view of the widespread distribution of adrenergic receptors in humans (see Chap. 2), it is perhaps not surprising that a large number of settings have been associated with changes in adrenergic receptor expression and function (Table 1). We will discuss some of the recent literature that relates to several of these settings, and we will emphasize the combined use of radioligand binding and other biochemical methods to examine receptors in parallel with more classic functional assays of these receptors. Interested readers may wish to consult other recent reviews that provide more detailed presentations of certain aspects of this topic, especially with respect to studies conducted with experimental animals and other models (1-3). In addition, other previous reviews summarize earlier literature in this area (4, 5).

The vast majority of information regarding adrenergic receptors and physiologic and pharmacologic factors that regulate these receptors has been derived from astute clinical observations on patients and from assays that have assessed changes in target cell response. Thus, clinicians and patients have learned to associate certain symptoms and signs—tachycardia, tremor, pallor, and sweating, for example— as evidence of stimulated adrenergic activity.

TABLE 1 Physiologic and Pharmacologic Settings with Altered
Adrenergic Receptors

Physiological
  Changes during development and aging
  Circadian changes
  Nutritional alterations
  Exercise
  Menstrual cycle/pregnancy

Pharmacological
  Glucocorticoids
  Thyroid hormone/antithyroid drugs
  Oral contraceptives/sex steroids
  Adrenergic agonists
  Adrenergic antagonists
  Drugs that alter plasma/tissue catecholamines
  Calcium channel blockers
  Miscellaneous drugs (e.g., indomethacin, penicillins, amiloride,
  quinidine)

In recent years, much attention has been focused on attempts to define mechanisms to explain changes in response and on the development of suitable model systems to examine human adrenergic receptors and response in a noninvasive manner. This has led to the growing use of peripheral blood cells as easily isolated and manipulable systems for studying adrenergic receptors. As discussed in more detail in Chapter 6, lymphocytes and granulocytes possess $beta_2$-adrenergic receptors and platelets contain $alpha_2$-adrenergic receptors. These receptors are linked through distinct guanine nucleotide binding (G) proteins that either promote stimulation (G_s, beta receptors) or inhibition (G_i, $alpha_2$ receptors) of adenylate cyclase, and it is also possible to examine more distal responses in these cells. Unfortunately, as summarized in Table 2, studies in peripheral blood cells have several drawbacks. Perhaps the most important one from a clinical perspective is that these cell types may not provide accurate insight into properties of receptors on internal tissues. While tissue-bound receptors are primarily exposed to synaptically released catecholamines, blood cells are exposed to catecholamines within the circulating blood. Thus, adrenergic receptors on blood cells probably undergo chronic stimulation and in turn, as will be discussed subsequently, feedback regulation. Moreover, since many factors are able to modulate plasma catecholamine levels, such as emotional stress, exercise, dietary Na^+, and methylxanthines, these changing levels probably contribute to

TABLE 2 Limitations of Using Blood Cells for Studying
Adrenergic Receptors

May not reflect properties of receptors in other tissues

Not innervated, only exposed to circulating catecholamines

Heterogeneous populations of cells

Assays available only for alpha$_2$ and beta$_2$ receptors

Platelets are unusual "cells": no nuclei, little protein synthesis

Variability among subjects due to genetic and environmental factors

---

the quite variable values reported for receptor numbers, affinities,
and responses in peripheral blood cells. The failure to control for
such factors probably contributes to the rather disparate results that
different investigators have found who have studied adrenergic
receptors in various physiologic and pharmacologic settings. As
methods are developed to assess receptors on other cell types and
as investigators prepare more purified cell populations, develop more
standardized methodology for assays, and control for factors that
contribute to variability in expression of receptors and response in
peripheral blood cells, we expect that more consistent results will be
obtained in the types of studies that we discuss here.

II. PHYSIOLOGICAL CHANGES IN ADRENERGIC RECEPTORS

A. Development and Aging

Response to catecholamines is highly dependent on the age of experimental animals and human subjects (1,6-8). For example, studies in
experimental animals document changes in receptors in the cardiovascular system, liver, and central nervous system during ontogeny
(1,7,8). Development of beta-adrenergic response can accompany
or follow the appearance of beta-adrenergic receptors that are
detected in radioligand binding assays. When receptors precede
tissue response, maturation of factors distal to receptors and/or of
domains of receptors other than the receptor binding site are likely
to explain the later appearance of response.

Much less information is available in humans. However, polymorphonuclear leukocytes and platelets obtained from neonates have
only half as many beta-adrenergic receptors and alpha$_2$-adrenergic
receptors, respectively, as do cells from adult subjects (9-11). In
one recent study, Reinhardt et al. (12) noted that lymphocyte beta-

adrenergic receptor number increased linearly during childhood and
was even higher in adults. In this study, platelet alpha$_2$-adrenergic
receptor number did not change with age. Other workers have found
that platelets from neonates have a striking and selective decrease in
aggregatory response to epinephrine, whereas leukocytes from neonates
show diminished beta-adrenergic-mediated cyclic adenosine monophosphate (AMP) generation. Whether these changes in neonates result
from developmental factors or effects secondary to the intrauterine
environment at the time of birth has not yet been determined. Interestingly, studies in animals indicate that cutting of the umbilical cord
prominently increases fetal catecholamine levels (13). If similar
increases occur in humans, desensitization/down-regulation of tissue
responses and receptors might result (as will be discussed subsequently).

Studies in animals indicate that fetal lung is a tissue in which
gestational increases in beta-adrenergic receptors and response are
temporally related to the plasma glucocorticoid concentration, that
glucocorticoid therapy can accelerate fetal lung development, and
that glucocorticoid-mediated increases in beta-adrenergic receptors
occur preferentially in alveoli (8 and 14 and references therein). We
imagine that steroids increase expression of beta-adrenergic receptors
(perhaps by induction of new receptors [15]) on type II pneumocytes,
the alveolar cells that are responsible for surfactant production.
Thus, these findings may relate to the beneficial effect of glucocorticoids in the treatment of neonatal respiratory distress syndrome.

Aging is accompanied by numerous alterations in adrenergic
response and receptors (6,7,16-18). For example, older ("aged")
experimental animals show decreased beta-adrenergic-mediated relaxation in aortae and decreased beta-adrenergic-mediated inotropic
response (16,18,19). Some investigators have also observed changes
in the number of adrenergic receptors that parallel these changes in
tissue responsivity, while other workers have emphasized the greater
importance of age-related changes in events distal to receptors (6,7,
19-21).

Much less information of this type is as yet available in humans,
although the data base is rapidly increasing regarding such alterations
(6,16,17,22,23). In the cardiovascular system, the data suggest
blunting of beta-adrenergic response and no consistent change discernible in alpha-adrenergic response (16,18).

Direct radioligand binding studies assessing adrenergic receptors
as a function of age have yielded ambiguous results. Initially, Schocken
and Roth reported an inverse correlation between age and number of
beta-adrenergic receptors on leukocyte membranes (24). Perhaps
because those workers used inappropriately high radioligand concentrations for their studies, subsequent investigators have, in general,
been unable to confirm this relationship (6,25-27). However, agonist

binding affinity (28) and beta-adrenergic-mediated cyclic AMP generation have been observed to be decreased in lymphocyte membranes prepared from elderly subjects (6,29,30). Such results suggest age-related changes in membrane components of the adenylate cyclase complex that are distal to receptors; one group has proposed that the catalytic component of adenylate cyclase may, in fact, be deficient or defective in cells from elderly subjects (29). One factor that could contribute to the results observed for leukocyte beta-adrenergic receptors is changes in receptors promoted by the increased levels of plasma catecholamines that occur in aging (26,28,29,31).

A few studies have also been reported on platelet alpha$_2$-adrenergic receptors as a function of age in human subjects, and again, the results are ambiguous (18): Some have found no relation between number of receptors and age, whereas others find an inverse relationship (e.g., see Refs. 6 and 26). Although one might hypothesize that decreases in receptor number result from increases in plasma catecholamines with aging, there is little convincing evidence that platelet alpha$_2$-adrenergic receptors are down-regulated by increases in catecholamines.

As of yet, there are little data available (32) regarding age-dependent expression of adrenergic receptors (as examined using radioligand binding techniques) on human tissues other than circulating blood cells. This is unfortunate because blood cells have short life spans relative to those of other cell types, and thus, it may not be appropriate to use such short-lived cells (platelets, 7-10 days; lymphocytes, a substantial portion of which live <14 days; polymorphonuclear cells, 3 days) as markers of age-dependent changes in other tissues. In other species, investigators have reported decreased beta-adrenergic sensitivity of fat cells from older animals, an effect resulting from changes distal to beta-adrenergic receptors (33), and enhanced alpha-adrenergic response of fat cells from older animals (34). Since fat cells in humans possess beta$_1$-, alpha$_1$-, and alpha$_2$-adrenergic receptors, further experiments in this tissue may prove of interest.

To summarize, aging is associated with prominent evidence of altered adrenergic response, especially within the cardiovascular system. Evidence obtained in humans indicates that receptor number on peripheral blood cells is probably not altered, although alterations in events that are distal to receptors has now been documented in several tissues. The precise definition of these alterations may contribute useful insights into changes that occur with aging in response to catecholamines and to other hormones and neurotransmitters as well.

B. Circadian Rhythm in Expression of Adrenergic Receptors

Some data in experimental animals have suggested that changes in expression of adrenergic receptors can occur in a circadian fashion (e.g., see Ref. 35). Data in humans suggest that $beta_2$-adrenergic receptors in circulating lymphocytes may (36,37) or may not (38) exhibit a circadian rhythm. Recent reports also came to opposite conclusions as to whether platelet $alpha_2$-receptors demonstrate such changes (38,39). The lack of consensus would suggest to us that circadian patterns probably do not exist on the cells that have been studied thus far in human subjects.

C. Nutritional Alterations in Receptor Expression

By nutritional alterations we refer to changes associated with altered caloric content or dietary composition, including electrolytes. Studies related to caloric content have largely focused on whether expression of adrenergic receptors in fat cells changes with obesity. Adipocytes possess an estimated 450,000 $beta_1$-adrenergic receptors per cell, which promote lipolysis, and an estimated 600,000 $alpha_2$-adrenergic receptors per cell, which exert antilipolytic effects (40-43). Numerous factors have been identified that can influence results in studies examining adrenergic response in adipocytes, including location from which cells are obtained, differential sensitivity of the various adrenergic receptor subtypes to regulation, generation of adenosine, and age of subjects. In spite of those potential problems, some data have been presented that document decreases in beta-adrenergic response of adipocytes when obese subjects lose weight (44). In addition, weight loss in obese subjects has also been reported to increase binding to platelet alpha-adrenergic receptors (45), and patients with anorexia nervosa show an enhanced number of platelet $alpha_2$ receptors and increased epinephrine-mediated platelet aggregation (46).

Other dietary manipulations have been less well studied, with the exception of changes in dietary sodium. Placing subjects on low sodium diets enhances plasma catecholamine concentration, an effect thought to be attributable to enhanced sympathoadrenal release of norepinephrine and epinephrine (47,48). As will be discussed subsequently, increases in circulating catecholamines can desensitize and down-regulate adrenergic receptors. Thus, to the extent that adrenergic receptors in a particular tissue are susceptible and sensitive to such desensitization, subjects on low sodium diets (typically less than 50 mEq Na^+) would be expected to demonstrate desensitization. Such changes have been reported for both granulocyte and lymphocyte membrane beta-adrenergic receptors (48).

One additional aspect of regulation of adrenergic receptors by sodium is the evidence that this ion can alter binding to $alpha_2$-adrenergic receptors on platelets, adipocytes, and other tissues (See Refs. 49 and 50 and references therein). Sodium appears to act at an intracellular site to decrease agonist binding to $alpha_2$-adrenergic receptors (49,50). Further studies are required to test whether dietary changes in sodium intake (or drug- or disease-induced changes) alter intracellular sodium enough to influence this regulation of agonist binding.

D. Exercise

Exercise, like a low sodium diet, prominently activates the sympathoadrenal axis and increases plasma catecholamines, albeit only for a short time during and following exercise. Evidence has been reported (51) that exercising individuals show reversible decreases in binding of agonists to platelet alpha$_2$-adrenergic receptors, but more detailed examination of this is warranted, since the reported changes were detected in intact platelets [in which agonist interaction is difficult to assess, since binding of agonists is regulated by intracellular guanosine triphosphate (GTP) and Na$^+$, (see Ref. 52)]. Exercise-induced increases in the number of beta-adrenergic receptors and in receptor-mediated cyclic AMP generation of lymphocyte membranes and intact lymphocytes have also been noted (53-55).

The resting bradycardia that occurs in trained individuals is thought to be related primarily to altered cholinergic and not adrenergic tone in the heart, although physical training results in a decrease in activity of the sympathetic nervous system (56). The state of physical conditioning has been proposed to determine the magnitude of increase in systolic blood pressure, but not that of chronotropic response, that occurs with infusion of epinephrine (56,57).

Several groups have examined whether beta-adrenergic responses and receptors are altered with exercise (as recently reviewed in Ref. 56). In one study, a striking inverse correlation between beta-adrenergic receptor number on lymphocyte membranes and degree of physical fitness (as assessed by maximal oxygen intake, max V_{O_2}) was found in men undergoing an intensive training program conducted over a 2-month period (58). In seven men completing the program, receptor number fell an average of about 60% in association with a 20% increase in max V_{O_2}. Since the number of subjects studied was small, these results need to be extended to a larger number of individuals and to include evaluation of other indices of receptor binding and function. Even so, the findings suggest that trained individuals have lower numbers of beta-adrenergic receptors on their lymphocytes. Were such changes also found on tissues like the heart, one might hypothesize that decreased numbers of beta-adrenergic receptors contribute to protective effects of exercise in decreasing the risks of myocardial ischemia and arrhythmias.

E. Menstrual Cycle/Pregnancy

Studies conducted in several animal species have indicated the importance of sex steroids in regulating expression of adrenergic receptors in the central nervous system (CNS) and in peripheral tissues (1,4,59). The influence of sex steroids on adrenergic receptors is complex and may involve effects on receptor metabolism as well as on the coupling of receptors to second messenger systems and to more distal events modulated by catecholamines. Moreover, different tissues responsive to sex steroids do not necessarily show concordant changes in adrenergic receptor expression (60,61).

Data in humans have been largely obtained on isolated blood cells, although several estrogen/progesterone-responsive tissues (e.g., fallopian tubes, uterine myometrium, and certain ovarian cells) also respond to catecholamines. Uterine contractility in response to cate-cholamines changes as a function of time during the menstrual cycle (62). Some recent data obtained with human myometrium indicate that alpha-adrenergic receptors can be identified using radioligand binding techniques (Table 3; see Refs. 63 and 64 and S. Bottari, personal communication). Bottari et al. used [^{3}H]prazosin to identify about 30 fmol/mg protein alpha$_1$-adrenergic receptors in myometrium from postmenopausal women (who have low estrogen and progesterone levels) and women in midfollicular (increased estrogen/low progesterone) or midluteal (very increased estrogen/increased progesterone) phase. By contrast, alpha$_2$-adrenergic receptor ([^{3}H]rauwolscine) number increased strikingly as levels of estrogen increased (from 47 fmol/mg in postmenopausal women to 120 and 262 fmol/mg in women in mid-follicular and midluteal phases, respectively). Beta-adrenergic receptor number was slightly higher at midfollicular phase than at midluteal phase, and both phases yielded higher numbers of receptors than in postmenopausal women. Thus, expression of myometrial alpha$_2$- and perhaps beta- but not alpha$_1$-receptors appears to be enhanced by estrogen in humans.

Other studies on alpha$_2$-adrenergic receptors during the menstrual cycle have been conducted with blood platelets; these studies have yielded discrepant results. In a longitudinal study, Jones et al. serially assayed platelet alpha$_2$-adrenergic receptors ([^{3}H]yohimbine sites) in nine women and found peak numbers of receptors at the onset of menses with a 20-25% decrease at midcycle (39); similar results were obtained by another group (26). A recent cross-sectional study involving 42 subjects undergoing weight control therapy for 6 weeks arrived at a different conclusion (65). Although weight loss was associated with about a 50% increase in alpha$_2$ receptor binding ([^{3}H]yohimbine binding assayed at a 10 nM concentration), women divided into four groups based on menstrual history showed equivalent values initially and similar increases with weight loss (i.e., at a time when they were 2 weeks out of phase compared to the initial determina-tion). Other groups have also reported no significant difference in platelet alpha$_2$-adrenergic receptor number in women assessed at different times during the menstrual cycle (66,67). From a functional viewpoint, there are no definitive data of which we are aware indicating changes in alpha-adrenergic responsivity of platelets during the menstrual cycle. Since platelets have no deoxyribonucleic acid (DNA) and limited capacity to synthesize proteins, effects of sex steroids on receptor expression likely result from changes produced in megakaryocytes rather than in circulating platelets. Beta-adrenergic receptor expression on mononuclear leukocytes appears similar during various phases of the menstrual cycle (67).

A more pharmacological approach to assessing the role of sex steroids in regulating adrenergic receptors is to evaluate changes induced by oral contraceptives. Peters et al. (66) used [^{3}H]dihydro-ergocryptine, which may not be an optimal radioligand for assessing platelet alpha$_2$ receptors, since it detects an excess number of sites (68), and found a 30% decrease in both site number and K_d of the radioligand during the menstrual cycle. Specifically, platelets obtained from 15 women on day 28 (the end of a cycle at the nadir of estrogen level) had lower values for B_{max} and K_d than did platelets obtained on day 21 (the last day of treatment; i.e., at the time of peak hormone levels). These women were receiving an oral contraceptive that contained 30 μg ethinylestradiol and 150 or 250 μg levonorgestrel. The platelets obtained at day 21 also showed enhanced aggregation in response to norepinephrine and serotonin. In two women taking preparations containing 50 μg ethinylestradiol and 50 μg norgestrel, Jones et al. found an opposite pattern—lower platelet alpha-receptor number (determined using a more specific alpha$_2$-receptor probe, [^{3}H]yohimbine) on days 18-21 than on days 23-28 (39). More women need to be studied before one can draw a definitive conclusion regarding effects of oral contraceptives on platelet alpha-adrenergic receptor binding and function.

During pregnancy large changes in the hormonal milieu occur. Uterine tone is altered by adrenergic influences, and this has led to the use of beta-adrenergic agonists to decrease uterine contractility in settings such as premature labor. Direct assays of adrenergic receptors on human myometrium (Table 3) indicate no change in alpha$_1$-adrenergic receptors number with pregnancy but a reduced number of alpha$_2$- and beta adrenergic receptors. This has been hypothesized to relate to an inhibitory effect of progesterone on alpha$_2$-adrenergic receptor expression, since values similar to those obtained in pregnancy were found in women receiving medroxyprogesterone acetate. By contrast, treatment with medroxyprogesterone acetate enhanced beta-adrenergic receptor number. Thus, perhaps increased levels of progesterone during pregnancy or with progestin therapy can decrease expression of beta-adrenergic receptors.

A related finding to these data on myometrium from pregnant women is the evidence (69) that platelet alpha$_2$-receptor number reportedly falls about 20% within 7-10 days during the postpartum

TABLE 3 Adrenergic Receptor Number in Human Myometrial Membranes

Setting	Plasma 17 β-estradiol (ng/ml)	Plasma progesterone (ng/ml)	[^{3}H]Prazosin (fmol/mg)	[^{3}H]Rauwolscine (fmol/mg)
Postmenopausal	0.01-0.02	<0.1	29 ± 6	47 ± 14
Midfollicular phase	0.025-0.075	<1.0	31 ± 4	120 ± 16
Midluteal phase	0.10-0.30	>0.20	23 ± 6	262 ± 34
Term pregnancy	10-45	75-250	25 ± 4	50 ± 16
Medroxyprogesterone acetate	0.025-0.075	5-20	39 ± 3	55 ± 7

Uteri were obtained from five to 12 women in each group, who were
anesthetized by general anesthesia and who were undergoing hysterec-
tomy. In the case of pregnant women, myometrial strips were obtained
during elective cesarean sections performed at term before onset of
labor. Subjects treated with medroxyprogesterone acetate received
150 mg/3 months. Binding data represent site numbers obtained from
Scatchard analysis of binding isotherms using [^{3}H] prazosin for
alpha$_1$-receptors and [^{3}H]rauwolscine for alpha$_2$-receptors, and are
expressed as fmol/mg myometrial membrane protein. Steroid values
were determined on plasma samples obtained preoperatively.
Source: Adapted from Ref. 63, used with permission.

period, in parallel with a decrease in plasma estrogen and progesterone
levels. Interestingly, women with less postpartum depression had
less of a decrease in platelet alpha-receptor number during this period.

III. PHARMACOLOGICAL REGULATION OF ADRENERGIC RECEPTORS

A. Adrenocortical Steroids (Glucocorticoids)

In Chapter 11, Cryer reviews evidence regarding regulation of
adrenergic receptors by adrenal cortical steroids. In addition, another
recent review of this topic has appeared (70). Thus, we will only
briefly discuss this subject.

Glucocorticoids appear to exert a variety of potentially important modulatory roles on catecholamine receptors and response, including changes in the number of beta-adrenergic receptors (15,71), enhanced coupling of beta-adrenergic receptors to cyclic AMP generation (72), inhibition of cyclic nucleotide phosphodiesterase activity (73), shared effects (with catecholamines) on distal responses in target cells (74), and reversal of agonist-mediated desensitization (75,76). Such effects may be responsible for the so-called "permissive" action of glucocorticoids in enhancing adrenergic responses. Although molecular details are as yet scanty, one imagines that glucocorticoids are able to increase receptor number by enhancing the rate of transcription of receptor genes, since the steroids have this effect on several other genes (77,78). However, the recent evidence that glucocorticoids can alter processing of membrane proteins (79) suggests that post-transcriptional mechanisms may be operative as well. Enhanced functional coupling of receptors to guanine nucleotide binding (G) proteins has been observed with beta-adrenergic receptors in human polymorpho-nuclear cell membranes (75,80); this enhanced coupling could result from changes in receptors, G proteins, the membrane environment in which these proteins reside, or some combination thereof. Thus, the regulation of adrenergic receptors and response by corticosteroids is a complex issue that likely involves multiple mechanisms. Moreover, tissue-specific changes in regulation of adrenergic receptors by glucocorticoids add even further complexity to this regulation (70).

B. Thyroid Hormone Derivatives

Hyperthyroidism and hypothyroidism and their effects on adrenergic receptors and response are also discussed elsewhere in this book by Cryer (Chap. 11). In addition, another recent article presents a detailed and comprehensive review of the influence of thyroid hormone on alpha- and beta-adrenergic receptors and responsiveness in experimental animals and tissues (81).

To date, data in humans fail to provide much evidence that basal levels of thyroid hormones are important for expression of adrenergic receptors or response—at least as assessed on circulating blood cells (67). However, triiodothyronine administration has been reported to increase the number of beta-adrenergic receptors on leukocytes in some (82,83) but not all (84) studies. In addition, antithyroid treatment can lower beta-adrenergic receptor number on mononuclear leukocytes from hyperthyroid patients (85). Since tissue-specific changes in adrenergic receptors occur in hyperthyroid animals, the possibility that treatment with thyroid hormone or ablation of thyroid function in man will alter adrenergic receptors in one or more tissues remains a strong possibility. Considerable data in animals indicate enhancement in beta-adrenergic response and perhaps decrease in alpha-adrenergic response with thyroid hormone treatment and opposite types of changes with ablation of thyroid function (81).

In view of the numerous clinical features of thyrotoxicosis that suggest increased adrenergic activity (e.g., tachycardia, increased cardiac output, increased lipid and glycogen mobilization, enhanced thermogenesis, tremor, and sweating) and the importance of beta-adrenergic blocking agents in the management of thyrotoxicosis, especially of the "hyperadrenergic" features, it seems almost certain that adrenergic responsiveness, and perhaps adrenergic receptors, are altered in this setting.

Hypothyroidism demonstrates an opposite clinical picture with respect to "adrenergic" symptoms and signs and reciprocal changes in receptors and response (relative to hyperthyroidism) have been observed. Recent data obtained with rat fat cells suggest that blunted beta-adrenergic response in hypothyroidism may represent a decreased ability of receptors (even though normal in number and affinity) to interact productively with the stimulatory guanine coupling protein (G_S) of adenylate cyclase (86). Adipocytes from human subjects would appear to represent a useful system to define effects of thyroid hormones on adrenergic receptors and response, but little detailed information on these cells is available.

C. Adrenergic Agonists

Feedback regulation of adrenergic receptors and response by adrenergic agonists is a well-documented phenomenon in a variety of model systems and experimental animals (1, 87, 88). Thus, settings in which agonist concentrations increase and responses are stimulated often demonstrate a subsequent blunting of response. This blunting of response, which has been termed desensitization, refractoriness, tachyphylaxis, and tolerance, can in theory be produced by a variety of mechanisms. Thus, agonist-mediated loss in response might occur because of changes in receptors or in postreceptor mechanisms. In general, agonist-mediated changes in receptors represent homologous (i.e., receptor-specific) desensitization, whereas changes in post-receptor events are typically heterologous (i.e., effecting several classes of receptors). Several recent reviews summarize current concepts regarding desensitization of beta-adrenergic receptors, which can be both homologous or heterologous (1, 3, 87).

Homologous desensitization appears to involve a series of discrete molecular changes in beta-adrenergic receptors (Table 4): (1) An initial "uncoupling" of receptors from functional response. This phase is characterized by no change in receptor number but instead by a decrease in affinity of receptors for agonists, and in studies with membrane preparations, in a loss in the ability of guanine nucleotides to modulate agonist binding. (2) In addition, the early (<15 min)

TABLE 4 Desensitization of Beta-Adrenergic Receptors

	Early (<15 min) events	Late (≥60 min) events
Receptor binding	Receptor number unchanged Decreased affinity for agonists Agonist binding insensitive to guanine nucleotides	Receptor number decreased
Receptor function	Decreased	Decreased
Location of receptors	Redistributed (sequestered) from cell surface	"Lost"
Reversibility	Rapid (minutes)	Slow (hr-days)
Molecular mechanism	Internalization of receptors? Receptor phosphorylation	Accelerated degradation of receptors

phase of desensitization has also been associated with a redistribution or sequestration of beta-adrenergic receptors, which may result from an internalization (endocytosis) of receptors from the plasma membrane into the cell interior. (3) After more prolonged incubation of cells with agonists (typically > 60 min), receptors are down-regulated; i.e., decreased in number. This down-regulation appears to result from an enhanced rate of turnover or clearance of receptors and not from suppression of receptor synthesis. Details of the molecular and cellular events mediating changes in desensitization are currently being actively investigated by many laboratories.

Heterologous desensitization of beta-adrenergic receptor systems has been less well studied, but it appears to result from a variety of different mechanisms, including covalent modification (e.g., phosphorylation) of receptors, changes in guanine nucleotide coupling proteins, and increases in cyclic nucleotide phosphodiesterase activity (the enzyme that cleaves cyclic AMP, the beta-adrenergic receptor second messenger).

Most of the current ideas regarding mechanisms mediating desensitization of adrenergic receptors have derived from studies of beta-adrenergic receptor systems, in particular various cultured and isolated cell systems. This is somewhat surprising, since much data indicate that agonists can also desensitize alpha-adrenergic receptors (e.g., see Ref. 89). Perhaps the difficulty in studying second messengers for alpha-adrenergic receptors has slowed research related to desensitization of these receptors. Recent findings indicate that, as for beta-adrenergic receptors, loss in agonist affinity can occur as an early event following agonist interaction with $alpha_1$-adrenergic receptors and that agonists can promote a more slowly occurring down-regulation of these receptors (89-92). Recent studies in experimental animals that are administered catecholamines or to which norepinephrine-producing pheochromocytomas are transplanted indicate that the extent of agonist-mediated down-regulation can vary among different adrenergic receptor types and subtypes and different tissues, and is dependent upon the agonist or stimulus used to promote down-regulation (93-97). $Alpha_2$-adrenergic receptors, at least in several tissues studied thus far, seem to be especially refractory to developing down-regulation (94,96,98-101).

In humans, certain responses to adrenergic agonists are known to desensitize rapidly. Perhaps the most well-known example is the blunting of vasoconstrictor responses in the upper airway following treatment with alpha-adrenergic agonists. In addition, the blood pressure of patients with pheochromocytoma is often comparatively low relative to the extremely high plasma catecholamine levels in these patients (102). For beta-adrenergic receptors, considerable debate continues as to whether $beta_2$-adrenergic-mediated bronchodilation desensitizes after chronic therapy with beta agonists, although tremor (another $beta_2$-adrenergic response) does appear to desensitize with such treatment (103,104).

Aside from those types of clinical observations, attempts to establish protocols to elicit desensitization in human subjects have, in general, met with much less success. For example, maneuvers that increase plasma catecholamines, such as assuming an upright from a supine posture, consuming a low sodium diet, or exercise, have yielded somewhat ambiguous data regarding "physiological" desensitization of adrenergic responses by endogenous catecholamines (Table 5). Thus, some studies have indicated that those types of maneuvers can blunt cardiovascular response to infused catecholamines (48) and can either decrease beta-adrenergic receptor number on leukocytes (48) or can "uncouple" receptors (i.e., decrease agonist-mediated cyclic AMP generation and also decrease receptor affinity for agonist in association with a loss in guanine nucleotide regulation of agonist binding) (e.g., see Refs. 28 and 105). Other workers have found that exercise and upright posture have no effect on beta-adrenergic receptor binding to intact lymphocytes, whether one looks at either total number of receptors or at redistributed (presumably desensitized) receptors (e.g., see Ref. 106). Thus, the evidence that beta-adrenergic recep-

TABLE 5 Desensitization of Beta Receptors on Mononuclear Leukocytes

Situation	Receptor changes	References
Upright posture plus ambulation	Decreased affinity for agonists	105
	No change in number	38,106
	Decreased adenylate cyclase activity	105
	increased cAMP accumulation	106
Exercise, 5-15 min	Increased number	54,55
	Increased cAMP accumulation	106,53-55
Agonist infusion <2 hr	Increased number	106,121
	No change in number	108
	Decreased cAMP accumulation	108
	Increased cAMP accumulation	106
Agonist infusion 4 hr exercise training	Decreased number	121
	Decreased number	58
	No change in number	56
beta-adrenergic agonists administered for many days	Decreased number	116
	Decreased cAMP accumulation	118

tors in peripheral blood mononuclear cells show rapid desensitization in vivo is not convincing. We are also aware of little firm evidence that tissue responses to catecholamines rapidly desensitize in vivo.

Nevertheless, in vitro incubation of mononuclear cells with beta-adrenergic agonists can produce a rapid (half-time of a few minutes) desensitization of cyclic AMP accumulation (107-109). In these types of studies, one can demonstrate receptor redistribution and functional uncoupling of beta-adrenergic receptors. Thus, even though agonists are able to bind to receptors, these receptors seem unable to promote activation of adenylate cyclase (107, 109).

For alpha-adrenergic receptors, desensitization of platelet alpha$_2$-adrenergic response (aggregation) can be produced by in vitro incubation of blood with catecholamines (110,111) or by agonist infusions (112), but this desensitization has not been consistently associated with altered properties of the receptors in receptor binding studies or with decreases in the ability of alpha-agonists to inhibit cyclic AMP accumulation in platelets or in platelet membranes. Protocols documenting rapid ("acute") desensitization of alpha$_1$ receptors or of alpha$_2$ receptors on other tissues (99,100) are not yet available.

Administration of adrenergic agonists to human subjects will down-regulate beta-adrenergic receptors (e.g., see Refs. 76 and 113-117) and will lead to decreased cyclic AMP generation in leukocytes (e.g., see Refs. 117 and 118). This more slowly occurring change (normally requiring many hours or days) has been shown by many laboratories (Table 5). Most studies have involved assays of receptors on peripheral blood cells. Some workers have shown that such treatment protocols also produce blunted responsiveness of the cardiovascular system to beta-adrenergic stimulation (114, 116, 119), although other investigators have questioned whether beta-adrenergic subsensitivity of bronchial airways develops even though leukocyte beta-adrenergic receptors are down-regulated (76). In addition, chronic administration of beta agonists also blunts metabolic and thermogenic responses to isoproterenol even though basal metabolic rate and triiodothyronine concentrations are increased by such treatment (120).

Taken together, the data from several studies clearly indicate that chronic administration of adrenergic agonists can down-regulate beta-adrenergic receptors in vivo. However, it is not yet resolved whether the changes in receptor number occur equally in all tissues or whether these changes in receptor number are of *major* importance in altering response to endogenous or exogenous catecholamines (76, 103, 104). When one considers the widespread use of beta-adrenergic agonists on a chronic basis in patients with asthma, for example, the ultimate resolution of these issues is likely to be of clinical importance.

There are two poorly understood phenomena that are somewhat related to agonist-mediated down-regulation of lymphocyte beta-adrenergic receptor in human subjects: (1) intravenous infusion of beta-adrenergic agonists (or acute exercise [53-55]) can produce transient increases in beta-adrenergic receptor number and in beta-adrenergic-stimulated cyclic AMP accumulation in these cells (106, 121); and (2) these same maneuvers also produce prominent changes in the number and subpopulations of circulating lymphocytes (122-125).

As noted above, it is not clear whether alpha$_2$-adrenergic receptors down-regulate in response to agonist administration. Data obtained from both in vitro and in vivo studies with platelets have led to non-uniform results (Table 6).

Methodologic differences, such as the use of probes like [^{3}H]clonidine (a partial agonist [26, 110, 126, 127]) or [^{3}H]dihydroergocryptine (which detects more than just alpha$_2$ receptors in platelets [68]) rather than "pure" alpha$_2$ antagonists, such as [^{3}H]yohimbine and [^{3}H]rauwolscine, may explain some of these different results. For example, a small and biologically insignificant decrease in affinity of agonist in binding to receptors could be detected as a decrease in number of [^{3}H]clonidine binding sites. Since [^{3}H]clonidine only recognizes a relatively small fraction (generally 25%) of the number of sites detected

TABLE 6 Studies Assessing Possible Desensitization of Alpha$_2$-Receptors on Platelets

Condition	Change	References
In vitro incubation with agonist	No change in number	101
	Decreased number	110
	Decreased epinephrine-initiated aggregation	51, 110, 111
	No change in epinephrine inhibition of adenylate cyclase	111
Infusion of epinephrine or upright posture, 2 hr	No change in number	112, 133
	Decreased agonist affinity	51
	Decreased epinephrine-initiated aggregation	112
Pheochromocytoma	Decreased epinephrine-initiated aggregation	131
	Decreased number	131
	No change in number	132, 133

with antagonist radioligands, the interpretation of changes in the number of these agonist sites is not necessarily straightforward. Similarly, since the significance of the "extra" sites detected with [^{3}H]dihydroergocryptine remains poorly understood (68), it may be difficult to precisely relate changes in number of binding sites detected with that radioligand with changes found with the "purer" alpha$_2$-antagonists. A further problem is that because platelets store substantial quantities of catecholamines (128), exposing platelets to elevated concentrations of catecholamines can result in a "pseudo-down-regulation" of platelet alpha$_2$ receptors attributable to retained agonist that interferes in binding assays (101).

Pathological conditions may also be associated with increases in catecholamines, either in particular tissues or within the circulation. Pheochromocytoma is the clearest example of such a condition. In rats with norepinephrine-producing pheochromocytomas, the high levels of plasma norepinephrine promote prominent down-regulation of beta$_1$ (but not beta$_2$) and alpha$_1$ (but not alpha$_2$)-adrenergic receptors in tissues such as kidney, heart, lung, and fat (94, 97); functional desensitization has also been demonstrated in such animals (97). The changes produced by these tumors may reflect influences of factors other than just the catecholamines because animals given infusions of norepinephrine show different patterns of down-regulation (96).

TABLE 7 Drugs Predicted to Alter Adrenergic Receptors through Changes in Levels of Catecholamines

Increased stimulation of receptors	Decreased stimulation of receptors
Adrenergic-receptor agonists	Adrenergic-receptor antagonists
Catecholamine uptake inhibitors (e.g., tricyclic depressants, cocaine)	Centrally acting alpha$_2$-receptor agonists
Monoamine oxidase inhibitors	Ganglion-blocking agents
Nicotine (or other ganglionic and presynaptic stimulants)	Adrenergic neuron blocking agents (e.g., guanethidine, bethanidine)
Methylxanthines	Reserpinelike drugs
Indirect-acting sympathomimetic amines (e.g., ephedrine)	Inhibitors of norepinephrine synthesis
	Presynaptic inhibitors of catecholamine release
	Compounds causing degenerative changes in adrenergic nerves

In humans, studies in patients with pheochromocytoma indicate that beta-adrenergic receptors on leukocytes and adipocytes are desensitized (118,129) and that platelet alpha$_2$-receptors may (130,131) or may not (112,132,133) be decreased in number. It has also been reported that patients with myocardial failure have decreases in the number and function of cardiac and lymphocyte beta receptors and in the number of platelet alpha$_2$ receptors (114,134-136); these decreases might result from increases in circulating catecholamines in such patients (137). Drugs (Table 7) and other settings that are associated with increases in catecholamines (e.g., low sodium intake, exercise, and emotional stress) would be likely to promote downregulation to the extent that a particular tissue was susceptible to this type of regulation.

D. Adrenergic Antagonists and Sympatholytics

The abrupt withdrawal of beta-adrenergic blockers can result in clinical symptoms and signs that suggest adrenergic hypersensitivity (138-140). Although the frequency and clinical importance of this syndrome is not yet clearly established, there seems little doubt that at least some patients can develop this clinical picture (propranolol withdrawal syndrome). Studies with experimental animals indicate that depletion of tissue catecholamines or treatment with adrenergic antagonists promotes supersensitivity to catecholamines and, in some cases, an up-regulation of beta-adrenergic receptors (1,4,88).

Prospective studies in normal individuals have shown that administration of propranolol will increase beta-adrenergic number (up to about 50%) in lymphocytes; after stopping treatment, plasma propranolol levels are undetectable within 1 day, whereas receptor number requires several days to decline to control values (141,142). Although not all investigators have found this up-regulation of beta-adrenergic receptors (143), the majority of studies have reported similar increases in receptors after propranolol treatment. Thus, abrupt withdrawal of propranolol can create a setting characterized by an increased number of "unblocked" beta-adrenergic receptors. During this period, the cardiovascular system has been shown by some (48,141,143), but not all (e.g., see Ref. 144), investigators to be more sensitive to catecholamines.

Recent work has suggested that $beta_1$-selective blockers may (145) or may not (146) induce up-regulation of lymphocyte $beta_2$ receptors. In addition, beta-adrenergic blocking agents that possess intrinsic sympathomimetic activity (partial agonist activity) do not appear to produce the withdrawal syndrome; instead, these partial agonists can induce down-regulation of beta-adrenergic receptors (142,147).

Such results emphasize the key role of intrinsic activity in determining whether down- or up-regulation of receptors (and in turn desensitization or supersensitization) will occur in response to drug therapy. Moreover, the findings also illustrate the likely importance of tonic stimulation produced by endogenous catecholamines in down-regulating beta-adrenergic receptors and in potentially dampening adrenergic response. Based on the up-regulation that is observed when subjects are treated with beta-adrenergic antagonists, humans must be chronically down-regulated to a certain extent by endogenous catecholamines.

$Alpha_2$-adrenergic agonists, like clonidine, lower blood pressure in large part owing to their inhibition of sympathetic nervous system through an effect in the CNS. Abrupt withdrawal of clonidine is known to be associated with a striking rebound in blood pressure and other evidence of adrenergic hypersensitivity (140). It seems likely that this syndrome is another example of supersensitivity secondary to withdrawal of normal sympathetic tone (148). A further example is that patients with various neurological lesions that lead to low concentrations of circulating catecholamines also show enhanced responses to administered catecholamines and, in some cases, reported increases in number of adrenergic receptors on blood cells (130,149,150).

A prediction would be that diseases or drugs that lower tissue or plasma catecholamines (or that block adrenergic receptors) have the potential to elicit enhanced receptor binding and response in target cells. Whether such changes will occur will depend on several factors, including the susceptibility of a particular receptor type or subtype to this regulation, relationship between receptor occupancy and response, kinetics of drug disappearance, and so forth. This prediction regarding up-regulation/supersensitivity is the converse of the principle that treatment with agonist or settings that increase catecholamines may result in desensitization/down-regulation (Table 7). Since certain types of adrenergic receptors may be located on presynaptic sites where these receptors may increase (e.g., $beta_2$) or decrease ($alpha_2$) norepinephrine release at synpathetic nerve endings, the ultimate change manifest in a particular target tissue will represent the integrated response of the several classes of receptors whose interaction impact on cells in that tissue. Table 7 lists drugs that are likely to be associated with altered levels of catecholamines and thus with the potential to change receptor expression. Although clinical studies are not yet available in patients who have received all of those agents, one can anticipate likely results of studies assessing up- and down-regulation of adrenergic receptors.

E. Calcium-Channel Blockers

A structurally and clinically diverse group of compounds block calcium entry via voltage-dependent and receptor-operated channels. Recent work indicates the existence of at least two classes of "receptors" for these agents, sites for phenylalkylamines (e.g., verapamil) and others for dihydropyridines (e.g., nifedipine and nitrendipine) (151-153). In addition, recent evidence indicates that these agents can also influence adrenergic receptors and response in several ways: competitive blockade of receptors (154,155), blockade of postreceptor events, decrease in the affinity of receptors for agonists (156), decrease in circulating levels of catecholamines (156), and perhaps blunting of desensitization of beta-adrenergic receptors (157). These effects may contribute to clinically observed responses to the calcium-channel blockers.

F. Miscellaneous Pharmacological Agents

An ever-growing literature documents that a widely diverse group of clinically useful drugs can interact with adrenergic receptors in humans. We will briefly discuss a few of these: penicillins, indomethacin, amiloride, and quinidine.

Carbenicillin and other penicillins can impair platelet function and promote a bleeding diathesis. Altered platelet alpha-receptor affinity for agonists and antagonists have been observed in platelets incubated in vitro with penicillins (158). It has been proposed (158) that this effect may represent a generalized change in platelet membranes produced by penicillins.

Administration of indomethacin is reported to increase the number of leukocyte beta-adrenergic receptors about twofold without altering platelet alpha-adrenergic receptors (159). Whether the changes in leukocyte receptors are related to anti-inflammatory effects of indomethacin or other cyclooxygenase inhibitors is not known.

The diuretic amiloride and certain of its analogs are able to block alpha-adrenergic receptors on platelets and other tissues, including rat renal cortical membranes (49,160). In addition, certain functional responses to alpha-adrenergic stimulation appear to be blunted independent of receptor blockade (161). The clinical significance of these effects must still be determined.

The antiarrhythmic agent quinidine (but not lidocaine or procainamide) can competitively and reversibly block alpha-adrenergic receptors in heart, kidney, and human platelets (162). This blockade may contribute to the idiosyncratic hypotensive reaction of patients given intravenous quinidine. In addition, patients receiving combined treatment with quinidine and verapamil may be particularly at risk for hypotension, perhaps secondary to additive, competitive blockade at alpha-adrenergic receptors (163).

IV. CONCLUSIONS

The recent availability of radioligand binding assays has provided a powerful means to begin to assess whether alterations in response to catecholamines occur because of changes in adrenergic receptors. We have discussed evidence documenting physiological and pharmacological regulation of these receptors. The most convincing data have accrued on homologous regulation of the number of beta-adrenergic receptors by catecholamines and receptor blocking drugs, while considerable inferential data indicate that regulation by heterologous hormones and other factors occurs as well.

Future studies will probably focus on two areas: (1) cellular and molecular mechanisms responsible for changes in receptor binding, and (2) comparison of receptor binding and function, in particular between circulating blood cells, on which it is relatively easy to study receptors and internal tissues, whose function is usually of more important concern to physicians. The mechanistic studies will require the use of affinity labels, antibodies, and other probes of receptor structure and function in order to develop a more detailed understanding of the "life cycle" of adrenergic receptors in human cells and tissues. In addition, it will be important to direct further attention at determining whether the various settings discussed in this chapter alter binding of agonists to receptors and, more importantly, how such alterations come about. The desirability of undertaking studies that assess not only radioligand binding but also functional activity of receptors is to us self evident.

The availability of appropriate cells to assess regulation of human adrenergic receptors is a continuing and vexing issue. Although blood cells have been the principal types of cells studied thus far, further investigation of adrenergic receptors on fat and muscle cells obtained by biopsy should be feasible. Ultimately, though, one will require noninvasive ways to assess adrenergic receptors on internal organs. We believe that a major challenge in clinical investigation of adrenergic receptors lies in the development of appropriate means for conducting those types of studies.

ACKNOWLEDGMENTS

Work in the authors' laboratories is supported by grants from the National Institutes of Health, National Science Foundation, American Heart Association, and American Heart Association, California Affiliate.

REFERENCES

1. Stiles, G. L., Caron, M. G., and Lefkowitz, R. J.: Beta-adrenergic receptors: Biochemical mechanisms of physiological regulation. Physiol. Rev. 64:661-743, 1984.

2. Kenakin, T. P.: The classification of drugs and drug receptors in isolated tissues. Pharmacol. Rev. 36:165-222, 1984.

3. Harden, T. K.: Agonist-induced desensitization of the beta-adrenergic receptor-linked adenylate cyclase. Pharmacol. Rev. 35:5-32, 1983.

4. Motulsky, H. J., and Insel, P. A.: Adrenergic receptors in man: Direct identification, physiologic regulation, and clinical alterations. N. Engl. J. Med. 307:18-29, 1982.

5. Nahorski, S. R., and Barnett, D. B.: Biochemical assessment of adrenoceptor function and regulation: New directions and clinical relevance. Clin. Sci. 63:97-105, 1982.

6. Kelly, J., and O'Malley, K.: Adrenoceptor function and ageing. Clin. Sci. 66:509-515, 1984.

7. Weiss, B., Clark, M. B., and Greenberg, L. H.: Modulation of catecholaminergic receptors during development and aging. In:

Handbook of Neurochemistry, 2nd Ed. (A. Lajtha, ed.). Plenum Press, New York, 1984, pp. 595-627.

8. Whitsett, J. A., Noguchi, A., and Moore, J. J.: Developmental aspects of alpha- and beta-adrenergic receptors. Semin. Perinatol. 6:125-141, 1982.

9. Roan, Y., and Galant, S. P.: Decreased neutrophil beta adrenergic receptors in the neonate. Pediatr. Res. 16:591-593, 1982.

10. Jones, C. R., McCabe, R., Hamilton, C. A., and Reid, J. L.: Maternal and fetal platelet responses and adrenoceptor binding characteristics. Thromb. Haemost. 53:95-98, 1985.

11. Corby, D. G., and O'Barr, T. P.: Decreased alpha-adrenergic receptors in newborn platelets: Cause of abnormal response to epinephrine. Dev. Pharmacol. Ther. 2:215-225, 1981.

12. Reinhardt, D., Zehmisch, T., Becker, B., and Nagel-Hiemke, M.: Age-dependency of alpha- and beta-adrenoceptors on thrombocytes and lymphocytes of asthmatic and nonasthmatic children. Eur. J. Pediatr. 142:111-116, 1984.

13. Padbury, J. F., Diakomanolis, E. S., Hobel, C. J., Perelman, A., and Fisher, D. A.: Neonatal adaptation: Sympatho-adrenal response to umbilical cord cutting. Pediatr. Res. 15:1483-1487, 1981.

14. Barnes, P., Jacobs, M., and Roberts, J. M.: Glucocorticoids preferentially increase fetal alveolar beta-adrenoreceptors: Autoradiographic evidence. Pediatr. Res. 18(11):1191-1194, 1984.

15. Fraser, C. M., and Venter, J. C.: The synthesis of beta-adrenergic receptors in cultured human lung cells: Induction by glucocorticoids. Biochem. Biophys. Res. Commun. 94(1):390-397, 1980.

16. Roth, G. S.: Hormone action during aging: Alterations and mechanisms. Mech. Aging Dev. 9:497-514, 1979.

17. Rowe, J. W., and Troen, B. R.: Sympathetic nervous system and aging in man. Endocr. Rev. 1:167-179, 1980.

18. Docherty, J. R., and O'Malley, K.: Ageing and alpha-adrenoceptors. Clin. Sci. 68 (Suppl. 10):133s-136s, 1985.

19. Guarnieri, T., Filburn, C. R., Zitnik, G. S., and Lakatta, E. G.: Contractile and biochemical correlates of beta-adrenergic stimulation of the aged heart. Am. J. Physiol. 239:H501-H508, 1980.

20. Kusiak, J. W., and Pitha, J.: Decreased response with age of the cardiac catecholamine-sensitive adenylate cyclase system. Life Sci. 33:1679-1686, 1983.

21. Narayanan, N., and Derby, J.: Alterations in the properties of beta-adrenergic receptors of myocardial membranes in aging: Impairments in agonist-receptor interactions and guanine nucleotide regulation accompanying diminished catecholamine responsiveness adenylate cyclase. Mech. Aging Dev. 19:127-139, 1982.

22. Fitzgerald, D., Doyle, V., Kelly, J. B., and O'Malley, K.: Cardiac sensitivity to isoprenaline, lymphocyte beta-adrenoceptors and aging. Clin. Sci. 66:697-699, 1984.

23. Kendall, M. J., Woods, K. L., Wilkins, M. R., and Worthington, D. J.: Responsiveness to beta-adrenergic receptor stimulation. The effects of age are cardioselective. Br. J. Clin. Pharmacol. 14:821-826, 1982.

24. Schocken, D. D., and Roth, G. S.: Reduced beta-adrenergic receptor concentrations in ageing man. Nature 267:856-858, 1977.

25. Landmann, R., Burgisser, E., and Buhler, F. R.: Human lymphocytes as a model for beta-adrenergic receptors in clinical investigation. J. Recept. Res. 3:71-88, 1983.

26. Brodde, O. E.: Endogenous and exogenous regulation of human alpha- and beta-adrenergic receptors. J. Recept. Res. 3:151-162, 1983.

27. Halper, J. P., Mann, J. J., Weksler, M. E., Bilezikian, J. P., Sweeney, J. A., Brown, R. P., and Golbourne, T.: Beta adrenergic receptors and cyclic AMP levels in intact human lymphocytes: Effects of age and gender. Life Sci. 35:855-863, 1984.

28. Feldman, R. D., Limbird, L. E., Nadeau, J., Robertson, D., and Wood, A. J. J.: Alterations in leukocyte beta-receptor affinity with aging. A potential explanation for altered beta-adrenergic receptors in the elderly. N. Engl. J. Med. 310:815-819, 1984.

29. Abrass, I. B., and Scarpace, P. J.: Catalytic unit of adenylate cyclase: Reduced activity in aged human lymphocytes. J. Clin. Endocrinol. Metab. 55:1026-1028, 1982.

30. Holdtke, R. D., and Climi, K. M.: Effects of aging on catecholamine metabolism. J. Clin. Endocrinol. Metab. 60:479-484, 1985.

31. Esler, M.: Assessment of sympathetic nervous function in humans from noradrenaline plasma kinetics. Clin. Sci. 62:247-254, 1982.

32. Maggi, A., Schmidt, M. J., Ghetti, B., and Enna, S. J.: Effect of aging on neurotransmitter receptor binding in rat and human brain. Life Sci. 24:367-374, 1979.

33. Hoffman, B. B., Chang, H., Farahbakshs, Z. T., and Reaven, G. M.: Age-related decrement in hormone-stimulated lipolysis. Am. J. Physiol. 247:E772-E777, 1984.

34. Lafontan, M.: Inhibition of epinephrine-induced lipolysis in isolated white adipocytes of aging rabbits by increased alpha-adrenergic responsiveness. J. Lipid Res. 20:208-216, 1979.

35. Wirz-Justice, A., Kafka, M. S., Naber, D., and Wehr, T. A.: Circadian rhythms in rat brain alpha- and beta-adrenergic receptors are modified by chronic imipramine. Life Sci. 27:341-347, 1980.

36. Titinchi, S., Shamma, M. A., Patel, K. R., Kerr, J. W., and Clark, B.: Circadian variation in number and affinity of beta$_2$-adrenoceptors in lymphocytes of asthmatic patients. Clin. Sci. 66:323-328, 1984.

37. Frohler, M., Saito, Y., Ackenheil, M., Bak, R., Bondy, B., Feistenauer, E., Hoftscuster, E., Vakis, A., and Welter, D.: Catecholaminergic binding sites of blood cells of healthy volunteers with special respect to circadian rhythms. Pharmacopsychiatria 18:147-148, 1985.

38. Sowers, J. R., Connelly-Fittinghoff, M., Tuck, M. L., and Krall, J. F.: Acute changes in noradrenaline levels do not alter lymphocyte beta-adrenergic receptor concentrations in man. Cardiovasc. Res. 17:184-188, 1983.

39. Jones, S. B., Bylund, D. B., Rieser, C. A., Shekim, W. O., Byer, J. A., and Carr, G. W.: Alpha$_2$-adrenergic receptor binding in human platelets: Alterations during the menstrual cycle. Clin. Pharmacol. Ther. 34:90-96, 1983.

40. Lafontan, M., Berlan, M., and Villeneuve, A.: Preponderance of alpha$_2$ over beta$_1$-adrenergic receptor sites in human fat cells is not predictive of the lipolytic effect of physiological catecholamines. J. Lipid Res. 24:429-440, 1983.

41. Fain, J. N., and Garcia-Sainz, J. A.: Adrenergic regulation of adipocyte metabolism. J. Lipid Res. 24:945-966, 1983.

42. Engfeldt, P., Arner, P., Kimura, H., Wahrenberg, H., and Ostman, J.: Determination of adrenoceptors of the alpha$_2$ subtype on isolated human fat cells. Scand. J. Lab. Invest. 43:207-213, 1983.

43. Engfeldt, P., Arner, P., Wahrenberg, H., and Ostman, J.: An assay for beta-adrenergic receptors in isolated human fat cells. J. Lipid Res. 23:715-719, 1982.

44. Kather, H., Weiland, E., Fischer, B., Wirth, A., and Schlierf, G.: Adrenergic regulation of lipolysis in abdominal adipocytes of obese subjects during caloric restriction: Reversal of catecholamine action caused by relief of endogenous inhibition. Eur. J. Clin. Invest. 15:30-37, 1985.

45. Sundaresan, P. R., Weintraub, M., Hershey, L. A., Kroening, B. H., Hasday, J. D., and Banerjee, S. P.: Platelet alpha-adrenergic receptors in obesity: Alteration with weight loss. Clin. Pharmacol. Ther. 33:776-781, 1983.

46. Luck, P., Mikhailidis, M. R., Dashwood, M. R., Barradas, M. A., Sever, P. S., Dandona, P., and Wakeling, A.: Platelet hyper-aggregability and increased alpha-adrenoceptor density in anorexia nervosa. J. Clin. Endocrinol. Metab. 57:911, 1983.

47. Esler, M., Jennings, G., Korner, P., Blombery, P., Sacharis, N., and Leonard, P.: Measurement of total organ-specific norepinephrine kinetics in humans. Am. J. Physiol. 247:E21-E28, 1984.

48. Fraser, J., Nadeau, J., Robertson, D., and Wood, A. J. J.: Regulation of human leukocyte beta receptors by endogenous catecholamines. Relationship of leukocyte beta receptor density to the cardiac sensitivity to isoproterenol. J. Clin. Invest. 67: 1777-1784, 1981.

49. Motulsky, H. J., and Insel, P. A.: Influence of sodium on the alpha$_2$-adrenergic receptor system of human platelets. Role for intraplatelet sodium in receptor binding. J. Biol. Chem. 258: 3913-3919, 1983.

50. Insel, P. A., and Motulsky, H. J.: A hypothesis linking intra-cellular sodium, membrane receptors, and hypertension. Life Sci. 34:1009-1113, 1984.

51. Hollister, A. S., Fitzgerald, G. A., Nadeau, J. H., and Robertson, D.: Acute reduction in human platelet alpha$_2$ adrenoceptor affinity for agonist by endogenous and exogenous catecholamines. J. Clin. Invest. 72:1498-1505, 1983.

52. Motulsky, H. J., Shattil, S. J., and Insel, P. A.: Characteriza-tion of alpha$_2$-adrenergic receptors on human platelets using [^{3}H]yohimbine. Biochem. Biophys. Res. Commun. 97:1562-1570, 1980.

53. Cundell, D., Danks, J., Phillips, M. J., and Davies, R. J.: Effect of exercise on isoprenaline-induced lymphocyte cAMP production in atopic asthmatics and atopic and non-atopic, non-asthmatic subjects. Clin. Allergy 14:433-442, 1984.

54. Butler, J., Kelly, J. G., O'Malley, K., and Pidgeon, F.: Beta-adrenergic adaptation to acute exercise. J. Physiol. 344:1131-117, 1983.

55. Brodde, O.-E., Daul, A., and O'Hara, N.: Beta-adrenoceptor changes in human lymphocytes induced by dynamic exercise. Naunyn Schmiedebergs Arch. Pharmacol. 325:190-192, 1984.

56. Williams, R. S.: Role of receptor mechanisms in the adaptive response to habitual exercise. Am. J. Cardiol. 55:68D-73D, 1985.

57. Mann, S. J., Krakoff, L. R., Felton, K., and Yeager, K.: Cardiovascular responses to infused epinephrine: Effect of the state of physical conditioning. J. Cardiovasc. Pharmacol. 6:339-343, 1984.

58. Butler, J., O'Brien, M., O'Malley, K., and Kelly, J. G.: Relationship of beta-adrenoceptor density to fitness in athletes. Nature 298:60-63, 1982.

59. Colucci, W. S., Gimbrone, M. A., Jr., McLaughlin, M. K., Halpern, W., and Alexander, R. W.: Increased vascular catecholamine sensitivity and alpha-adrenergic receptor affinity in female and estrogen-treated male rats. Circ. Res. 50:805-811, 1982.

60. Roberts, J. M., Goldfien, R. D., Tsuchiya, A. M., Godfien, A., and Insel, P. A.: Estrogen treatment decreases alpha-adrenergic binding sites on rabbit platelets. Endocrinology 104:722-728, 1979.

61. Mishra, N., Hamilton, C. A., Jones, C. R., Leslie, C., and Reid, J. L.: Alpha-adrenoceptor changes after oestrogen treatment in platelets and other tissues in female rabbits. Clin. Sci. 69: 235-238, 1985.

62. Garret, W. J.: The effects of adrenalin and noradrenalin on the intact non-pregnant human uterus. J. Obstet. Gynecol. Br. Empire 62:876, 1955.

63. Bottari, S. P., Vokaear, A., Kaivez, E., Lescrainier, P., and Vauquelin, G.: Differential regulation of the alpha-adrenergic receptor subclasses by gonadal steroids in human myometrium. J. Clin. Endocrinol. Metab. 57:937-943, 1983.

64. Bottari, S. P., Lescrainier, J.-P., Kaivez, E., Severne, Y., Vauquelin, G., and Vokaer, A.: Regulation of adrenergic receptor subtypes by gonadal steroids in human myometrium. Arch. Int. Physiol. Biochim. 91:B86, 1983.

65. Sundaresan, P. R., Madan, M. K., Kelvie, S. L., and Weintraub, M.: Platelet alpha$_2$ adrenoceptors and the menstrual cycle. Clin. Pharmacol. Ther. 37(3):337-342, 1985.

66. Peters, J. R., Elliot, J. M., and Grahame-Smith, D. G.: Effect of oral contraceptives on platelet noradrenaline and 5-hydroxytryptamine receptors and aggregation. Lancet 2:933-936, 1979.

67. Rosen, S. G., Berk, M. A., Popp, D. A., Serusclat, P., Smith, E. B., Shah, S. D., Ginsberg, A. M., Clutter, W. E., and Cryer, P. E.: Beta$_2$ and alpha$_2$-adrenergic receptors and receptor coupling

to adenylate cyclase in human mononuclear leukocytes and platelets
in relation to physiological variations of sex steroids. J. Clin.
Endocrinol. Metab. 58:1068-1075, 1984.

68. Motulsky, H. J., and Insel, P. A. [³H]Dihydroergocryptine
    binding to alpha-adrenergic receptors of human platelets. A
    reassessment using the selective radioligands [³H]prazosin,
    [³H]yohimbine, and [³H]rauwolscine. Biochem. Pharmacol. 31:
    2591-2597, 1982.

69. Metz, A., Stump, K., Cowen, P. J., Elliot, J. M., Gelder, M. G.,
    and Grahame-Smith, D. G.: Changes in platelet $alpha_2$ adrenoceptor
    binding post-partum: Possible relation to maternity blues. Lancet
    1:495-498, 1983.

70. Davies, A. O., and Lefkowitz, R. J.: Regulation of beta-adrenergic
    receptors by steroid hormones. Ann. Rev. Physiol. 46:119-130,
    1984.

71. Cotecchia, S., and De Blasi, A.: Glucocorticoids increase beta-
    adrenoceptors on human intact lymphocytes in vitro. Life Sci.
    35:2359-2364, 1984.

72. Marone, G., Lichtenstein, L. M., and Plaut, M.: Hydrocortisone
    and human lymphocytes: Increases in cyclic adenosine 3':5'-
    monophosphate and potentiation of adenylate cyclase-activating
    agents. J. Pharmacol. Exp. Ther. 215(2):469-477, 1980.

73. Lee, T. P., and Reed, C. E.: Effects of steroids on the regulation
    of the levels of cyclic AMP in human lymphocytes. Biochem.
    Biophys. Res. Commun. 78(3):998-1004, 1977.

74. Insel, P. A., and Honeysett, M.: Glucocorticoid-mediated inhibition
    of ornithine decarboxylase activity in S49 lymphoma cells. Proc.
    Natl. Acad. Sci. U.S.A. 78:5669-5672, 1981.

75. Davies, A. O., and Lefkowitz, R. J.: In vitro desensitization of
    beta adrenergic receptors in human neutrophils. J. Clin. Invest.
    71:565-571, 1983.

76. Tashkin, D. P., Conolly, M. E., Deutsch, R. I., Hui, K. K.,
    Littner, M., Scarpace, P., and Abrass, I.: Subsensitization of
    beta-adrenoceptors in airways and lymphocytes of healthy and
    asthmatic subjects. Am. Rev. Respir. Dis. 125:185-193, 1982.

77. O'Malley, B. W.: Steroid hormone action in eucaryotic cells.
    J. Clin. Invest. 74:307-312, 1984.

78. Ringold, G. M.: Steroid hormone regulation of gene expression.
    Ann. Rev. Pharmacol. Toxicol. 25:529-566, 1985.

79. Firestone, G. L., Payvar, F., and Yamamoto, K. R.: Gluco-corticoid regulation of protein processing and compartmentalization. Nature 300: 221-225, 1982.

80. Davies, A. O., and Lefkowitz, R. J.: Agonist-promoted high affinity state of the beta-adrenergic receptor in human neutrophils: Modulation by corticosteroids. J. Clin. Endocrinol. Metab. 53(4): 703-708, 1981.

81. Bilezikian, J. P., and Loeb, J. N.: The influence of hyperthyroidism and hypothyroidism on alpha- and beta-adrenergic receptor systems and adrenergic responsiveness. Endocr. Rev. 4:378-388, 1983.

82. Andersson, R. G. G., Nilsson, O. R., and Kuo, J. F.: Beta-adrenoceptor-adenosine 3'-5'-monophosphate system in human leukocytes before and after treatment for hyperthyroidism. J. Clin. Endocrinol. Metab. 56:42-45, 1983.

83. Ginsberg, A. M., Clutter, W. E., Shah, S. D., and Cryer, P. E.: Triiodothyronine-induced thyrotoxicosis increases mononuclear leukocyte beta-adrenergic receptor density in man. J. Clin. Invest. 67:1785-1791, 1981.

84. Hui, K. K. P., Wolfe, R. N., and Conolly, M. E.: Lymphocyte beta-adrenergic receptors are not altered in hyperthyroidism. Clin. Pharmacol. Ther. 32:161-165, 1982.

85. Cognini, G., Piantanelli, L., Paolinelli, E., Orlandoni, P., Pelligrini, A., and Masera, N.: Decreased beta-adrenergic receptor density in mononuclear leukocytes from thyroidectomized patients. Acta Endocrinol. 103:1-5, 1983.

86. Malbon, C. C., Graziano, M. P., and Johnson, G. L.: Fat cell beta-adrenergic receptor in the hypothyroid rat: Impaired inter-action with the stimulatory regulatory component of adenylate cyclase. J. Biol. Chem. 259(5):3254-3260, 1984.

87. Hertel, C., and Perkins, J. P.: Receptor-specific desensitization of beta-adrenergic receptor function. Mol. Cell. Endocrinol. 37: 245-256, 1984.

88. Insel, P. A., and Sanda, M.: Temperature-dependent binding to beta-adrenergic receptors of intact S49 cells: Implications for the state of the receptor that activates adenylate cyclase under physio-logical conditions. J. Biol. Chem. 254:6554-6559, 1979.

89. Colucci, W. S., Brock, T. A., Gimbrone, M. A., Jr., and Alexander, R. W.: Regulation of alpha$_1$-adrenergic receptor-coupled calcium flux in cultured vascular smooth muscle cells. Hypertension 6(2):I19-I24, 1983.

90. Schwarz, K. R., Carter, E. A., Homcy, R. M., and Graham, R. M.: Regulation of adrenergic agonist affinity states. Fed. Proc. 44:1628, 1985.

91. Hughes, R. J., and Insel, P. A.: An intermediate state of the alpha$_1$-adrenergic receptor along the pathway to down-regulation in BC3H-1 cells. Fed. Proc. 44:1796, 1985.

92. Hughes, R. J., and Insel, P. A.: Agonist-mediated regulation of α_1- and β_2-adrenergic receptor metabolism in a muscle cell line, BC3H-1. Mol. Pharmacol. 29:521-530, 1986.

93. Hasegawa, M., and Townley, R. G.: Difference between lung and spleen susceptibility of beta-adrenergic receptors to de-sensitization by terbutaline. J. Allergy Clin. Immunol. 71:230-238, 1983.

94. Snavely, M. D., Mahan, L. C., O'Connor, D. T., and Insel, P. A.: Selective down-regulation of adrenergic receptor sub-types in rats with pheochromocytoma. Endocrinology 113:354-361, 1983.

95. Snavely, M. D., Ziegler, M. G., and Insel, P. A.: A new approach to determine rates of receptor appearance and disappearance in vivo: Agonist-mediated down-regulation of rat renal cortical beta$_1$- and beta$_2$-adrenergic receptors. Mol. Pharmacol. 27:19-26, 1985.

96. Snavely, M. D., Ziegler, M. G., and Insel, P. A.: Subtype-selective down-regulation of rat renal cortical alpha- and beta-adrenergic receptors by catecholamines. Endocrinology, 117:2182-2189, 1985.

97. Tsujimoto, G., Manger, W. M., and Hoffman, B. B.: Desensitiza-tion of beta-adrenergic receptors by pheochromocytoma. Endo-crinology 114(4):1271-1278, 1984.

98. Pecquery, R., Leneveu, M.-C., and Giudicelli, Y.: In vivo desensitization of the beta, but not the alpha$_2$-adrenoreceptor-coupled-adenylate cyclase system in hamster white adipocytes after administration of epinephrine. Endocrinology 114(5):1576-1583, 1984.

99. Villeneuve, A., Carpene, C., Berlan, M., and Lafontan, M.: Lack of desensitization of alpha$_2$-mediated inhibition of lipolysis in fat cells after acute and chronic treatment with clonidine. J. Pharmacol. Exp. Ther. 233:433-440, 1985.

100. Burns, T. W., Langley, P. E., Terry, B. E., and Bylund, D. B.: Studies on desensitization of adrenergic receptors of human adipocytes. Metabolism 31(3):288-293, 1982.

101. Karliner, J. S., Motulsky, H. J., and Insel, P. A.: Apparent "down-regulation" of human platelet alpha$_2$-adrenergic receptors is due to retained agonist. Mol. Pharmacol. 21:36-43, 1982.

102. Bravo, E. L., Tarazi, R. C., Gifford, R. W., and Stewart, B. H.: Circulating and urinary catecholamines in pheochromocytoma. Diagnostic and pathophysiologic implications. N. Engl. J. Med. 301:682-686, 1979.

103. Jenne, J. W.: Whither beta-adrenergic tachyphylaxis? J. Allergy Clin. Immunol. 70:413-416, 1982.

104. van den Berg, W., Leferink, J. G., Fokkens, J. K., Kreukniet, J., Maes, R. A. A., and Bruynzeel, P. L. B.: Clinical implications of drug-induced desensitization of the beta receptor after continuous oral use of terbutaline. J. Allergy Clin. Immunol. 69(5):410-417, 1982.

105. Feldman, R. D., Limbird, L. E., Nadeau, J., FitzGerald, G. A., Robertson, D., and Wood, A. J. J.: Dynamic regulation of leukocyte beta adrenergic receptor-agonist interactions by physiological changes in circulating catecholamines. J. Clin. Invest. 72:164-170, 1983.

106. De Blasi, A., Maisel, A., Feldman, R. D., Ziegler, M. G., Fratelli, M., Di Lallo, M., Smith, D., Lai, C.C., Motulsky, H.: In vivo regulation of beta adrenergic receptors on human mononuclear leukocytes: Assessment of receptor number, function, and location following posture change, exercise, and infusion of isoproterenol. J. Clin. Endocrinol. Metab., in press, 1986.

107. Motulsky, H. J., Cunningham, E. M. S., DeBlasi, A., and Insel, P. A.: Agonists promote rapid desensitization and redistribution of beta adrenergic receptors on intact human mononuclear leukocytes. Am. J. Physiol. 250:E583-E590, 1986.

108. Krall, J. F., Connelly, M., and Tuck, M. L.: Acute regulation of beta adrenergic catecholamine sensitivity in human lymphocytes. J. Pharmacol. Exp. Ther. 214:554-560, 1980.

109. De Blasi, A., Lipartiti, M., Motulsky, H. J., Insel, P. A., and Fratelli, M.: Agonist-induced redistribution of beta-adrenergic receptors on intact human mononuclear leukocytes. Redistributed receptors are nonfunctional. J. Endocrinol. Clin. Metab. 61:1081-1088, 1985.

110. Cooper, B., Handin, R. I., Young, L. H., and Alexander, R. W.: Agonist regulation of the human platelet alpha-adrenergic receptor. Nature 274:703-706, 1978.

111. Motulsky, H. J., Shattil, S. J., Ferry, N., Rozansky, D., and Insel, P. A.: Desensitization of epinephrine-initiated platelet

aggregation does not alter binding to the alpha$_2$-adrenergic receptor or its coupling to adenylate cyclase. Mol. Pharmacol., 29:1-6, 1986.

112. Jones, C. R., Hamilton, C. A., Whyte, K. F., Elliott, H. L., and Reid, J. L.: Acute and chronic regulation of alpha(2)-adrenoceptor number and function in man. Clin. Sci. 68 (Suppl. 10):129s-132s, 1985.

113. Galant, S. P., Duriseti, L., Underwood, S., and Insel, P. A.: Decreased beta-adrenergic receptors on polymorphonuclear leukocytes after adrenergic therapy. N. Engl. J. Med. 299:933-936, 1978.

114. Colucci, W. S., Alexander, R. W., Williams, G. H., Rude, R. E., Holman, B. L., Konstam, M. A., Wynne, J., Mudge, G. H., Jr., and Braunwald, E.: Decreased lymphocyte beta-adrenergic-receptor density in patients with heart failure and tolerance to the beta-adrenergic agonist pirbuterol. N. Engl. J. Med. 305: 185-190, 1981.

115. Sano, Y., Watt, G., and Townley, R. G.: Decreased mononuclear cell beta-adrenergic receptors in bronchial asthma: parallel studies of lymphocyte and granulocyte desensitization. J. Allergy Clin. Immunol. 72:495-503, 1983.

116. Aarons, R. D., Nies, A. S., Gerber, J. G., and Molinoff, P. B.: Decreased beta adrenergic receptor density on human lymphocytes after chronic treatment with agonists. J. Pharmacol. Exp. Ther. 224(1):1-6, 1983.

117. Galant, S. P., Duriseti, L., Underwood, S., Allred, S., and Insel, P. A.: Beta-adrenergic receptors of polymorphonuclear particulates in bronchial asthma. J. Clin. Invest. 65:577-586, 1980.

118. Greenacre, J. K., and Conolly, M. E.: Desensitization of the beta-adrenoceptor of lymphocytes from normal subjects and patients with phaeochromocytoma: Studies in vivo. Br. J. Clin. Pharmacol. 5:191-197, 1978.

119. Kaywin, P., McDonough, M., Insel, P. A., and Shattil, S. J.: Platelet function in essential thrombocythemia. Decreased epinephrine responsiveness associated with a deficiency of platelet alpha-adrenergic receptors. N. Engl. J. Med. 299:505-509, 1978.

120. Scheidegger, K., O'Connell, M., Robbins, D. C., and Danforth, E., Jr.: Effects of chronic beta-receptor stimulation on sympathetic nervous system activity, energy expenditure, and thyroid hormones. J. Clin. Endocrinol. Metab. 58(5):895-903, 1984.

121. Tohmeh, J. F., and Cryer, P. E.: Biphasic adrenergic modulation of beta-adrenergic receptors in man. Agonist-induced early increment and late decrement in beta-adrenergic receptor number. J. Clin. Invest. 65:836-840, 1980.

122. Hedfors, E., Holm, G., Ivansen, M., and Wahren, J.: Physiological variation of blood lymphocyte reactivity: T-cell subsets, immunoglobulin production, and mixed-lymphocyte reactivity. Clin. Immunol. Immunopathol. 27:9-14, 1983.

123. Edwards, A. J., Bacon, T. H., Elms, C. A., Verardi, R., Felder, M., and Knight, S. C.: Changes in the populations of lymphoid cells in human peripheral blood following physical exercise. Clin. Exp. Immunol. 58:420-427, 1984.

124. Soppi, E., Varjo, P., Eskola, J., and Laitinen, L. A.: Effect of strenuous physical stress on circulating lymphocyte number and function before and after training. Clin. Lab. Immunol. 8: 43-46, 1982.

125. Crary, B., Hauser, S. L., Borysenko, M., Kutz, I., Hoban, C., Ault, K. A., Weiner, H. L., and Benson, H.: Epinephrine-induced changes in the distribution of lymphocyte subsets in peripheral blood of humans. J. Immunol. 131(3):1178-1181, 1983.

126. Garcia, S. J. A., Zis, A. P., Hollingsworth, P. J., Greden, J. F., and Smith, C. B.: Platelet alpha$_2$-adrenergic receptors in major depressive disorder. Binding of tritiated clonidine before and after tricyclic antidepressant drug treatment. Arch. Gen. Psychiatry 38:1327-1333, 1981.

127. Shattil, S. J., McDonough, M., Turnbull, J., and Insel, P. A.: Characterization of alpha-adrenergic receptors in human platelets using [^{3}H]clonidine. Mol. Pharmacol. 19:179-183, 1981.

128. Zweifler, A. J., and Julius, S.: Increased platelet catecholamine content in pheochromocytoma. N. Engl. J. Med. 306:890-894, 1982.

129. Smith, U., Sjostrom, L., Stenstrom, G., Isaksson, O., and Jacobsson, B.: Studies on the catecholamine resistance in fat cells from patients with phaeochromocytoma. Eur. J. Clin. Invest. 7:355-361, 1977.

130. Davies, I. B., Mathias, C. J., Sudera, D., and Sever, P. S.: Agonist regulation of alpha-adrenergic receptor responses in man. J. Cardiovasc. Pharmacol. 4:S139-S144, 1982.

131. Brodde, O.-E., and Bock, K. D.: Changes in platelet alpha$_2$-adrenoceptors in human phaeochromocytoma. Eur. J. Clin. Pharmacol. 26:265-267, 1984.

132. Snavely, M. D., Motulsky, H. J., O'Connor, D. T., Ziegler, M. G., and Insel, P. A.: Adrenergic receptors in human and experimental pheochromocytoma. Clin. Exp. Hypertens. (A) 4:829-848, 1982.

133. Pfeifer, M. A., Ward, K., Malpass, T., Stratton, J., Halter, J., Evans, M., Beiter, H., Harker, L. A., and Porte, D., Jr.: Variations in circulating catecholamines fail to alter human platelet alpha$_2$-adrenergic receptor number or affinity for [^{3}H]yohimbine or [^{3}H]dihydroergocryptine. J. Clin. Invest. 74:1063-1072, 1984.

134. Weiss, R. J., Tobes, M., Wertz, C. E., and Smith, B.: Platelet alpha$_2$ adrenoreceptors in chronic congestive heart failure. Am. J. Cardiol. 52:101-105, 1983.

135. Bristow, M. R., Ginsburg, R., Minobe, W., Cubicciotti, R. S., Sageman, W. S., Lurie, K., Billingham, M. E., Harrison, D. C., and Stinson, E. B.: Decreased catecholamine sensitivity and beta-adrenergic-receptor density in failing human hearts. N. Engl. J. Med. 307:205-211, 1982.

136. Minakuchi, K., Ogawa, K., Ban, M., and Satake, T.: Decreased generation of cyclic AMP in lymphocytes by beta-adrenergic stimulation in heart failure. Jpn. Heart J. 22:585-592, 1981.

137. Viquerat, C. E., Daly, P., Swedberg, K., Evers, C., Curran, D., Parmley, W. W., and Chatterjee, K.: Endogenous catecholamine levels in chronic heart failure: Relation to the severity of hemodynamic abnormalities. Am. J. Med. 78:455-460, 1985.

138. Wood, A. J. J.: Beta-blocker withdrawal. Drugs 25 (Suppl. 2): 318-321, 1983.

139. Prichard, B. N. C., Tomlinson, B., Walden, R. J., and Bhattacharjee, P.: The beta-adrenergic blockade withdrawl phenomenon. J. Cardiovasc. Pharmacol. 5:S56-S62, 1983.

140. Hart, G. R., and Anderson, R. J.: Withdrawal syndromes and the cessation of antihypertensive therapy. Arch. Intern. Med. 141:1125-1127, 1981.

141. Aarons, R. D., Nies, A. S., Gal, J., Hegstrand, L. R., and Molinoff, P. B.: Elevation of beta-adrenergic receptor density in human lymphocytes after propranolol administration. J. Clin. Invest. 65:949-957, 1980.

142. Brodde, O.-E., Daul, A., Stuka, N., O'Hara, N., and Borchard, U.: Effects of beta-adrenoceptor antagonist administration on

beta$_2$-adrenoceptor density in human lymphocytes. The role of "intrinsic sympathomimetic activity". Nauyn Schmiedebergs Arch. Pharmacol. 328:417-422, 1985.

143. Goldstein, R. E., Corash, L. C., Tallman, J. F., Jr., Lake, C. R., Hyde, J., Smith, C. C., Capurro, N. L., and Anderson, J. C.: Shortened platelet survival time and enhanced heart rate responses after abrupt withdrawal of propranolol from normal subjects. Am. J. Cardiol. 47:1115-1122, 1981.

144. Kiyingi, K. S., and Shaw, K.: The phenomenon of beta-adrenergic hypersensitivity following propranolol withdrawal studied in normal subjects. Eur. J. Clin. Pharmacol. 27:423-428, 1984.

145. Piantanelli, L., Giunta, S., Basso, A., Cognini, G., Andreoni, A., and Paciaroni, E.: Atenolol-induced regulation of leukocyte beta$_2$-adrenoceptors in hypertension. Pharmacology 29:210-214, 1984.

146. De Blasi, A., Cortellaro, M., and Costantini, C.: Effects of chronic metoprolol and sulphinpyrazone on human lymphocyte beta-adrenoceptors. Br. J. Clin. Pharmac. 18:45-50, 1984.

147. Molinoff, P. B., and Aarons, R. D.: Effects of drugs on beta-adrenergic receptors on human lymphocytes. J. Cardiovasc. Pharmacol. 51:S63-S67, 1983.

148. Hamada, S., Yamamura, H. I., and Roeske, W. R.: An increase in cardiac alpha$_1$-adrenoceptors following chronic clonidine treatment. Naunyn Schmiedebergs Arch. Pharmacol. 320:115-118, 1982.

149. Brodde, O.-E., Anlauf, M., Arroyo, J., Wagner, R., Weber, F., and Buck, K. D.: Hypersensitivity of adrenergic receptors and blood-pressure response to oral yohimbine in orthostatic hypotension. N. Engl. J. Med. 308(17):1033, 1983.

150. Hui, K. K. P., and Conolly, M. E.: Increased numbers of beta receptors in orthostatic hypotension due to autonomic dysfunction. N. Engl. J. Med. 304(24):1473-1476, 1981.

151. Snyder, S. H., and Reynolds, I. J.: Calcium-antagonist drugs: Receptor interactions that clarify therapeutic effects. N. Engl. J. Med. 313:995-1002, 1985.

152. Janis, R. A., and Triggle, D. J.: New developments in Ca^{2+} channel antagonists. J. Med. Chem. 26(6):775-785, 1983.

153. Glossmann, H., Ferry, D. R., Goll, A., and Rombusch, M.: Molecular pharmacology of the calcium channel: Evidence for

subtypes, multiple drug-receptor sites, channel subunits, and
the development of a radioiodinated 1,4-dihydropyridine calcium
channel label, [^{125}I]iodipine. J. Cardiovasc. Pharmacol. 6:
S608-S621, 1984.

154. Barnathan, E. S., Addonizio, V. P., and Shattil, S. J.:
     Interaction of verapamil with human platelet alpha-adrenergic
     receptors. Am. J. Physiol. 242:H19-H23, 1982.

155. Motulsky, H. J., Snavely, M. D., Hughes, R. J., and Insel,
     P. A.: Interaction of verapamil and other calcium channel blockers
     with alpha$_1$- and alpha$_2$-adrenergic receptors. Circ. Res. 52:
     226-231, 1983.

156. Feldman, R. D., Park, G. D., and Lai, C.-Y. C.: The Inter-
     action of verapamil and norverapamil with beta-adrenergic
     receptors. Circulation 72:547-554, 1985.

157. Borst, S. E., Hui, K. K., and Conolly, M. E.: Verapamil inhibits
     agonist induced desensitization of human lymphocyte beta-
     adrenergic receptors. Fed. Proc. 44:1621, 1985.

158. Shattil, S. J., Bennett, J. S., McDonough, M., and Turnbull, J.:
     Carbenicillin and penicillin G inhibit platelet function in vitro
     by impairing the interaction of agonists with the platelet surface.
     J. Clin. Invest. 65:329-337, 1980.

159. Gullner, H.-G., Kafka, M. S., and Bartter, F. C.: Indomethacin
     increases leukocyte beta-adrenoceptors in man. Clin. Sci. 59:
     397-400, 1980.

160. Insel, P. A., Snavely, M. D., Healy, D. P., Munzel, P. A.,
     Potenza, C. L., and Nord, E. P.: Radioligand binding and func-
     tional assays demonstrate postsynaptic alpha$_2$-receptors on
     proximal tubules of rat and rabbit kidney. J. Cardiovasc. Pharma-
     col. 7 (Suppl. 8):S9-S17, 1985.

161. Sweatt, J. D., Johnson, S. L., Cragoe, E. L., and Limbird,
     L. E.: Inhibitors of Na$^+$/H$^+$ exchange block stimulus-provoked
     release in human platelets. J. Biol. Chem. 260:12910-12919,
     1985.

162. Motulsky, H. J., Maisel, A. S., Snavely, M. D., and Insel, P. A.:
     Quinidine is a competitive antagonist at alpha$_1$- and alpha$_2$-
     adrenergic receptors. Circ. Res. 55:376-381, 1984.

163. Maisel, A. S., Motulsky, H. J., and Insel, P. A.: Hypotension
     after quinidine plus verapamil: Possible additive competition at
     alpha-adrenergic receptors? N. Engl. J. Med. 312:167-170, 1985.

8

Adrenergic Receptors in Cardiovascular Disease

R. WAYNE ALEXANDER   *Brigham and Women's Hospital, Harvard Medical School, Boston, Massachusetts*

I. INTRODUCTION

The sympathetic nervous system is critically important for normal cardiovascular homeostasis. Norepinephrine released from sympathetic nerve endings and circulating epinephrine released from the adrenal medulla increase heart rate and force of contraction and regulate the tone of blood vessels. The realization by early investigators of the central role of the sympathetic nervous system in the control of normal cardiovascular function stimulated intense interest in the potential role of disordered sympathetic function in the pathogenesis of cardiovascular diseases such as congestive heart failure and hypertension. Initially, attention was focused on putative alterations in the intensity of the catecholamine stimulus to target organs such as the failing heart or hypertensive vasculature. Subsequently, it was realized that alterations in target organ responsiveness to catecholamines might be pathophysiologically important. The development of radioligand binding assays for initially beta-adrenergic (1-3) and subsequently alpha-adrenergic receptors (4) facilitated the development of the concept that the adrenergic receptor was a discrete entity which by physiological regulation or pathological alteration could be a central locus of control of cardiovascular responsiveness to catecholamines (5,6). Putative abnormalities of adrenergic receptor function in cardiovascular diseases have been the subject of intense and accelerating interest in recent years. The purpose of this chapter is to review the available data concerning disorders of sympathetic nervous system function in cardiovascular diseases with particular emphasis on abnormalities of adrenergic receptor function.

II. APPROACH TO THE PROBLEM OF STUDYING ADRENERGIC RECEPTORS IN CARDIO-VASCULAR DISEASES

The difficulties in directly assessing adrenergic receptor function in man have been discussed in detail elsewhere in this book. A major limitation has been tissue availability. Human blood vessels have not been available in suitable form for ligand binding studies, although in recent years human myocardial specimens have become available with the increasing frequency of cardiac transplantation and trans-venous biopsy. While direct analysis of receptor function in relevant cardiovascular tissues is likely to result in the most definitive answers to questions concerning receptor function in disease states, this approach has obvious practical limitations. An alternative approach to the problem is to evaluate receptor function in more readily available tissues such as circulating, blood-formed elements (7,8). As discussed elsewhere, this approach is predicated upon the assumption that putative receptor abnormalities in target tissues of interest such as the heart will be manifested generally in other more accessible cell types. This assumption has proven to be valid in a number of instances. Thus, analysis of receptors for insulin on mononuclear leukocytes and for LDL on fibroblasts cultured from skin has yielded important insights into receptor pathophysiology in diabetes mellitus (9) and familial hyper-cholesterolemia, respectively (10). When using this approach, it is important that the underlying assumptions be considered and that the validity of these assumptions be tested insofar as is possible.

The issue is not resolved as to whether physiological or patho-physiological changes in adrenergic receptors in cardiovascular tissues are reflected with reasonable fidelity by changes in receptors on blood cells. The most definitive approach to the problem requires direct comparison between adrenergic receptors on formed elements and in the tissue of interest. This strategy has been used by two groups and reported in preliminary form. Thus, Gordon et al. compared lymphocyte and myocardial beta-adrenergic receptor density and affinity in humans with normal ventricular function or with severe biventricular failure (11). Over this wide range of ventricular function a good correlation was found between myocardial beta-adrenergic receptor density and ventricular function and between receptor density in myocardium and in lymphocytes. In contrast, over a narrow range of sympathetic activity in patients with normal ventricular function a correlation has not been found between lymphocyte and myocardial beta-adrenergic receptor density (12). These results appear to support the notion that when considered broadly a correlation between beta-adrenergic receptor characteristics in myocardium and lymphocytes can be found but that the techniques are not sufficiently precise to detect changes in receptor function over a narrow range of sympathetic activity.

The study of alpha-adrenergic receptors in cardiovascular disease is fraught with even more difficulty than is the study of beta-adrenergic receptors. There are several reasons for this difficulty. First is the fact that of the two alpha-adrenergic subtypes, $alpha_1$ and $alpha_2$, only the latter appears to be represented on a circulating formed element— the platelet. Secondly, the physiology of the two alpha-receptor subtypes in mediating events in the cardiovascular system is less completely understood than in the case of beta-adrenergic receptors. For example, $alpha_2$-adrenergic receptors may affect cardiovascular function by inhibiting norepinephrine release at a presynaptic location on sympathetic nerve terminals (13), or by inducing contraction by virtue of a postsynaptic location on certain vascular smooth muscle (14,15). In contrast, $alpha_1$-adrenergic receptors are thought to be located exclusively postsynaptically. Finally, the most clearly defined cardiovascular effects of alpha-adrenergic receptors are to mediate vascular contraction. Arterial tissue from man is particularly difficult to obtain for pharmacological and physiological studies.

Insights into adrenergic receptor function in disease can be gained using classic pharmacological techniques. Thus, by measuring, for example, chronotropic or pressure responses to various concentrations of catecholamines one can provide evidence for or against the existence of alterations in adrenergic sensitivity (16). The limitations of this approach are that it is difficult to define the mechanism responsible for any alterations in sensitivity which are observed, and that it frequently is difficult to distinguish true changes in sensitivity from apparent changes due to shifts in baseline.

III. CONGESTIVE HEART FAILURE

The pathophysiology of congestive heart failure has been the subject of intense interest among basic and clinical investigators for many years. The discovery of the potent role of the catecholamines and the sympathetic nervous system in modulating the inotropic and chronotropic state of the normal myocardium naturally led to early speculation that disordered function of the sympathetic nervous system could play an important role in the syndrome of congestive heart failure.

Clinically, patients in heart failure appear to be in a hyperadrenergic state with tachycardia and frequent diaphoresis. Early data tended to support this conclusion. Compared to normal controls, subjects with congestive heart failure had increased plasma catecholamine levels at rest and these levels increased dramatically with exercise, suggesting that the sympathetic nervous system plays an important role in supporting the circulation in heart failure (17). More recently, circulating norepinephrine levels have been found to correlate with the severity of the congestive heart failure (18). Although the evidence for a generalized hyperactivity of the sympathetic nervous system in congestive heart failure, especially with some degree of decompensation, is incontrovertible, findings with respect to sympathetic activity in the heart were seemingly paradoxical. Thus, cardiac norepinephrine levels in experimental congestive heart failure were markedly decreased (19). Furthermore, this decrease was functionally significant, since stimulation of cardiac sympathetic nerves resulted in a marked diminution of inotropic responsiveness relative to normal (20). Depletion of cardiac catecholamines has been observed in human congestive heart failure (21). The mechanisms for the depletion of catecholamine stores selectively in the heart in congestive heart failure have not been clearly defined, but the depletion is experimentally associated with decreased activity of tyrosine hydroxylase (22), the rate-limiting enzyme for catecholamine synthesis.

The increases in circulating catecholamines affect cardiovascular homeostasis in congestive heart failure in at least two ways. First the catecholamines provide inotropic support and chronotropic stimulation (23). A manifestation of the dependence of the failing myocardium on circulating catecholamines is the worsening of heart failure that is precipitated by administering beta-adrenergic receptor blockers (24). This dependence upon catecholamines of the failing heart contrasts with the situation in normal heart where there is little evidence that sympathetic stimulation is required for normal function at rest (25). A second function of the high circulating catecholamine levels in congestive heart failure is to increase peripheral resistance by stimulating vascular alpha-adrenergic receptors (26).

The fact that in spite of increased stimulation from high circulating catecholamines the heart in failure has depressed contractility raised the possibility that desensitization of the myocardium to beta-adrenergic receptor stimulation contributed to the overall reduction in inotropic state. Evidence supporting this notion was derived initially from studies on lymphocytes (27). In patients with severe heart failure and elevated levels of catecholamines, cyclic adenosine monophosphate (AMP) generation in blood lymphocytes in response to isoproterenol was decreased. Direct evidence for depressed myocardial beta-adrenergic function in heart failure has been obtained (28). Thus, patients with heart failure had about a 50% decrease in beta-adrenergic receptor density and maximum isoproterenol-stimulated adenylate activity and up to a 70% reduction in maximum isoproterenol-stimulated contraction relative to normals. The hyporesponsiveness appeared to be specific for beta-adrenergic receptor-mediated responses, since histamine and fluoride stimulation of adenylate cyclase activity was unchanged. The seeming paradox of postulating that agonist-induced desensitization exists in the heart under circumstances in which endogenous stores of catecholamines are depleted (as in heart failure) may be resolved by assuming that the stimulus from the elevated circulating catecholamines is sufficient to induce desensitization. These data are potentially of much significance in enhancing the understanding of aspects of the pathophysiology of congestive heart failure.

For both theoretical and practical clinical reasons it is important to consider carefully the implications of beta-adrenergic desensitization in congestive heart failure. Major issues include the need to define the regulatory mechanisms involved and to relate these mechanistic insights to problems of therapeutics. The regulation of beta-adrenergic receptor number and sensitivity is the subject of an extensive literature (reviewed in Ref. 6). The most widely studied phenomenon is that of agonist regulation of beta-adrenergic receptors by beta-adrenergic catecholamines capable of activating the receptors. Although the mechanism by which cells decrease their beta-adrenergic responsiveness after exposure to catecholamines is incompletely understood, there are some rather general features in various cell types. First, there may be decreased responsiveness of adenylate cyclase without a loss of receptor number. Subsequently there may be loss of receptors from the cell surface, a phenomenon known as down-regulation. The decrease in beta-adrenergic receptor number also is associated with a diminished capacity for catecholamines to stimulate adenylate cyclase.

It is likely for a number of reasons that the decreased beta-adrenergic responsiveness to catecholamines of the failing myocardium is owing to agonist-induced desensitization. First, as noted, the hyporesponsiveness is apparently specific for beta-adrenergic agonists (28). Second, as mentioned above, there is probably sufficient stimulus for agonist-induced desensitization even with markedly reduced catecholamine stores in the failing heart, since one of the major features of the altered metabolism of patients in congestive heart failure is the substantially elevated circulating catecholamine levels (17). Third, experimental evidence suggests that exposure of myocardium to beta-adrenergic catecholamines results in a rapid attenuation of the initial inotropic response. This phenomenon has been demonstrated in a model system utilizing embryonic chick hearts in which the velocity of contraction is measured using optical techniques (29). As shown in Figure 1, exposure to isoproterenol results in an increase in velocity of contraction. After 30 min of exposure to isoproterenol (1 μM) control contractility after removal of the drug is unchanged, whereas rechallenge with isoproterenol results in an attenuated contractile response. This partial physiological desensitization is associated with diminished capacity of catecholamines to stimulate adenylate cyclase (Fig. 2), which was interpreted, in this instance, to result from uncoupling of the receptor from the enzyme. Finally, there are additional

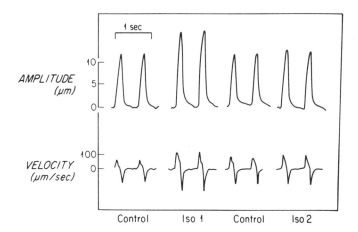

FIGURE 1 Desensitization of beta-adrenergic receptor-mediated increases in contractility in chick embryo ventricle. The tracings show
the amplitude and velocity of contraction measured optically. Iso 1
denotes responses to 0.1 μM isoproterenol. The tissue was then
exposed to 1 μM isoproterenol for 30 min. The second control represents the unchanged response after removing isoproterenol. Iso 2
depicts the diminished response resulting from desensitization of beta-
adrenergic-mediated effects. The inotropic response to increasing
calcium concentrations was unchanged after isoproterenol incubation.
(From Ref. 29, used with permission.)

clinical data which are consistent with the possibility that agonist-
induced desensitization of myocardial beta-adrenergic responsiveness
in congestive heart failure may contribute to decreased contractility
in response to catecholamines. Dobutamine infusion in patients with
congestive heart failure resulted in an initial increase in cardiac output
and heart rate but the response became attenuated with continued
infusion of dobutamine over many hours (30). In another study, the
synthetic beta-adrenergic agonist pirbuterol was given to patients
with congestive heart failure and resulted in an increase in left ventricular ejection fraction acutely (31) (Fig. 3). When the hemodynamic
response was reassessed after chronic therapy for 1 month, the effects
of pirbuterol were markedly attenuated (Fig. 4). Concomitantly there
was a decrease in lymphocyte beta-adrenergic receptor number, suggesting that receptor down-regulation might account for the tolerance
to the drug. Thus, both of these clinical studies are consistent with
the hypothesis that exogenous catecholamines result in at least partial
desensitization of the heart to the inotropic effects of catecholamines
although a contribution of altered preload or afterload resulting from
changing vascular sensitivity to beta-adrenergic stimulation cannot be
excluded.

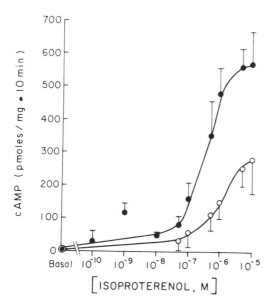

FIGURE 2 Desensitization of embryonic chick heart beta-adrenergic receptor-coupled adenylate cyclase after incubation of intact tissue with isoproterenol (1 μM) for 30 min. Closed circles show control response and open circles represent the response in desensitized hearts. (From Ref. 29, used with permission.)

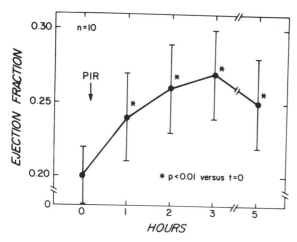

FIGURE 3 Time course of the effect of the initial administration of pirbuterol (PIR) on the left ventricular ejection fraction measured by equilibrium (gated) radionuclide angiocardiogram. (From Ref. 31, used with permission.)

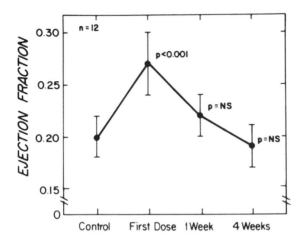

FIGURE 4 Desensitization of effects of the beta-adrenergic agonist pirbuterol on left ventricular ejection fraction in patients with congestive heart failure. Although the initial dose gave significant improvement, this effect was attenuated after treatment for 1 and 4 weeks. (From Ref. 31, used with permission.)

A. Clinical and Pathophysiological Implications of Beta-Adrenergic Receptor Regulation in Congestive Heart Failure

The available data suggesting that the heart in congestive heart failure is dependent for inotropic support upon the elevated circulating catecholamine levels and that beta-adrenergic responsiveness of the failing myocardium is decreased because of agonist regulation of receptor sensitivity has interesting implications. Thus, a situation can be conceived in which a person with well-compensated congestive heart failure and high-normal or only slightly increased levels of circulating catecholamines might exhibit a relatively high level of sensitivity to inotropic effects of catecholamines, either basal levels or transiently increased levels brought about by brief periods of exercise. These circumstances would be compatible with a relatively comfortable clinical situation. On the other hand, prolonged, high levels of circulating catecholamines brough about by excessive levels and duration of exercise or by the stress resulting from decompensation from other causes could desensitize myocardial beta-adrenergic responsiveness, resulting in further increases of heart failure. Thus, according to this scenario, low circulating catecholamine levels would be associated with relatively high myocardial beta-adrenergic responsiveness and clinical stability.

Conversely, high circulating catecholamine levels would be associated with decreased beta-adrenergic responsiveness and a worsening of the clinical situation. These speculations are summarized in Figure 5. Thus, in spite of the fact that cardiac catecholamine levels are low in congestive heart failure, high circulating catecholamine levels may be counterproductive by leading to enhanced myocardial beta-adrenergic desensitization.

B. Alpha-Adrenergic Receptors in Congestive Heart Failure

As alluded to previously, the prototypic alpha-adrenergic receptor-mediated response in the cardiovascular system is vasoconstriction. It has become apparent that the myocardium also possesses alpha-adrenergic receptors primarily of the alpha$_1$ subtype (32). Relative to the extensive literature on beta-adrenergic receptor function in congestive heart failure, there are many fewer data concerning the contribution of cardiovascular alpha-adrenergic receptors to the pathophysiological changes in congestive heart failure.

Elevated circulating catecholamine levels may contribute to the increased peripheral vascular resistance and decreased venous capacitance in congestive heart failure through the stimulation of vascular alpha-adrenergic receptors. Furthermore, in contrast with the catecholamine depletion found in the myocardium in heart failure, norepinephrine released by tyramine is increased in the limbs in congestive heart failure in man (33), suggesting that vascular catecholamine stores may actually be increased but certainly are not decreased. The contribution of alpha receptors to the increase in peripheral resistance frequently seen in congestive heart failure is manifested by the decrease in afterload brought about by the alpha$_1$ antagonist prazosin (34). Prazosin also decreases preload, suggesting that venous tone is enhanced by alpha$_1$-adrenergic contraction in congestive heart failure (34). On balance there appears to be little evidence that the peripheral vascular alpha-adrenergic receptors are hypersensitive (33). Blockade of the effects of increased circulating and locally released catecholamines on the receptors, however, can have salutory effects on congestive heart failure by augmenting cardiac output and decreasing pulmonary capillary wedge pressure by decreasing afterload and preload through arteriolar and venular dilatation respectively (35,36).

There are few data concerning the role of changes in myocardial alpha-adrenergic receptors in congestive heart failure. The physiological role of these receptors is not even clearly defined. In animal models, alpha-adrenergic agonists are capable of increasing contractility (37). In a guinea pig model of congestive heart failure, ligand binding

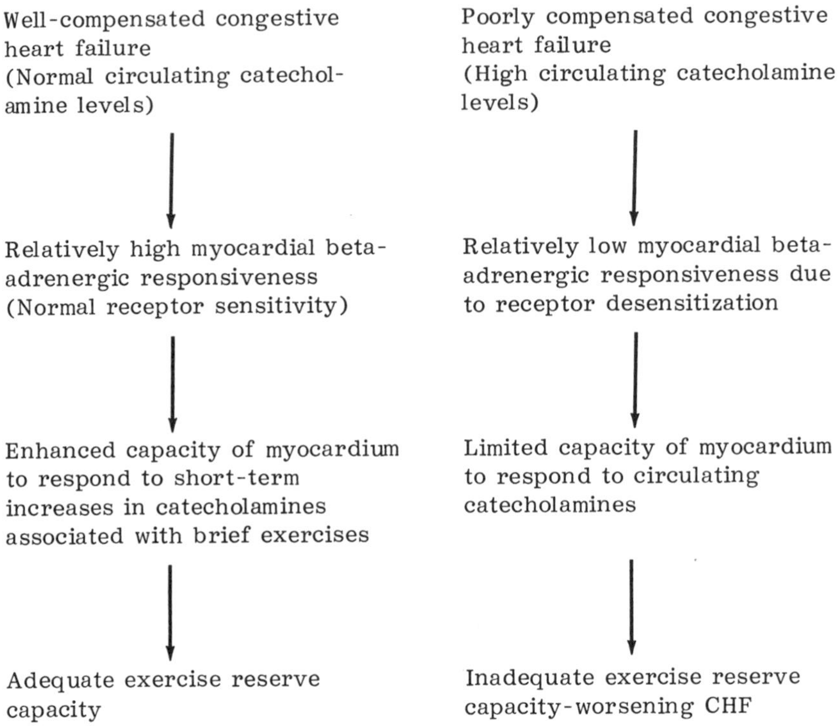

Well-compensated congestive
heart failure
(Normal circulating catecholamine levels)

Poorly compensated congestive
heart failure
(High circulating catecholamine
levels)

Relatively high myocardial beta-
adrenergic responsiveness
(Normal receptor sensitivity)

Relatively low myocardial beta-
adrenergic responsiveness due
to receptor desensitization

Enhanced capacity of myocardium
to respond to short-term
increases in catecholamines
associated with brief exercises

Limited capacity of myocardium
to respond to circulating
catecholamines

Adequate exercise reserve
capacity

Inadequate exercise reserve
capacity-worsening CHF

FIGURE 5 Possible implications of myocardial beta-adrenergic receptor
regulation in congestive heart failure.

studies have shown an increase in myocardial $alpha_1$-adrenergic receptors (38). The pathophysiological implications of these observations
remain to be determined as does their relevance to the disease in man.

IV. HYPERTENSION

The role of the sympathetic nervous system in the pathogenesis of
systemic hypertension has been the focus of intense investigation for
many years. Interest was initiated as increased understanding of the
central role of catecholamines in cardiovascular control was developing,
and was sustained by the early demonstration of efficacy of a variety
of sympatholytic drugs in ameliorating the elevated blood pressure.
As in the case of congestive heart failure, initial attention focused
on the intensity of the stimulus—the level of sympathetic nervous
system activity—with less consideration being given to the possibility

of altered target organ responsiveness. Although indirect evidence
for the participation of the sympathetic nervous system in human
essential hypertension is substantial, direct evidence for a general
increase in activity is not available. Some subsets of patients appear
to have elevated plasma catecholamine levels, but this finding is not
consistent in all studies (reviewed in Refs. 39 and 40). This lack of
consistency in demonstrating increased sympathetic activity likely
relates to the heterogeneity of essential hypertension with respect to
etiology and to the inadequacy of currently available tools for assessing
sympathetic nervous system function.

In view of the rather inconclusive observations concerning the
level of sympathetic activity in hypertension, it was logical to consider
whether there might be altered end organ sensitivity to catecholamines
in this condition. Thus, whether there was decreased beta-adrenergic
vasodilator responsiveness or increased alpha-adrenergic vasoconstrictor
responsiveness have been issues of wide and continuing interest. Early
evidence from an experimental model of hypertension—the spontaneously
hypertensive rat (SHR)—suggested that both hypotheses might have
some validity. Thus aortic (41,42) and heart (43) homogenates from
the SHR have been reported to have decreased activity of beta-
adrenergic receptor-stimulated adenylate cyclase activity, although
conflicting data have appeared (summarized in Ref. 44). Moreover,
a number of studies have suggested that there is increased vascular
contractile reactivity to catecholamines in the SHR. Thus, under
certain conditions, increased sensitivity to infused norepinephrine
has been seen in SHR hindquarter (45) and in the renal vascular beds
(46). In contrast, studies with nerve stimulation in SHR have frequently not shown evidence of vascular alpha-adrenergic hypersensitivity, although maximum extent of contraction may be affected (47).
Insights into this apparent paradox were derived from a study by
Halpern and his colleagues (48). Thus, in small isolated mesenteric
arteries from SHR and control rats with unperturbed innervation,
concentration-response curves for norepinephrine were quite similar.
In contrast, when neuronal catecholamine uptake was inhibited with
cocaine, the artery showed increased sensitivity to norepinephrine-
induced contraction. Thus, these data and those of others (47)
suggest that the sympathetic innervation of the SHR vasculature has
enhanced uptake which may mask, at least at certain ages, an intrinsic
alpha-adrenergic hyperresponsiveness of the vasculature. An increase
relative to control in the norepinephrine-induced membrane depolarization of the SHR caudal artery also provides evidence of a specific
increase in alpha-adrenergic vascular sensitivity in this model (49).

The molecular mechanisms underlying the apparent changes in
alpha- and beta-adrenergic responsiveness in at least some vascular
beds of the SHR are incompletely understood. Decreased numbers

of beta-adrenergic receptors have been reported in SHR myocardial
membranes (50) and brain cerebral microvessels (51). Molecular mechanisms underlying the apparent increased alpha-adrenergic sensitivity
in the SHR vasculature demonstrated by several investigators have
not been defined. An increase in alpha$_2$-adrenergic receptor number
has been reported in SHR kidney, but was speculated to represent
changes in the tubules (52). Thus, the tentative evidence for alteration of alpha-adrenergic as well as beta-adrenergic cardiovascular
sensitivity in this animal model having some features of the human
disease provides a precedent suggesting that similar mechanisms may
obtain in human essential hypertension.

A. Adrenergic Responsiveness in Human Essential Hypertension

1. Alpha-Adrenergic Responses

There are a number of studies which suggest that the resistance
vasculature in human essential hypertension shows enhanced contractile
responsiveness when alpha-adrenergic agonists are administered (53-55).
It is not clear, however, whether this hyperresponsiveness is due to true
alpha-adrenergic hypersensitivity or whether it can be explained on the
basis of structural factors. Thus, the medial hypertrophy seen in hypertension would result in a higher resistance at any level of tone (56,57).
This fact makes it difficult to attribute hyperresponsiveness to supersensitivity. Indeed, it has been argued that mechanical factors can
explain virtually all of the increased responsiveness (58). In contrast,
other carefully performed clinical studies in essential hypertensives have
concluded that there may be a contribution of alpha-adrenergic supersensitivity to the observed vascular hyperresponsiveness (59,60).
However, no evidence of altered alpha-adrenergic responsiveness
was observed in temporal artery strips from human essential hypertensives (61). Although these muscular arteries may not be representative of the resistance vasculature, the findings suggest that
any putative alpha-adrenergic hypersensitivity is not a general feature
of all arteries in essential hypertensives. Thus, on balance there
appear to be few compelling data that specific alpha-adrenergic vascular
supersensitivity is a general feature in human hypertension.

2. Beta-Adrenergic Responses

A broad range of beta-adrenergic receptor-mediated functions are
important in blood pressure control. Cardiac rate and contractility, renal
renin release, and vasodilatation are modulated by beta-adrenergic receptors and derangements in any of these or several other beta-mediated
functions could contribute to the pathogenesis of essential hypertension.
Subjects with labile, hyperkinetic hypertension have been reported
to have an increase in cardiac rate response and in vasodilator responses
to isoproterenol infusion (62). In contrast, in subjects with more estab-

lished hypertension, a blunted chronotropic response to infused iso-proterenol has been observed (63,64). Similarly, a tendency has been noted for normal age-related decreases in beta-adrenergic receptor-mediated chronotropic and vasodilator responses to be enhanced in hypertensives (65). Thus, there is considerable evidence for abnormal beta-adrenergic responsiveness in essential hypertension.

B. Receptor Binding Studies in Essential Hypertension

Technical and practical limitations have thus far precluded the direct assessment of adrenergic receptor function in cardiovascular target tissues from hypertensive patients. There are data from bind-ing studies on circulating formed elements, and these results potentially provide mechanistic insights into the pathophysiological abnormalities, especially those involved in beta-adrenergic receptor function, as noted above.

Lymphocyte beta-adrenergic receptor density has been reported not to be decreased in essential hypertension (66). However, very recently other particularly intriguing findings have been reported concerning lymphocytic beta-adrenergic receptors in essential hyper-tension. In membranes of lymphocytes from normal controls isopro-terenol competition curves with the antagonist [^{125}I]iodocyanopindolol are shallow, and computer analysis of these data show that they fit best to a model consisting of two states of high and low affinity (67). The high-affinity binding state is thought to result from the formation of a complex between the agonist, the beta-adrenergic receptor, and a guanine nucleotide regulatory protein involved in coupling the receptor to the catalytic unit of adenylate cyclase. Thus, the propor-tion of receptors in the high-affinity binding state (with respect to agonist) is related to the ability to stimulate adenylate cyclase. As shown in Figure 6, when the subjects were supine at rest, the propor-tion of lymphocyte beta-adrenergic receptors in the high-affinity binding state (R_h) was decreased from 42% in normal controls to 25% in essential hypertensives. This decrease in high-affinity binding was associated with a diminished capacity to stimulate adenylate cyclase.

These data should be viewed in the context of other data, includ-ing previous observations from the same group on the regulation of lymphocyte beta-adrenergic receptors in normal subjects by endogenous catecholamines (68). Thus, the increase in plasma catecholamines that occurs when subjects assume an upright position was associated with a decrease in high-affinity receptors and, as described in hyper-tensive subjects, with attenuation of stimulation of adenylate cyclase. It might be concluded that the changes in high-affinity binding in lymphocytes from hypertensive subjects could be related to elevated circulating catecholamines. Although differences in circulating cate-cholamines between normals and hypertensives were not detected by

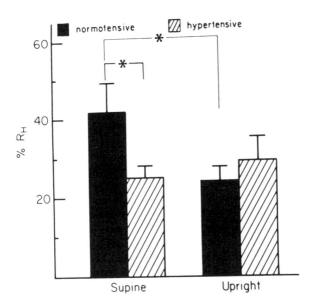

FIGURE 6 Proportions of lymphocyte beta-adrenergic receptors in high-affinity binding state in normals and hypertensives at rest and after assuming the upright posture. *, $P < 0.05$ compared with supine normotensives. (From Ref. 67, used with permission.)

Feldman et al. (67), Buhler and his colleagues (69) have reported consistent increases in hypertensives of circulating epinephrine, the endogenous catecholamine interacting most potently with beta$_2$-adrenergic receptors, such as those on lymphocytes (and vascular smooth muscle). Thus, it is possible (as was considered by Feldman et al. [67]) that the changes in hypertensive lymphocytes result from regulation of the receptors by elevations of endogenous catecholamines not detected by their methods.

There are potentially important implications of the disordered beta-adrenergic receptor function for understanding the pathophysiology of essential hypertension. Thus, if as the clinical physiological studies suggest, the decreased lymphocyte beta-adrenergic sensitivity extends to the vasculature, then beta-adrenergic receptor-mediated vasodilatation will be attenuated. This situation may result in a predominance of vasoconstrictor tone.

It is interesting to consider the potential role of small increases in circulating catecholamines, and in particular epinephrine, in initiating the state of beta-adrenergic desensitization and the postulated predominance of vascular vasoconstrictor influences. It is important to note that beta-adrenergic receptors exhibit rapid regulatory and desensitization responses (67), whereas alpha-adrenergic receptors, and in particular the $alpha_1$-adrenergic receptors of vascular smooth muscle, may be much more resistant to agonist-induced down-regulation of receptors, at least with respect to the relatively prolonged time of catecholamine exposure which is required (70). Thus, elevated circulated epinephrine in essential hypertensives could differentially desensitize beta-adrenergic as opposed to alpha-adrenergic responses. This potential scenario is depicted in Figure 7.

There are relatively few ligand binding studies on alpha-adrenergic receptors in essential hypertension. A major problem has been the fact that, insofar as is known, none of the circulating formed elements express $alpha_1$-adrenergic receptors. Human platelets possess an $alpha_2$-adrenergic receptor, and there are a limited number of studies which have been conducted on this receptor in essential hypertensives. Platelet $alpha_2$-adrenergic receptor number and affinity are similar in normals and hypertensives (71). Receptor number did not change with treatment with an $alpha_2$ agonist nor did receptor number correlate with urinary catecholamines (71). In contrast, platelet $alpha_2$-receptor number has been reported to be decreased in pheochromocytoma in

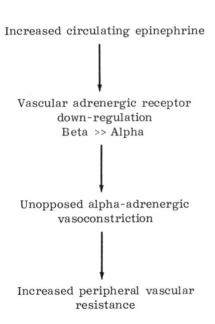

Increased circulating epinephrine

Vascular adrenergic receptor
down-regulation
Beta >> Alpha

Unopposed alpha-adrenergic
vasoconstriction

Increased peripheral vascular
resistance

FIGURE 7 Possible consequences of the putative differential sensitivity of vascular alpha- and beta-adrenergic receptors to regulation by elevated catecholamines seen in subsets of essential hypertensive.

which catecholamine levels are very high (72), although this finding has not been confirmed by others (71). These observations tend to support the proposition that alpha-adrenergic receptors are more resistant to agonist regulation than are beta-adrenergic receptors.

V. ISCHEMIC HEART DISEASE

Alpha-adrenergic-mediated responses have been considered to be involved in two aspects of ischemic heart disease—coronary vasospasm and postmyocardial infarction arrhythmia. It is logical to speculate that coronary vasospasm could result either from general or localized increases in catecholamine release or localized coronary alpha-adrenergic supersensitivity. There is no evidence of generalized increases in sympathetic tone in variant angina (73). Initial clinical reports implicated a role for $alpha_1$-adrenergic-mediated contraction in Printzmetal's variant angina, since alpha-adrenergic blockers appeared to be efficacious in ameliorating the spasm (74, 75). A very recent report has failed to confirm these findings and could demonstrate no efficacy of prazosin in vasospastic angina, thus mitigating against an important role for coronary $alpha_1$ receptors in contributing to this syndrome (76). This view is shared by others (77). Thus, the data on balance do not appear to support a major role for alpha-adrenergic receptors in mediating coronary vasospasm in the majority of patients. Beta-adrenergic blockade may worsen vasospastic angina, however, presumably by removing basal beta-adrenergic vasodilator stimulus (78).

A putative role for myocardial alpha-adrenergic receptors in mediating postmyocardial infarction arrhythmias has been considered based on studies in cats. In these studies, coronary occlusion with or without reperfusion was associated with an increase in postocclusion ventricular arrhythmias which were markedly attenuated by alpha-adrenergic receptor blockade with prazosin or phentolamine (79). More recently, ligand binding studies have shown an increase in $alpha_1$-adrenergic receptor density in the ischemic area after coronary occlusion (80). These data thus suggest that after myocardial occlusion there is an appearance of cryptic alpha-adrenergic receptors which are involved in mediating postocclusion arrhythmia. The relevance of these observations to postinfarction arrhythmias in humans has not yet been defined. A rapid increase in beta-adrenergic receptors after coronary occlusion in dogs has also been observed (81), but the pathophysiological importance is not known.

VI. BETA-ADRENERGIC BLOCKER WITHDRAWAL SYNDROME

Beta-adrenergic blocking drugs play a central role in the treatment of angina pectoris because of their capacity to block catecholamine-mediated increases in chronotropy and inotropy, thus minimizing increases in oxygen demand. Severe exacerbation of angina after withdrawal of propranolol was reported a number of years ago and was speculated to be the consequence of beta-adrenergic supersensitivity (82,83). Subsequent studies have demonstrated that chronic propranolol treatment results in the appearance of increased beta-adrenergic receptors in human lymphocytes (84) and in rat heart (85). Although there appears to be little doubt that chronic beta-blocker therapy can result in an up-regulation of beta-adrenergic receptors, it remains controversial whether true beta-adrenergic supersensitivity is a general feature of abrupt propranolol withdrawal after chronic therapy in man. There are clinical data which suggest that beta-adrenergic supersensitivity exists after beta-blocker withdrawal (86), and there are other carefully performed studies which raise doubts as to whether true beta-adrenergic supersensitivity is generally observed under these conditions (87). It seems likely that at least some patients are at risk after propranolol withdrawal. Consideration should be given to more prolonged withdrawal using low doses of beta-blockers.

ACKNOWLEDGMENTS

I should like to acknowledge the excellent assistance of Ms. Joan Macauley with preparation of the manuscript. The work in this manuscript was partially supported by the National Institutes of Health grant HL-2-9763.

REFERENCES

1. Aurbach, G. D., Fedak, S. A., Woodard, C. J., Palmer, J. S., Hauser, D., and Troxler, F.: Science 186:1223, 1974.

2. Levitzki, A., Atlas, D., and Steer, M. L.: Proc. Natl. Acad. Sci. U.S.A. 71:2773, 1974.

3. Mukherjee, C., Caron, M. G., Coverstone, M., and Lefkowitz, R. J.: J. Biol. Chem. 250:4869, 1975.

4. Williams, L. T., and Lefkowitz, R. J.: Science 192:791, 1976.

5. Lefkowitz, R. J.: N. Engl. J. Med. 295:323, 1976.

6. Lefkowitz, R. J., Caron, M. G., and Stiles, G. L.: N. Engl. J. Med. 310:1570, 1984.

7. Williams, L. T., Snyderman, R., and Lefkowitz, R. J.: J. Clin. Invest. 57:149, 1975.

8. Alexander, R. W., Cooper, B., and Handin, R. I.: J. Clin. Invest. 61:1137, 1978.

9. Olefsky, J. M., and Reaven, G. M.: J. Clin. Invest. 54:1323, 1974.

10. Brown, M. S., and Goldstein, J. L.: J. Biol. Chem. 249:5153, 1974.

11. Gordon, E. P., Bristow, M. R., Laser, J. A., Minobe, W. A., Fowler, M. B., and Savin, W. M.: Circulation 68:III-99, 1983.

12. Hausen, M., Kramer, B., and Kubler, W.: Circulation 68:III-100, 1983.

13. Langer, S. Z.: Biochem. Pharmacol. 23:1793, 1974.

14. Berthelsen, S., Pettinger, W. A.: Life Sci. 21:595, 1973.

15. Drew, G. M., and Whiting, S. B.: Br. J. Pharmacol. 67:207, 1984.

16. Lindenfield, J., Crawford, M. H., O'Rourke, R. A., Levine, S. P., Montiel, M. M., and Horwitz, L. D.: Circulation 62:704, 1980.

17. Chidsey, C. A., Harrison, D. C., and Braunwald, E.: N. Engl. J. Med. 267:650, 1962.

18. Thomas, J. A., and Marks, B. H.: Am. J. Cardiol. 41:233, 1978.

19. Chidsey, C. A., Kaiser, G. A., Sonnenblick, E. H., Spann, J. F., Jr., and Braunwald, E.: J. Clin. Invest. 43:2386, 1964.

20. Covell, J. W., Chidsey, C. A., and Braunwald, E.: Circ. Res. 19:51, 1966.

21. Chidsey, C. A., Braunwald, E., and Morrow, A. G.: Am. J. Med. 39:442, 1965.

22. Pool, P. E., Covell, J. W., Levitt, M., Gibb, J., and Braunwald, E.: Circ. Res. 20:349, 1967.

23. Gaffney, T. E., Braunwald, E.: Am. J. Med. 34:320, 1963.

24. Epstein, S. E., and Braunwald, E.: Ann. Intern. Med. 65:20, 1966.

25. Epstein, S. E., Robinson, B. F., Kahler, R. L., and Braunwald, E.: J. Clin. Invest. 44:1745, 1965.

26. Packer, M., Meller, J., Gorlin, R., and Herman, M.: Circulation 59:531, 1979.

27. Thomas, J. A., and Marks, B. H.: Am. J. Cardiol. 41:233, 1978.

28. Bristow, M. R., Ginsberg, R., Minobe, W., Cubiccioti, R. S., Sageman, W. S., Lurie, K., Billingham, M. E., Harrison, D. C., and Stinson, E. B.: N. Engl. J. Med. 307:205, 1982.

29. Marsh, J. D., Barry, W. H., Neer, E. J., Alexander, R. W., and Smith, T. W.: Circ. Res. 47:493, 1980.

30. Unverferth, D. V., Blanford, M., Kates, R. E., and Leier, C. V.: Am. J. Med. 69:262, 1980.

31. Colucci, W. S., Alexander, R. W., Williams, G. H., Rude, R. E., Holman, B. L., Konstam, M. A., Wynne, J., Mudge, G. H., and Braunwald, E.: N. Engl. J. Med. 305:185, 1981.

32. Karliner, J. S., Barnes, P., Hamilton, C. A., and Dollery, C. T.: Biochem. Biophys. Res. Commun. 90:142, 1979.

33. Kramer, R. S., Mason, D. T., and Braunwald, E.: Circulation 38:629, 1968.

34. Miller, R. R., Awan, N. A., Maxwell, K. S., and Mason, D. T.: N. Engl. J. Med. 297:303, 1977.

35. Packer, M., Meller, J., Gorlin, R., and Herman, M. V.: Circulation 59:531, 1979.

36. Colucci, W. S.: Ann. Intern. Med. 97:67, 1983.

37. Perez, J. E., Sobel, B. E., and Henry, P. D.: Am. J. Cardiol. 43:381, 1979.

38. Karliner, J. S., Barnes, P., Brown, M., and Dollery, C. T.: Eur. J. Pharmacol. 67:115, 1980.

39. Kopin, I. J., Goldstein, D. S., and Feuerstein, G. I.: In: *Frontiers in Hypertension Research.* J. H. Laragh, F. R. Buhler, and D. W. Seldin (eds.). New York, Springer-Verlag, p. 283, 1981.

40. Kuchel, O.: In: *Hypertension.* J. Genest, O. Kuchel, P. Hamet, and M. Cantin (eds.). New York, McGraw-Hill, p. 140, 1983.

41. Amer, M. S.: Science 179:807, 1973.

42. Klenerova, V., Albrecht, I., and Hynie, S.: Pharmacol. Res. Commun. 7:453, 1975.

43. Bhalla, R. C., Sharma, R. V., and Ashley, T.: Biochem. Biophys. Res. Commun. 82:273, 1978.

44. Hamet, P.: In: *Hypertension.* J. Genest, O. Kuchel, P. Hamet, and M. Cantin (eds.). New York, McGraw-Hill, p. 408, 1983.

45. Lais, L. T., and Brody, M. J.: Circ. Res. (Suppl. I) 36:I-216, 1975.

46. Berecek, K. H., Schwertschlog, V., and Gross, F.: Am. J. Physiol. 238:H287, 1980.

47. Vanhoutte, P. M., Webb, R. C., and Collis, M. G.: Clin. Sci. 59:2115, 1980.

48. Whall, C. W., Myers, M. M., and Halpern, W.: *Blood Vessels* 17: 1, 1980.

49. Hermsmeyer, K., and Walton, S.: Circ. Res. (Suppl. I) 40:I-53, 1977.

50. Limas, C., and Limas, C. J.: Biochem. Biophys. Res. Commun. 83:710, 1978.

51. Magnoni, M. S., Kobayaski, H., Cazzaniga, A., Isumi, F., Spano, P. F., and Trabucchi, M.: Circulation 67:610, 1983.

52. Graham, R. M., Pettinger, W. A., Sagalovsky, A., Brabson, J., and Gandler, T.: *Hypertension* 4:881, 1982.

53. Mendlowitz, M.: Am. Heart J. 85:252, 1973.

54. Doyle, A. E., Fraser, J. R. E.: Circ. Res. 9:755, 1961.

55. Sivertsson, R., and Olander, R.: Life Sci. (Part I) 7:1291, 1968.

56. Folkow, B., Grimby, G., and Thulesius, O.: Acta Physiol. Scand. 44:255, 1958.

57. Folkow, B.: Clin. Sci. 41:1, 1971.

58. Sivertsson, R.: Acta Physiol. Scand. (Suppl.) 343, 1970.

59. Mendlowitz, M., Torosdag, S. M., and Sharney, L.: J. Appl. Physiol. 10:436, 1957.

60. Mendlowitz, M., Naftchi, N. E., Gitlow, S. E., and Wolf, R. L.: Geriatrics 20:797, 1965.

61. Horwitz, D., Clineschmidt, B. V., Van Buren, J. M., and Ommaya, A. K.: Circ. Res. 34&35 (Suppl. I):I109, 1974.

62. Messerli, F. H., Kuchel, O., Hamet, P., Tolis, G., Guthrie, G. P., Fraysse, J., Nowaczynski, W., and Genest, J.: Circ. Res. (Suppl. II) 39:II43, 1976.

63. McAllister, R. G., Love, D. W., Guthrie, G. P., Dominick, J. A., and Kotchen, T. A.: Arch. Intern. Med. 139:879, 1979.

64. Volpe, M., Trimarco, B., Ricciardetti, B., Sacca, L., Ungaro, B., Rengo, F., and Condorelli, M.: Cardiovasc. Res. 16:732, 1982.

65. Buhler, F. R., Kiowski, W., Van-Brummelen, P., Amann, F. W., Bertel, O., Landman, R., Lutold, B. E., and Bolli, P.: Clin. Exp. Hypertens. 2:409, 1980.

66. Doyle, V., O'Malley, K., and Kelly, J. G.: J. Cardiovasc. Pharmacol. 4:738, 1982.

67. Feldman, R. D., Limbird, L. E., Nadeau, J., Robertson, D., and Wood, A. J. J.: J. Clin. Invest. 73:648, 1984.

68. Feldman, R. L., Limbird, L. E., Nadeau, J., Fitzgerald, G. A., Robertson, D., and Wood, A. J. J.: J. Clin. Invest. 72:164, 1983.

69. Buhler, F. R., Amann, F. W., Bolli, P., Hulthen, L., Kiowski, W., Landmann, R., and Burgisser, E.: J. Cardiovasc. Pharmacol. (Suppl. I) 4:5134, 1982.

70. Colucci, W. S., Gimbrone, M. A., Jr., and Alexander, R. W.: Circ. Res. 48:104, 1981.

71. Motulsky, H. J., O'Connor, D. T., and Insel, P. A.: Clin. Sci. 64:265, 1983.

72. Davies, I. B., Sudera, D., and Sever, P. S.: Clin. Sci. (Suppl.) 61:2075, 1981.

73. Robertson, D., Robertson, R. M., Nies, A. S., Oates, J. A., and Friesinger, G. C.: Am. J. Cardiol. 43:1080, 1979.

74. Levene, D. L., and Freeman, M. R.: J.A.M.A. 236:1018, 1976.

75. Tzivoni, D., Keren, A., Benharin, J., Gottlieb, S., Atlas, D., and Stern, S.: Am. Heart J. 105:262, 1983.

76. Winniford, M. D., Filipchuk, N., and Hillis, L. D.: Circulation 67:1185, 1983.

77. Chierchia, S., Davies, G. J., Berkeuboom, G., Crea, F., Crean, P., and Maseri, A.: Circulation 69:8, 1984.

78. Robertson, R. M., Wood, A. J. J., Vaughn, W. K., and Robertson, D.: Circulation 65:281, 1982.

79. Sheridan, D. J., Penkoske, P. A., Sobel, B. E., and Corr, P. B.: J. Clin. Invest. 65:161, 1980.

80. Corr, P. B., Shayman, J. A., and Kramer, J. B.: J. Clin. Invest. 67:1232, 1981.

81. Mukherjee, A., Wong, T. M., Buja, L. M., Lefkowitz, R. J., and Willerson, J. T.: J. Clin. Invest. 64:1423, 1979.

82. Miller, R. R., Olson, H. G., Amsterdam, E. A., and Mason, D. T.: N. Engl. J. Med. 293:416, 1975.

83. Shand, D. G., and Wood, A. J.: Circulation 58:202, 1978.

84. Aarons, R. D., Nies, A. S., and Gal, J.: J. Clin. Invest. 65:949, 1980.

85. Glaubiger, G., and Lefkowitz, R. J.: Biochem. Biophys. Res. Commun. 78:720, 1977.

86. Boudoulas, H., Lewis, R. P., Kates, R. E., and Dalamangas, G.: Ann. Intern. Med. 87:433, 1977.

87. Lindenfeld, J., Crawford, M. H., O'Rourke, R. A., Levine, S. P., Montiel, M. M., and Horwitz, L. D.: Circulation 62:704, 1980.

9

Adrenergic Receptors in Allergic Disorders

DIANA L. MARQUARDT and STEPHEN I. WASSERMAN   *University of California, San Diego Medical Center, San Diego, California*

I. INTRODUCTION

Current concepts of the role of adrenergic receptors in allergic disease have been derived from a variety of observations. Clinical experience with adrenergic agonists and antagonists in patients with allergic disorders provided initial insights into adrenergic processes in allergy. Subsequent functional investigations into adrenergic processes in the key allergic target tissues in normal individuals have been supplemented more recently by direct analysis of adrenergic receptors in health and disease.

Adrenergic responsiveness in allergy is dependent upon a complex series of interrelationships achieved within a specific target tissue. These interrelationships reflect the presence of direct adrenergic innervation, the content of nonneural adrenergic receptors, as well as tissue and blood levels of neurohormones, and exogenously administered therapeutic drugs. In a single individual, these interrelationships may vary between specific tissues, may display regional differences within a particular tissue, and may be altered dramatically in disease.

This chapter will briefly discuss the important physiological adrenergic mechanisms in tissues most affected by allergic disease, and focus upon the current state of our understanding of the pathobiology of adrenergic processes in specific allergic conditions. Data obtained from the study of human tissues will be emphasized.

II. ADRENERGIC REGULATION OF CUTANEOUS PROCESSES

The tissues most pertinent to the study of allergic disorders are the
respiratory tract (upper and lower) and the skin. It is important to
recognize that these tissues are complex and contain within them
numerous cell types and structures. The majority of the information
regarding the role of receptors is based upon physiological studies
utilizing adrenergic agonists and antagonists with some recent information based upon direct radioligand binding techniques.

In the epidermis, adrenergic receptors have been identified by
both physiological and radioligand binding techniques. A catecholamine-
sensitive adenylate cyclase activity, thought to regulate cell proliferation, has been demonstrated in this tissue (1). In one study, this
adenylate cyclase activity was shown to respond to the beta$_2$ agonist
salbutamol, whereas in others, isoproterenol was demonstrated to be
more active than epinephrine or norepinephrine (2,3). In addition,
stimulatory effects of beta-adrenergic drugs in epidermal proliferation
are blocked by propranolol and butoxamine (a beta$_2$ antagonist) but
not by practolol, suggesting this receptor function is primarily beta$_2$
(4). A similar receptor specificity utilizing radioligand binding of
[^{125}I]iodohydroxybenzylpindolol has been obtained in newborn mouse
skin (5). The cell which bears such a beta$_2$ receptor, however, has
yet to be identified, although the basal cell is most sensitive to
epinephrine-induced increases in cyclic adenosine monophosphate
(cAMP) (6). Pharmacological studies are inconclusive regarding the
presence of alpha-adrenergic receptors in human epidermis (1).

The microcirculation appears to be the site most responsive to
adrenergic regulation in the dermis. Classic studies have demonstrated
constriction to be mediated via alpha-adrenergic receptors and vasodilation by beta-adrenergic processes (7). Studies in human skin
have been hampered by technical factors and by regional differences
in cutaneous reactivity (8). Acral areas primarily demonstrate constriction and central areas vasodilation. Catecholamine-dependent
constriction has been surmised from physiological studies of human
papillary loop vessels. Other tissues of the dermis also respond to
adrenergic stimuli. Fibroblast proliferation is regulated by beta$_2$-
adrenergic receptors (9), whereas eccrine sweating, primarily a
cholinergic response, has been shown both to be augmented (10) and
inhibited (11) by beta-adrenergic stimulation. Catecholamine-containing
nerve endings, however, are present around eccrine sweat glands
(12), and both phenylephrine and isoproterenol can augment secretion
from isolated palmar sweat glands from monkeys (10). As such secretory responses are inhibited by phentolamine and propranolol, respectively, it is likely that eccrine sweat glands possess both alpha- and
beta-adrenergic receptors. Similarly, human apocrine sweat glands
secrete in response to circulating catecholamines, especially epinephrine (13).

III. CLINICAL EVIDENCE OF ADRENERGIC ABNORMALITIES IN SKIN IN ALLERGIC DISEASE

Atopic dermatitis, a pruritic, erythematous, cutaneous disorder, is characterized by microvesicles, excoriation, and lichenification (14). It occurs prominently in individuals with a personal or family history of atopic disorders and is associated in most cases with elevations in circulating levels of immunoglobulin E (IgE) antibody (15). Patients with atopic dermatitis manifest pallor rather than the usually observed erythema after stroking of the skin. This response is felt to be due to an abnormally brisk vasoconstriction which has been demonstrated in cutaneous vessels and digital arteries of patients with atopic dermatitis (16-18). A similar blanch can be demonstrated in patients with atopic dermatitis upon the injection of epinephrine or norepinephrine in amounts insufficient to blanch the skin of normal individuals (19). Further evidence for enhanced alpha-adrenergic tone in atopic dermatitis is the piloerection so consistently noted in these patients (20).

Beta-adrenergic responsiveness has not been fully elucidated in skin of patients with atopic dermatitis. In this tissue, normal basal levels of adenylate cyclase and phosphodiesterase activities have been reported (21,22), but in vitro epidermal mitosis appears to be less sensitive to catecholamine-mediated inhibition (23) in atopic as compared to normal skin. Norepinephrine comprises 95% of cutaneous catecholamines and levels of this hormone appear normal in atopic dermatitis. Injection of [^{14}C]norepinephrine into children with this disorder has been associated with delayed urinary excretion of the radiolabeled material and enhanced binding to skin when compared to normal children (24). Increased levels of catechol-o-methyltransferase also are reported in the skin of patients with atopic dermatitis (25).

IV. PHYSIOLOGICAL ADRENERGIC REGULATION OF THE UPPER AIRWAY

In the nasal airway, the major effects of stimulation of adrenergic receptors are the regulation of blood flow and mucus production and/or secretion (26). Sympathetic stimulation of nasal mucosa leads to an alpha-adrenergic-mediated vasoconstriction and a decrease in mucus production. Beta-adrenergic responses are less prominent. When added in vitro, beta agonists augment nasal mucus production, whereas topical application of the beta agonist fenoterol in vivo increases nasal airway resistance, probably owing to vasodilation (27).

Exercise and the concomitant release of neurohormones into the circulation are associated with an increase in nasal patency due to decreases in nasal blood flow, presumably mediated by alpha-adrenergic responses (28). Similarly, the nasal congestion associated with use of reserpine and α-methyldopa suggests the importance of adrenergic influences on nasal blood flow.

V. CLINICAL EVIDENCE OF ADRENERGIC ABNORMALITIES IN ALLERGIC NASAL DISORDERS

In allergic rhinitis, clinical evidence for alpha- and beta-adrenergic mechanisms has developed primarily from therapeutic observations and experimental antigen provocation of allergic nasal disease. Although not particularly prominent in the clinical armamentarium for the treatment of allergic rhinitis, the topical application of fenoterol has been shown to decrease nasal responses to aerosolized antigen in sensitized subjects (29). The effect of fenoterol was apparent in decreased reactivity to antigen as well as by alleviation of allergen-induced sneezing and rhinorrhea. Alpha-adrenergic agonists are popular as nasal decongestants in a variety of disorders, but their continued use is associated with rebound vasodilation, nasal obstruction, and eventually rhinitis medicamentosa. No specific role for alpha-adrenergic agonists or antagonists has been demonstrated in allergic rhinitis.

VI. PHYSIOLOGICAL ADRENERGIC REGULATION OF THE LOWER AIRWAY

The adrenergic regulation of pulmonary physiology is exceedingly complex and responses to alpha- and beta-adrenergic agonists may be reflected by smooth muscle, by mucus-secreting cells and glands, by mediator-generating cells (mast cells), and by the pulmonary and/or bronchial circulations as well as by a variety of other cell types located throughout the lung.

A. Beta–Adrenergic Responses

The function of human airway smooth muscle does not appear to be regulated by direct adrenergic innervation, a finding confirmed histologically utilizing immunocytochemistry, electron microscopy, and fluorescent techniques (30). Thus, airway smooth muscle responds primarily to circulating adrenergic agonists (31). The response of human airway smooth muscle to beta-adrenergic stimulation is relaxation. Utilizing selective $beta_1$ and $beta_2$ agonists and antagonists, it has been demonstrated that the relaxation of the smooth muscle of the human trachea and the smaller intrapulmonary airways is mediated by $beta_2$-adrenergic receptors (32). Inhibitors of neuronal and extraneuronal uptake of beta agonists do not alter concentration effect relationships to these agents, consistent with the absence of significant direct beta-adrenergic innervation and further supporting morphological interpretations. Direct radioligand binding studies confirm this beta$_2$ predominance in human airway smooth muscle (33), whereas beta$_1$-adrenergic receptors mediate some sympathetic nerve responses in directly innervated animal lungs. Beta-adrenergic receptors on mammalian pulmonary smooth muscle, quantitated with an autoradiographic technique, are increased at peripheral locations (34), suggesting increased importance of beta-adrenergic receptor-mediated relaxation in the smaller airways of the human lung.

The role of beta-adrenergic mechanisms in human mucus glycoprotein production and secretion is unclear. Some reports claim beta agonists enhance mucus production in human tracheal tissue in vitro (35,36), a finding supported by stimulation of cat tracheal mucus production by these agents (37). Other studies employing isoproterenol as the beta agonist fail to demonstrate such a beta-adrenergic stimulation, as assessed by release of radiolabeled mucus glycoprotein from human bronchial explants in tissue culture (38). It has also been suggested that beta agonists enhance mucociliary transport in humans (39).

Another potentially key role of beta-adrenergic receptors in human allergic disease is the regulation of mast cell secretory responses. In chopped human lung tissue (40), in enzyme digested and dispersed human lung mast cells (41), and in human skin in vivo (42), beta-adrenergic agonists inhibit antigen-induced release of mediators in a concentration-dependent fashion. Isoproterenol or norepinephrine (in the absence of propranolol) inhibits histamine release from human lung tissue; fenoterol prevents mediator release from isolated human lung mast cells; and terbutaline prevents cutaneous mast cell degranulation. In dogs, intra-arterial injections of compound 48-80 induce histamine release from pulmonary tissue, and this is prevented by isoproterenol (43). Rat serosal mast cells possess approximately 40,000 high-affinity ($K_d \sim 1$ nM) beta-adrenergic receptors, predominantly (85%) of the beta$_2$ subtype. These receptors are linked to adenylate cyclase and stimulation of the cells with beta agonists induces rapid, transient increases in cyclic AMP. However, in the rat mast cell, beta agonists do not affect antigen-induced mediator release (44). To date, there is no radioligand binding information regarding the presence, number, or affinity of beta-adrenergic receptors on human mast cells.

B. Alpha-Adrenergic Responses

Alpha-adrenergic receptors mediate bronchoconstriction of airway smooth muscle in humans and animals (45). Norepinephrine in the

presence of propranolol can cause bronchoconstriction of tracheal
smooth muscle strips, and this response is inhibited by phentolamine.
The functional relevance of this in normal airways is unclear. In some
in vivo studies of normal humans, airway constriction could be induced
by methoxamine and reversed by an alpha-adrenergic antagonist (46),
whereas histamine-induced bronchospasm could be reversed by the
alpha antagonist thymoxamine (47). However, others have failed to
demonstrate such in vivo alpha-adrenergic-mediated bronchoconstriction in normal subjects (48). In animal studies, radioligand binding
employing yohimbine and prazosin reveal a predominance of alpha$_2$-
adrenergic receptors (5:1) in large airway smooth muscle and a
predominance of alpha$_1$ (10:1) receptors in peripheral lung tissue
(49). Field stimulation-induced tracheal contractions were inhibited
best by a combination of alpha$_1$ and alpha$_2$ inhibitors, and intermediate
inhibition was achieved with an alpha$_2$ inhibitor, with lesser inhibition
in the presence of an alpha$_1$ antagonist alone.

In studies of mucus glycoprotein secretion, alpha-adrenergic
agonists are potent secretagogues (38). Norepinephrine (1-10 μM)
in the presence of propranolol augments (by 25-50%) mucus secretion
from cultured explants of human bronchial epithelium, a response
mimicked by phenylephrine. In such tissue preparations, submucosal
acinar glands far outnumber goblet cells, but the precise target of
alpha stimulation is not known.

Regulation of mast cell function by alpha agonists was first
demonstrated in chopped human lung fragments in which phenylephrine,
or norepinephrine in the presence of propranolol, enhanced antigen-
mediated histamine release (50). Subsequent studies with isolated
rat and human mast cells have not substantiated this effect, and its
relevance remains uncertain.

VII. CLINICAL EVIDENCE OF ADRENERGIC ABNORMALITIES
IN LUNG IN ALLERGIC DISEASE

In asthmatic airways, abnormal responses to both alpha- and beta-
adrenergic regulation have been identified. Blockade of beta-adrenergic
receptors of asthmatic individuals by propranolol is associated with
increased airway resistance and deterioration of pulmonary function
at rest (51). The fall in pulmonary function induced by a bronchoconstrictive stimulus is exacerbated by propranolol (52). In contrast,
in normal humans at rest or after exercise, or in monkeys after histamine inhalation (53), propranolol is without effect on airways responses.
These findings suggest that in asthmatic individuals, an on-going
beta-adrenergic stimulus for bronchodilation is present. This point
is supported by clinical observations of the direct correlation of the

circadian rhythms of bronchoconstriction and circulating catecholamines
in asthmatics, in which the most severe bronchospasm occurs con-
currently with the nadir in plasma epinephrine (54). This relationship
has long been exploited therapeutically in asthma with first epinephrine
and now a series of selective beta$_2$-adrenergic agonists used to treat
bronchospasm. Pretreatment of asthmatic subjects with beta$_2$ agonists
such as fenoterol, terbutaline, or salbutamol is associated with blunted
bronchoconstrictive responses to antigen, exercise, or cold air, whereas
the use of these agents after experimental or spontaneous induction
of bronchospasm is associated with prompt bronchodilation (55). Also,
recovery form exercise-induced bronchospasm is temporally associated
with increases in circulating levels of catecholamines (56). Although
there has been some concern that chronic beta-adrenergic therapy
in asthmatics may result in an attenuated bronchodilatory response
to beta-adrenergic agents, 11 asthmatics after a month of regular
oral terbutaline therapy demonstrated no changes in the bronchodilation
induced by inhaled isoproterenol or subcutaneous terbutaline and no
change in the beneficial effect of terbutaline subcutaneously on
histamine-induced bronchospasm. This stable clinical responsiveness
occurred despite a marked decrease in lymphocyte beta-adrenergic
receptor numbers (57). A similar study of 2 weeks of oral terbutaline
therapy in asthmatics showed that their pulmonary function improved
with further beta-adrenergic stimulation in spite of a pharmacological
down-regulation of beta-adrenergic receptor numbers (58).

A potential role for alpha-adrenergic mechanisms in asthma was
first suggested by the ability of alpha agonists to cause bronchoconstriction in canine respiratory smooth muscle pretreated with
beta-blockers and indomethacin. This hypothesis was made particularly
pertinent with the demonstration that airway smooth muscle from
Ascaris-sensitive dogs with hyperreactive airways did not require
pharmacological manipulation to manifest an alpha agonist-induced
contraction. This response was most marked in central airways (59).
However, when human subjects with asthma were pretreated with
inhaled atropine and propranolol to blunt cholinergic and beta-
adrenergic responses, there was no significant difference in forced
expiratory volume in 1 sec between those receiving phenylephrine
and those receiving saline, suggesting that asthmatic bronchial hyperresponsiveness was not secondary to increased bronchial smooth muscle
alpha-adrenergic receptor activity (60). Because these patients did
not receive methacholine or antigen bronchoprovocation at the time
of the phenylephrine, it is possible that at times of bronchoconstriction,
augmented alpha-adrenergic responsiveness is present in asthmatics.
In general, studies of pulmonary functional changes with alpha-
adrenergic blockers have yielded conflicting results, ranging from
increased airway conductance and flow rates in asthmatics after inhala-

tion of alpha blockers (61) and prevention of exercise-induced asthma
after alpha blockade (62) to no consistent change in spirometry after
inhalation of these agents (63,64). In a group of 15 youth asthmatics,
inhalation of albuterol was associated with greater improvement in
pulmonary function than was noted with the inhalation of phentolamine.
The inhalation of the alpha-blocker was accompanied by increased
bronchoconstriction in two and some bronchodilation in three subjects,
but for the entire group phentolamine-induced changes were equivalent
to those produced by placebo (65). Thus, although the existence of
alpha-adrenergic receptors in human bronchial smooth muscle is clear,
their role in bronchial hyperreactivity remains uncertain.

VIII. ASSESSMENT OF ALPHA- AND BETA-ADRENERGIC
PROCESSES IN ALLERGIC DISEASE

Because many of the physiological abnormalities in asthma and atopic
disease seemed to mimic models of beta-receptor blockade, Szentivanyi,
in 1968, postulated that atopic disease was associated with (and perhaps caused by) an abnormal pharmacological state, that of beta-
adrenergic blockade (66). This theory of beta-adrenergic blockade
has subsequently been expanded to the concept that an abnormality
in the ratio of alpha- to beta-adrenergic receptors exists in atopic
disease, and that this abnormality may predate the onset of clinical
manifestation of allergic disorders.

This concept of adrenergic receptor abnormalities in allergic
disease has proven extremely heuristic, and a large body of functional,
biochemical, and direct radioligand binding data of alpha- and beta-
adrenergic receptors in allergic patients have accumulated in the past
15 years. The overwhelming preponderance of these studies has
suggested that patients with allergic disease demonstrate a variety
of abnormalities either in beta- or alpha-adrenergic receptor number,
in physiological responses to adrenergic agents, or in cellular biochemical responses to adrenergic stimuli. Even now, however, it remains
elusive as to whether such abnormalities are causative of allergic disease, are concomitants of allergic disease, or are secondary responses
to the allergic disorders or their therapy (Table 1).

IX. PHYSIOLOGICAL PARAMETERS OF
ADRENERGIC RESPONSIVENESS

Clinical observations of patients with allergic disorders, particularly
asthmatic patients, repeatedly have demonstrated abnormal beta-
adrenergic responsiveness. For example, it is a standard clinical

TABLE 1 Role of Adrenergic Receptors in Tissues Central to Allergic Diseases

Tissue	Alpha receptors	Beta receptors
Lung	Bronchoconstriction Enhanced mast cell mediator release Vasoconstriction	Bronchial relaxation Inhibited mast cell mediator release Increased airway mucus (?) Improved mucociliary clearance (?)
Nose	Vasoconstriction Decreased mucus production (?)	Increased nasal airways resistance Vasodilation Inhibited mast cell-mediator release (?)
Skin	Vasoconstriction Increased pilomotor smooth muscle tone	Epidermal mitosis inhibition

axiom that the bronchodilator response to inhaled or oral beta-adrenergic agonists may be impaired or absent in some patients with severe asthma at the height of their disease. Other abnormal responses to adrenergic agonists in asthmatics have been known since 1963. Such abnormalities include diminished rises in pulse rate, free fatty acids, lactate, pyruvate, glucose, and cyclic AMP levels, and an impaired eosinopenic response to beta-adrenergic stimuli (67,68). Unfortunately, since normal individuals who receive inhaled or oral beta-adrenergic agonists develop tachyphylaxis to these agents (as assessed by both airway responses to inhaled bronchodilators and metabolic responses to beta stimulation [69]) and drugs were not

TABLE 2 Potential Adrenergic Receptor Alterations in Asthma

	References
1. Hyperresponsiveness to alpha-adrenergic stimuli	70, 72
2. Hyporesponsiveness to beta-adrenergic stimuli	71, 72
3. Increased sensitivity to beta-adrenergic blockade	51, 52
4. Autoantibodies to beta-adrenergic receptors	111, 112

withheld from patients studied, these early observations of adrenergic
hyporesponsiveness in asthmatics are difficult to interpret (Table 2).
In an attempt to clarify the responsiveness of patients to adrenergic stimuli in a series of studies, Kaliner and colleagues have examined
the physiological responses to alpha- and beta-adrenergic agonists in
normal subjects, allergic asthmatics, patients with allergic rhinitis,
and asymptomatic individuals with positive immediate hypersensitivity
skin tests, whom they termed "preallergic" (70-72). All studied
individuals abstained from medications for 30 days. Criteria for
alpha-adrenergic responses were the concentrations of phenylephrine
required to reduce cutaneous blood flow by 50% or to dilate the pupil
by more than 0.5 mm, whereas a beta-adrenergic response was defined
by the concentration of isoproterenol required to increase the pulse
pressure by 22 mm Hg. Patients with asthma required significantly
less phenylephrine and more isoproterenol than did normals to manifest
the criteria responses to alpha- and beta-adrenergic stimuli. Interestingly, while allergic rhinitis and preallergic individuals were less
sensitive to beta-adrenergic stimuli than were normal individuals,
they did not demonstrate evidence of hyperresponsiveness to alpha
agonists. These findings were interpreted to reflect a generalized
diminished beta-adrenergic responsiveness in atopy with a superimposed
increase in alpha-adrenergic reactivity in asthma. It is important to
note, however, certain limitations of the work: The responses studied
were not those pertinent to the physiological and clinical alterations
manifested by the patient groups; the measured responses were but
indirect measures of drug action; and because specific blocking agents
were not employed, the alterations in response could not unequivocally
be attributed to direct beta- or alpha-adrenergic actions of the employed drugs. This last issue is made more significant by the demonstration of isoproterenol-induced pulmonary vascular responses which
could be prevented by indomethacin but not propranolol (73). A
study which demonstrated decreased sensitivity to the bronchodilator
properties of aerosolized salbutamol in asthma as compared to normals
attributed the differences to difficulty in achieving tissue penetration
in asthma due to airway obstruction (74), raising further methodological
questions in the interpretation of studies comparing drug effects between normals and asthmatics.

X. PHARMACOLOGICAL PARAMETERS OF ADRENERGIC RESPONSIVENESS

In physiological or biochemical studies involving isolated cells to
further explore the "alpha-beta imbalance" theory of atopy, investigations have proceeded along two lines: an assessment of the differences

in generation of cyclic AMP and/or leukocyte enzyme secretion, or in receptor number or affinity as assessed by radioligand binding to cells or tissues from atopics as compared to normals.

A. Cyclic AMP Studies

Parker and Smith examined resting and isoproterenol-stimulated cyclic AMP levels in leukocytes and lymphocytes from asthmatics and normals (75) and reported a reduction in cyclic AMP accumulation in response to beta-adrenergic agonists in cells from asthmatic subjects. Furthermore, the reduction observed generally correlated with the severity of asthmatic symptoms. However, they found little evidence for beta-adrenergic blockade in those chronic asthmatics without a recent exacerbation of symptoms. Unfortunately, a large percentage of the asthmatic patients were exposed to elevated blood catecholamines during the time of the study either because of administered beta-adrenergic drugs or owing to the effect of methylxanthines to elevate endogenous catecholamine levels (76,77). Such increases in circulating levels of catecholamines could potentially down-regulate beta-adrenergic receptors and thereby produce a temporary insensitivity to beta-adrenergic stimulation which might have no relationship to asthma per se. Other subsequent data emphasized the contribution of therapy with adrenergic agonists to a desensitized state. For example, isoproterenol-stimulated cyclic AMP levels in lymphocytes of asthmatics receiving beta sympathomimetics were markedly reduced compared to those of normal individuals or asthmatics on nonadrenergic therapy, the latter two groups exhibiting similar lymphocyte cyclic AMP accumulation (78). Other investigators reported no differences in the leukocyte cyclic AMP response to isoproterenol between untreated asthmatics and normals, and that during terbutaline therapy both normals and asthmatics develop a functional desensitization of the leukocyte beta-adrenergic receptor (79).

Similar studies of leukocyte function have been undertaken in patients with atopic dermatitis. Although it has been found that basal concentrations of cyclic AMP in lymphocytes and granulocytes from normal subjects and subjects with atopic eczema are nearly identical, the lymphocyte cyclic AMP response to both epinephrine and isoproterenol (but not prostaglandin E_1 [PGE_1]) was attenuated in cells from atopic dermatitis patients (80). Additionally, the inhibition by isoproterenol of lysosomal enzyme release from granulocytes after zymosan stimulation was significantly less in eczema, demonstrating a decreased beta-adrenergic response in peripheral blood cells in atopic eczema, in the absence of cyclic AMP alterations in resting or PGE_2-stimulated cells (80). Another group has performed similar studies in patients with severe atopic dermatitis and noted a decreased cyclic AMP response

to both isoproterenol and prostaglandin E_1, suggesting a defect in
cyclic AMP generation that occurs distal to the beta-adrenergic
receptor (81). A more detailed consideration of this question was
made by Safko and colleagues (82). In their studies, exposure of
mononuclear leukocytes from eczema patients to isoproterenol, histamine,
and prostaglandin E_1 resulted in a depressed cyclic AMP response for
each agonist. Moreover, following exposure of normal cells to low
concentrations of these same agents, the cells become refractory to
stimulation of cAMP by any of the agonists employed. These studies
suggest that depressed cyclic AMP responses to beta-adrenergic
agonists need not reflect a primary event at the level of the beta-
adrenergic receptor. Thus, considerable uncertainty remains as to
whether cAMP generation in peripheral blood leukocytes differs between allergic patients and normal individuals, and even if such
differences do exist, whether they reflect an alteration in beta-
adrenergic receptors in the studied tissue, and more importantly,
if such alterations occur in target tissues pertinent to the disorder
in question.

B. Radioligand Binding Studies

With the advent of radioligand binding methods, a more direct approach
to the study of cell and tissue adrenergic receptors became available.
However, access to relevant human tissues has remained a problem.
Because of their accessibility, peripheral blood leukocytes have most
often been chosen for analysis, and numerous investigations have
been undertaken to compare beta-adrenergic receptors on these cells
in normals and patients with allergic disease.

In a small study of 10 control subjects and 11 stable asthmatics,
Brooks et al. reported a decrease in specific binding of [^{3}H]dihydro-
alprenolol ([^{3}H]DHA) in lymphocytes from asthmatics and a positive
correlation between the amount of specific [^{3}H]DHA binding and
FEV /FVC% (83,84). In this study, usage of beta-adrenergic drugs
was not completely avoided. In a larger study, Galant et al. found
no significant differences in [^{3}H]DHA binding (site number, K_d, or
agonist affinity) to polymorphonuclear leukocyte membranes of 31
control individuals and 30 asthmatics taking no adrenergic drugs (85).
Wolfe and colleagues reported no differences in [^{3}H]DHA binding to
lymphocytes between aspirin-sensitive asthmatics and controls (86).
Neutrophils from 17 asthmatic individuals currently taking adrenergic
drugs bound 70% less [^{3}H]DHA than those from the normal individuals
or untreated asthmatic patients, suggesting the defect in ligand binding
in asthma is not intrinsic to the disease but an alteration in beta-
adrenergic receptors acquired through usage of medication. This
concept was supported by the finding that after a 6-day course of

oral terbutaline therapy, [³H]DHA binding to polymopholeukocytes was decreased by approximately 85% in both normals and asthmatics, but this decrease could be reversed by withholding the drug for 6 days (87).

Townley and colleagues (88) have analyzed the binding of the beta-adrenergic ligand [¹²⁵I]iodohydroxybenzylpindolol ([²⁵I]HYP) to membranes of mixed leukocytes, mononuclear leukocytes, and neutrophilic polymorphonuclear leukocytes in normals, and in patients with asthma, both those taking medications regularly and those using no oral beta agonists and using other medications rarely. This group found a decrease in specific [¹²⁵I]HYP binding only to mononuclear cell preparations in asthmatics not using regular medications, while binding to both cell types was abnormal in asthmatics taking oral beta agonists. They also found that [¹²⁵I]HYP binding to mononuclear leukocytes was impaired in normal individuals for at least a week after discontinuing oral terbutaline. Taken together, these data suggest the possibility that beta-adrenergic binding defects exist in asthma, and that such defects may be identified only in selected cell populations. However, the persistence of down-regulation in mononuclear cells after discontinuation of oral beta agonist suggests these cells may be uniquely sensitive in this regard and that even intermittent drug usage by asthmatic patients may cause prolonged alteration in receptor binding.

Leukocytes from patients with atopic dermatitis have also been used in radioreceptor assays to determine beta-adrenergic receptor characteristics. In a study of [³H]DHA binding to polymorphonuclear leukocyte membranes of eight normal individuals, six patients with mild atopic eczema and nine patients with moderate to severe atopic eczema (89), Galant et al. noted no significant differences in receptor number or affinity between any of the groups. Similar results were reported by another group who studied granulocytes from 12 patients with atopic dermatitis and 12 controls and who used both [³H]DHA and [¹²⁵I]HYP to assess beta-adrenergic receptors (90). This information, taken together with the finding of decreased cyclic AMP responsiveness to a variety of agonists, including isoproterenol, in leukocytes from subjects with atopic eczema (see above), suggests that the number of beta-adrenergic receptors on cells from patients with atopic dermatitis is normal, and that it is the functional coupling of the beta-adrenergic receptors to adenylate cyclase that may be altered in atopic disease. As yet, no direct proof or refutation of this hypothesis is available. Szentivanyi (91,92) has presented two brief and as yet unsubstantiated reports evaluating adrenergic receptors on peripheral blood lymphocytes and on lung membranes of patients with allergic disorders. Utilizing [³H]DHA and [³H]dihydroergocryptine ([³H]DHE), a nonselective alpha-adrenergic antagonist, as ligands to assess alpha- and beta-adrenergic receptors in membranes of lymphocytes from 10 patients with nonallergic skin disease, eight with atopic dermatitis, and 70 normal individuals, Szentivanyi's group found that atopic dermatitis patients bound significantly less (36 fmol/mg protein) [^{3}H]DHA than did normals (64 fmol/mg protein) or those with non-allergic skin disease (58 fmol/mg protein) (91). Additionally, it was found that membrane preparations from atopic dermatitis patients bound significantly more [^{3}H]DHE (69 fmol/mg protein) than the normals (12 fmol/mg protein) or nonallergic skin disease patients (11 fmol/mg protein). Similar results were obtained by this group, utilizing the same techniques to study lymphocyte membranes of patients with reversible airway disease (asthma) as compared to fixed airway disease and normal individuals. In this preliminary study (92), membrane preparations from asthmatics bound 48 fmol [^{3}H]DHA/mg protein and 59 fmol [^{3}H]DHE/mg protein, values significantly different from those of normals (79 fmol [^{3}H]DHA, 11 fmol [^{3}H]DHE). These two studies are particularly difficult to interpret as the cell prepara-tions were not well described, the clinical criteria were not given, and the binding data were obtained using single ligand concentrations (11 nM [^{3}H]DHA, 4 nM [^{3}H]DHE). As most workers (93) find that alpha-adrenergic receptors detectable in peripheral blood cells are confined to alpha$_2$ receptors on platelets, the [^{3}H]DHE binding data in these studies may merely reflect differing platelet contamination of lymphocyte preparations. In the only study of platelet alpha-adrenergic receptors in allergic disease, no differences were identified in alpha-adrenergic inhibition of prostaglandin-induced cyclic AMP increases in the platelets of normal and asthmatic patients (94).

The desirability of examining the target tissue appropriate to asthma has prompted numerous investigations of mammalian lung adrenergic receptors. Lung exhibits a higher density of beta-adrenergic receptors than any other human tissue (95), with the number of these receptors increasing from the trachea to the terminal bronchioles (34). Human lung obtained at the time of surgery in patients with chronic lung disorders possesses similar numbers of alpha- and beta-adrenergic receptors in radioligand binding studies (96). Beta-adrenergic receptors generally represent the majority of adrenergic receptors in other mammalian lung (97).

Only one study of human lung membrane adrenergic receptor binding in asthma has been published (92). This study, as noted above, is a preliminary report in which single concentrations of the ligands [^{3}H]DHA and [^{3}H]DHE were employed. Membrane preparations of lung specimens obtained from asthmatic individuals bound 160 fmol [^{3}H]DHA/mg protein and 172 fmol/mg protein [^{3}H]DHE as compared to normal lung which bound 212 fmol/mg protein [^{3}H]DHA and 38 fmol/mg protein [^{3}H]DHE, or to tissue from nonasthmatic pulmonary disease patients (194 fmol [^{3}H]DHA, 62 fmol [^{3}H]DHE/mg protein).

The data for adrenergic binding to asthmatic lung presented in this report closely resemble those reported for lung from nonasthmatic patients observed by Barnes and co-workers, while the preponderance of beta receptors in nonasthmatic lung resembles data in animal models rather than previous human data (96). Because clinical criteria for assessing asthma were not reported and the binding data presented do not reflect saturation binding experiments, the studies of Szenti-vanyi remain inconclusive (91,92).

As lung tissue is difficult to obtain from humans, particularly those with asthma or allergy, other workers have attempted to assess beta- and alpha-adrenergic binding to lung tissues from animals immunized to express IgE antibody or experimental asthma.

To examine the effects of experimental asthma on lung adrenergic receptors, Barnes et al. sensitized guinea pigs to ovalbumin aerosol for 4 weeks (98). The nature of the antibody response engendered by this protocol was not identified, but the animals manifested marked respiratory distress and occasional cyanosis when challenged with aerosolized ovalbumin (98). Sensitized animals demonstrated an increase in the number of alpha-adrenergic receptors and a slight decrease in beta-adrenergic receptor number in membranes prepared from homogenates of lungs. Further analysis of adrenergic binding in experimental airways disease in guinea pigs was undertaken in animals sensitized in three different ways: intraperitoneal injection of ovalbumin or saline; 10-day aerosol administration of ovalbumin or saline; 21-day low-dose aerosol administration of ovalbumin or saline (99). No differences were seen in alpha- or beta-adrenergic receptor numbers or binding affinities in lung membranes from guinea pigs receiving either ovalbumin or saline intraperitoneally. The short sensitization protocol with ovalbumin aerosol produced dramatic asthmatic symptoms as well as a small reduction in lung membrane beta-adrenergic receptors and a modest increase in alpha-adrenergic receptors, whereas the prolonged sensitization was characterized by only a mild degree of asthmatic symptoms and no change in adrenergic receptor characteristics compared to control animals. The interpretation most consistent with these two studies is that alterations in receptor binding reflect the degree of physiological insult and resultant hypoxia and are perhaps secondary to changes related to altered catecholamine release.

In a preliminary report, however, Halonen and colleagues suggest that receptor abnormalities may occur as a consequence of an immune response per se (100). Utilizing an immunization protocol known to induce IgE synthesis in newborn rabbits, these workers found that the concentration of isoproterenol required to affect pulse pressure and heart rate were greater in the immunized rabbits as compared to control animals. No change in phenylephrine-mediated (alpha-

adrenergic) pupillary responses could be discerned in these same
animals. These preliminary studies thus suggest that sensitization
to produce IgE antibody may in itself be sufficient to alter physiological
responses to beta agonists and suggest defects in beta-adrenergic
receptor properties and/or function.

At the present time, it is impossible to state with confidence
what the effects of either sensitization, experimental antigen challenge,
or spontaneous allergic disease may have on adrenergic receptor numbers, binding characteristics, or function. Until studies are performed
in which defined antibody responses are elicited and characterized,
and adrenergic receptors characterized both functionally and by
radioligand techniques in well-controlled experimental models, it will
be difficult to clarify this problem.

XI. PHYSIOLOGICAL MANIPULATION OF ADRENERGIC RECEPTORS IN ALLERGIC DISEASE

In addition to studies directed at comparing adrenergic receptors on
cells and tissues from normals and those with allergic disease, investigation has also focused upon the pharmacological manipulation of
receptor characteristics by administration of sympathomimetic agents
or corticosteroids in allergic patients.

The topic of desensitization (or refractoriness) to beta-adrenergic
agonists in normal and allergic individuals has been reviewed (69).
It would appear that this phenomenon occurs relatively slowly, is
somewhat more difficult to demonstrate in asthmatics as compared to
normal individuals (especially when large airways are studied), is
easier to demonstrate when oral rather than inhaled agonists are
utilized, and is characterized by a great deal of individual variability
in the expression of this phenomenon when assessed physiologically.
Similarly, prolonged exposure of normal subjects to high doses of
beta-adrenergic agonists reduced lymphocyte cyclic AMP responses
(78), but the susceptibility to desensitization appears to be quite
different between lymphocytes and bronchial smooth muscle (101,102).
In fact, there is some evidence that tolerance to the side effects of
beta-adrenergic drugs may develop during chronic therapy, but the
beneficial effects of these agents on lung function may persist (57,58,
103). It is possible that beta-adrenergic desensitization may be most
prominent in extrapulmonary tissues and that lung, by virtue of its
larger number of "spare receptors," may be less readily desensitized
(104). It has also been shown that allergen-induced asthmatic attacks
may lead to an impaired cyclic AMP response to isoproterenol in lymphocytes which was not present prior to the asthmatic episode (105),
presumably owing to augmentation of circulating catecholamines but

perhaps on account of other mediators (histamine, prostaglandins) capable of cross desensitizing cells to the effects of beta agonists (see above).

Numerous observations indicate that chronic adrenergic therapy also induces abnormalities at the receptor level termed down-regulation. In a report of normals and asthmatics receiving oral terbutaline therapy, cyclic AMP responsiveness is diminished by day 7 of treatment and returns to normal within 1 week after therapy is discontinued (79). Comparable data for polymorphonuclear leukocytes in normals and asthmatics have also been observed (85). Thus, agonist-mediated alterations in beta-adrenergic receptors have been noted in both normal and asthmatic patients. In no study has the mechanism of this change been addressed, and it remains unclear whether synthesis, degradation, or occupancy of receptors or other undefined events are responsible for the observed effects.

The numerous biochemical effects of corticosteroids are beyond the scope of this discussion, but it has been observed that in many cells and tissues, glucocorticoids augment beta-adrenergic receptor numbers. Clinically, beta-adrenergic agonists and corticosteroids work synergistically in the treatment of asthma (106), and in normal individuals and asthmatic patients (107,108) unresponsive to adrenergic therapy, corticosteroid administration may restore the bronchodilator effects of isoproterenol. Hydrocortisone induces (within 24 hr) an increase in the number of beta-adrenergic receptors on cultured human lung cells and doubles the rate of receptor synthesis (109), whereas rat lungs display significantly greater numbers of beta receptors 24 hr after an in vivo dexamethasone injection (44). A 4-hr incubation of granulocytes with cortisone acetate in vitro results in an increased number of beta-adrenergic receptors ($[^3H]$DHA sites), but lymphocytes under the same conditions demonstrate a decreased number of beta-adrenergic receptors; after 16 hr incubation with hydrocortisone, $[^3H]$DHA binding increases in these cells (110). These findings underline the care which must be taken in interpreting studies in which different cells and/or tissues are the subjects of analysis of adrenergic receptor biology.

One controversial point about the action of corticosteroids on beta receptors has been whether steroids will augment "resting" beta receptor numbers or whether a prior down-regulation of beta adrenoceptors is required before corticosteroids will have an enhancing effect. In a study employing neutrophils incubated in vitro with isoproterenol for 3 hr, a 40% reduction in $[^3H]$DHA binding and an 86% decrease in cAMP responsiveness to isoproterenol were noted (111). When hydrocortisone was present during the isoproterenol incubation, the desensitization of cAMP generation was attenuated, but the receptor number still declined. Of interest, incubation with the corticosteroid

also partially restored the ability of agonists to demonstrate high-
affinity binding to the membranes. This suggests that changes in
agonist affinity may be as important for steroid effects as are previously
reported increases in receptor number. However, the limited time
course in these experiments prevents drawing a firm conclusion about
the basis of the steroid effect. Suffice it to say that both functionally
and pharmacologically, corticosteroids appear to enhance beta-
adrenergic responsiveness, and their useful role in the treatment
of allergic diseases may be partially based on this activity.

XII. AUTOANTIBODIES

The possibility that asthmatics or atopics could have autoantibodies
to tissue beta-adrenergic receptors stems from the identification of
pathophysiologically relevant antibodies to cell surface receptors in
myasthenia gravis, Graves' disease, and in an infrequently encountered
form of diabetes. The possibility of antireceptor antibodies would be
consistent with the adrenergic imbalance theory of allergy and asthma,
since an antibody could functionally block beta-adrenergic receptors
and, if tightly bound, might alter radioligand binding data and suggest
a reduced number of receptors. Autoantibodies to beta$_2$-adrenergic
receptors, which were initially demonstrated in serum from a patient
with allergic rhinitis and two patients with asthma, were detected by
an indirect immunoprecipitation assay and the ability of these sera to
inhibit [^{125}I]IHYP binding to dog lung membrane receptors (112).
In a larger follow-up study, Fraser et al. screened 17 subjects with
allergic asthma, eight with allergic rhinitis, nine asymptomatic subjects
with positive immediate hypersensitivity skin tests, termed preallergic,
19 controls, and seven cystic fibrosis patients for beta-adrenergic
autoantibodies by the ability to affect [^{125}I]protein A binding to calf
lung membrane, to inhibit radioligand binding to calf lung beta-
adrenergic receptors, and to precipitate solubilized calf lung beta-
adrenoceptors by indirect immunoprecipitation (113). Three of 19
normals, four of 17 asthmatics, one preallergic patient, and one with
cystic fibrosis demonstrated autoantibodies to beta$_2$-adrenergic receptors
by these criteria. The presence of autoantibodies in these patients
correlated with a physiological alpha-adrenergic hyperresponsiveness
as assessed by pupillary reaction to phenylephrine and a beta-
adrenergic hyporesponsiveness as assessed by the concentration of
isoproterenol required to increase pulse pressure more than 22 mm Hg.
In spite of the intriguing possibilities offered by these observations,
there was no correlation between the presence of autoantibodies and
clinical disease severity, and a similar percentage of normals and
asthmatics possessed the antibodies. Moreover, it is possible that

protease contamination of serum may degrade receptors, thereby falsely suggesting the presence of autoantibodies to beta-adrenergic receptors in patient samples. Therefore, the importance of autoantibodies to the $beta_2$-adrenergic receptor to any disease state has not been established. Until additional data are presented, we conclude that it appears unlikely that these antibodies explain the observed changes in adrenergic receptors in disease.

XIII. CONCLUSIONS

Adrenergic receptors are important in the physiological function of target tissues of allergic disease. The importance of these receptors is underscored by the clinical usefulness of adrenergic drugs in the therapy of allergic disease and in the finding of abnormalities in functional responses and in radioligand binding of adrenergic agents to receptors in allergic patients. Despite a large body of studies, it remains uncertain as to whether such identified abnormalities in receptor-mediated processes are primary and causative of allergic disorders or whether these findings are secondary to disease processes or even to the agents used to treat the disordered physiology seen in allergic disease.

The answer to these questions will come only from well-designed and carefully executed experimental studies in which adrenergic drugs are avoided, catecholamine levels are monitored, appropriate tissues investigated, and state of the art receptor analysis by both functional and ligand binding techniques is undertaken. Without such careful analysis the role of adrenergic receptors in allergic disease will remain an intriguing mystery.

ACKNOWLEDGMENT

Supported in part by a grant from the National Institutes of Health AI 17268.

REFERENCES

1. Tharp, M. D.: In: *Biochemistry and Physiology of the Skin*. L. A. Goldsmith (ed.). New York, Oxford University Press, 1983, p. 1210.

2. Yoshikawa, K., Adachi, K., Halprin, K. W., and Levine, B.: Br. J. Dermatol. 92:619, 1975.

3. Orenberg, E. K., Pfendt, E. A., and Wilkinson, D. I.: J. Invest. Dermatol. 80:503, 1983.

4. Duell, E. A.: Biochem. Pharmacol. 29:97, 1980.

5. Solanki, V., and Murray, A.: J. Invest. Dermatol. 71:344, 1978.

6. Grommans, J., Bergers, M., Van Erp, P., Van Den, Hark J., Van de Kerkhof, P., Mier, P., and Roelfzema, H.: Br. J. Dermatol. 101:413, 1979.

7. Ahlquist, R. P.: Am. J. Physiol. 153:586, 1948.

8. Fox, R. M., Goldsmith, R., and Kidd, D. J.: J. Phys. (Lond.) 161:298, 1962.

9. Rao, G. J. S., Del Monte, M., and Nadler, H. L.: Nature 232: 253, 1971.

10. Sato, K.: In: Biochemistry and Physiology of the Skin. L. A. Goldsmith (ed.). New York, Oxford University Press, 1983, p. 596.

11. Ogawa, T.: Jpn. J. Physiol. 26:517, 1976.

12. Uno, H.: J. Invest. Dermatol. 69:111, 1977.

13. Robertshaw, D.: J. Invest. Dermatol. 63:160, 1974.

14. Hanafin, J. M.: Semin. Dermatol. 2:5, 1983.

15. Stone, S. P., Muller, A., and Gleich, G. J.: Arch. Dermatol. 108:806, 1973.

16. Ramsay, C.: Br. J. Dermatol. 81:37, 1969.

17. Black, W.: Arch. Klin. Exp. Dermatol. 203:63, 1958.

18. Johnson, L. A., and Winkelman, R. K.: Arch. Dermatol. 92:621, 1965.

19. Juhlin, L.: J. Invest. Dermatol. 37:257, 1961.

20. Juhlin, L.: J. Invest. Dermatol. 37:201, 1961.

21. Flaxman, B. A., and Harper, R. A.: J. Invest. Dermatol. 65:52, 1975.

22. Marcello, C. L., Duell, E. A., Stawiski, M. A., Anderson, T. F., and Voorhees, J. J.: J. Invest. Dermatol. 72:20, 1979.

23. Carr, R. H., Busse, W. W., and Reed, C. E.: J. Allergy Clin. Immunol. 51:255, 1973.

24. Solomon, L. M., Wentzel, H. E., and Tulsky, E.: J. Invest. Dermatol. 43:193, 1964.

25. Bamshad, J.: J. Invest. Dermatol. 52:100, 1969.

26. Ritter, F. N.: In: *Allergy, Principles and Practice*. E. Middleton, C. E. Reed, and E. F. Ellis (eds.). St. Louis, Mosby, 1978, p. 359.

27. Schumacher, M. J.: J. Allergy Clin. Immunol. 66:33, 1980.

28. Mygind, N., and Weeke, B.: In: *Allergy, Principles and Practice*. E. Middleton, C. E. Reed, and E. F. Ellis (eds.). St. Louis, Mosby, 1983, p. 1101.

29. Borum, P., and Mygind, N.: J. Allergy Clin. Immunol. 66:25, 1980.

30. Richardson, J. B., and Ferguson, C. C.: Fed. Proc. 38:202, 1979.

31. Davis, C., Kannan, M. S., Jones, T. R., and Daniel, E. E.: J. Appl. Physiol. 53:1080, 1982.

32. Zaagsma, J., Van der Heijden, P. J. C. M., Van der Schaar, M. W. G., and Bank, C. M. C.: J. Recept. Res. 3:89, 1983.

33. Engel, G.: Postgrad. Med. J. (Suppl. 1) 57:77, 1981.

34. Barnes, P. J., Basbaum, C. B., Nadel, J. A., and Roberts, J. M.: Nature 299:444, 1982.

35. Boat, T. F., and Kleineman, J. I.: Chest (Suppl.) 67:32, 1975.

36. Phipps, R. J., Williams, I. P., Richardson, P. S., Pell, J., Pach, J., and Wright, N.: Clin. Sci. 63:23, 1982.

37. Liedtke, C. M., Rudolph, S. A., and Boat, T. F.: Am. J. Physiol. 244:C391, 1983.

38. Shelhamer, J., Marom, Z., and Kahner, M. A.: J. Clin. Invest. 66:1400, 1980.

39. Pavia, D., Bateman, J. R. M., and Clarke, S. W.: Bull. Eur. Physiopathol. Respir. 16:335, 1980.

40. Orange, R. P., Austen, W. G., and Austen, K. F.: J. Exp. Med. 134:136S, 1971.

41. Peters, S. P., Schulman, E. S., Schleimer, R. P., MacGlashan, D. W., Jr., Newhall, M. H., and Lichtenstein, L. M.: Am. Rev. Respir. Dis. 126:1034, 1982.

42. Ting, S., Zweiman, B., and Lavker, R.: J. Allergy Clin. Immunol. 71:437, 1983.

43. Brown, J. K., Leff, A. P., Frey, M. J., Reed, B. R., Lazarus, S. C., Shields, R., and Gold, W.: Am. Rev. Respir. Dis. 126:842, 1982.

44. Marquardt, D. L., and Wasserman, S. I.: J. Immunol. 129:2122, 1982.

45. Kneussl, M. P., and Richardson, J. B.: J. Appl. Physiol. 45: 307, 1978.

46. Anthracite, R. J., Vachon, J., and Knapp, P. M.: Psychosom. Med. 33:481, 1971.

47. Gaddie, J., Legge, J. S., Petrie, G., and Palmer, K. N. V.: Br. J. Dis. Chest 66:141, 1972.

48. Pael, K. R., and Kerr, J. W.: Clin. Allergy 3:439, 1973.

49. Barnes, P. J., Skoogh, B. E., Nadel, J. A., and Roberts, J. M.: Mol. Pharmacol. 23:570, 1983.

50. Kaliner, M. A., Orange, R. P., and Austen, K. F.: J. Exp. Med. 136:556, 1972.

51. McNeill, R. S., and Ingram, C. G.: Am. J. Cardiol. 18:473, 1966.

52. Dorow, P.: Respiration 43:359, 1982.

53. Pare, P. D., and Nicholls, I.: J. Allergy Clin. Immunol. 69:213, 1982.

54. Barnes, P., Fitzgerald, G., Brown, M., and Dollery, C.: N. Engl. J. Med. 303:263, 1980.

55. Nelson, H. S.: In: *Allergy, Principles and Practice*. E. Middleton, C. E. Reed, and E. F. Ellis (eds.). St. Louis, Mosby, 1983, p. 511.

56. Larsson, K., Jhemdahl, P., and Martinsson, A.: Chest 82:560, 1982.

57. Tashkin, D. P., Conolly, M. E., Deutsch, R. I., Hui, K. K., Littner, M., Scarpace, R., and Abrass, I.: Am. Rev. Respir. Dis. 125:185, 1982.

58. van den Berg, W., Leferink, J. G., Fokkens, J. K., Kreukniet, J., Maes, R. A. A., and Bruynzeel, P. L. B.: J. Allergy Clin. Immunol. 69:410, 1982.

59. Malo, P. E., and Wasserman, M. A.: Eur. J. Pharmacol. 86:27, 1982.

60. Thomson, N. C., Daniel, E. E., and Hargreave, F. E.: Am. Rev. Respir. Dis. 126:521, 1982.

61. Bianco, S., Griffin, J. P., Kamburoff, P. L., and Prince, F. J.: Br. Med. J. 4:18, 1974.

62. Biel, M., and de Koch, M. A.: Respiration 35:78, 1978.

63. Campbell, S. C.: Ann. Allergy 49:135, 1982.

64. Barnes, P. J., Ind, P. W., and Dollery, C. T.: Thorax 36:378, 1981.

65. Shiner, R. J., and Molhorn, I.: Chest 83:603, 1983.

66. Szentivanyi, A.: J. Allergy 42:203, 1968.

67. Cookson, D. U., and Reed, C. E.: Am. Rev. Respir. Dis. 88: 636, 1963.

68. Lochey, S. D., Jr., Glennon, J. A., and Reed, C. E.: J. Allergy 40:349, 1967.

69. Jenne, J. W.: J. Allergy Clin. Immunol. 70:413, 1982.

70. Henderson, W. R., Shelhamer, J. H., Reingold, D. B., Smith, L. J., Evans, R., and Kaliner, M. A.: N. Engl. J. Med. 300: 642, 1979.

71. Shelhamer, J. H., Metcalfe, D. D., Smith, L. J., and Kaliner, M.: J. Allergy Clin. Immunol. 66:52, 1980.

72. Kaliner, M. A., Shelhamer, J. H., Davis, P. B., Smith, L. J., Venter, J. C.: Ann. Intern. Med. 96:349, 1982.

73. Rubin, L. J., and Lazar, J. D.: J. Clin. Invest. 71:1366, 1983.

74. Barnes, P. J., and Pride, N. B.: Br. J. Clin. Pharmacol. 15: 677, 1983.

75. Parker, C. W., and Smith, J. W.: J. Clin. Invest. 52:48, 1973.

76. Higbee, M. D., Kumar, M., and Galant, S. P.: J. Allergy Clin. Immunol. 70:377, 1982.

77. Berkowitz, B. A., and Spector, S.: Eur. J. Pharmacol. 13:193, 1971.

78. Conolly, M. E., and Greenacre, J. K.: J. Clin. Invest. 58:1307, 1976.

79. Bruijnzeel, P. L. B., van den Berg, W., Hamelink, M. L., van den Bagaard, W., Houben, L. A. M. J., and Kreukniet, J.: Ann. Allergy 43:105, 1979.

80. Busse, W. W., and Lee, T.: J. Allergy Clin. Immunol. 58:586, 1976.

81. Parker, C. W., Kennedy, S., and Eisen, A. Z.: J. Invest. Dermatol. 60:302, 1977.

82. Safko, J. M., Chan, S., Cooper, K. D., and Hanafin, J. M.:
    J. Allergy Clin. Immunol. 68:218, 1981.

83. Brooks, S. M., McGowan, K., and Altenan, P.: Chest 75:232,
    1979.

84. Brooks, S. M., McGowan, K., Bernstein, I. L., Altenan, P.,
    and Peagler, J.: J. Allergy Clin. Immunol. 63:401, 1979.

85. Galant, S. P., Duriseti, L., Underwood, S., Allred, S., and
    Insel, P. A.: J. Clin. Invest. 65:577, 1980.

86. Wolfe, R. N., Hui, K. K., Conolly, M. E., Tashkin, D. P., and
    Fisher, H. K.: J. Allergy Clin. Immunol. 69:46, 1982.

87. Galant, S. P., Duriseti, L., Underwood, S., and Insel, P. A.:
    N. Engl. J. Med. 299:933, 1978.

88. Sano, Y., Watt, G., Townley, R. G.: J. Allergy Clin. Immunol.
    72:495, 1983.

89. Galant, S. P., Underwood, S., Allred, S., and Hanafin, J. M.:
    J. Invest. Dermatol. 72:330, 1979.

90. Ruoho, A. A. E., De Clrque, J. L., and Bussee, W. W.:
    J. Allergy Clin. Immunol. 66:46, 1980.

91. Szentivanyi, A., Heim, O., Schultze, P., and Szentivanya, J.:
    Acta Derm. Venererol. (Suppl.) (Stockh.) 92:19, 1980.

92. Szentivanyi, A., Heim, O., and Schultze, D.: N.Y. Acad. Sci.
    332:295, 1979.

93. Casale, T., and Kaliner, M. A.: J. Allergy Clin. Immunol. 74:
    812, 1984.

94. Davis, P. B., and Lieberman, P.: J. Allergy Clin. Immunol. 69:
    35, 1982.

95. Rugg, E. L., Barnett, D. B., and Nahorski, S. R.: Mol. Pharma-
    col. 14:996, 1978.

96. Barnes, P. J., Karliner, J. S., and Dollery, C. T.: Clin. Sci.
    58:457, 1980.

97. Marquardt, D. L., Motulsky, J. H., and Wasserman, S. I.:
    J. Appl. Physiol. 53:731, 1982.

98. Barnes, P. J., Dollery, C. T., and MacDerme, A.: Nature 285:
    569, 1980.

99. Meta, H., Yui, Y., Yasueda, H., and Sheda, T.: Int. Arch.
    Allergy Appl. Immunol. 70:225, 1983.

100. Halonen, M., Lohman, I. C., and Kaliner, M. A.: Fed. Proc.
     42:6159, 1983.

101. Fleisch, J. H., and Titus, E.: J. Pharmacol. Exp. Ther. 181: 425, 1972.

102. Svedmyr, N., Larsson, S., and Thiringer, G.: Br. Med. J. 2: 668, 1974.

103. Larsson, S., Svedmyr, N., and Thiringer, G.: J. Allergy Clin. Immunol. 59:93, 1977.

104. Avner, B. P., and Jenne, J. W.: J. Allergy Clin. Immunol. 68: 51, 1981.

105. Kaeter, G. H., Neurs, H., Kauffman, J. H., and De Vries, K.: Eur. J. Resp. Dis. 63:72, 1982.

106. Shenfield, G. M., Hodson, M. E., Clarke, S. W., Paterson, J.: Thorax 30:430, 1975.

107. Ellul-Micallef, R., Borthwick, R. C., and McHardy, G. J. R.: Clin. Sci. Mol. Med. 47:105, 1974.

108. Ellul-Micallef, R., and Fenech, F. F.: Lancet 2:1269, 1975.

109. Fraser, C. M., and Venter, J. C.: Biochem. Biophys. Res. Commun. 94:390, 1980.

110. Davies, A. O., and Lefkowitz, R. J.: J. Clin. Endocrinol. Metab. 51:500, 1980.

111. Davies, A. O., and Lefkowitz, R. J.: J. Clin. Invest. 71:565, 1983.

112. Venter, J. C., Fraser, C. M., and Harrison, L. C.: Science 207:1361, 1980.

113. Fraser, C. M., Venter, J. C., and Kaliner, M. A.: N. Engl. J. Med. 305:1165, 1981.

10

Adrenergic Receptors in Endocrine and Metabolic Diseases

PHILIP E. CRYER   *Washington University in St. Louis, St. Louis, Missouri*

I. INTRODUCTION

There is considerable evidence that the catecholamines, including the adrenomedullary hormone epinephrine, and several nonagonist hormones can modify adrenergic receptors and responsiveness to catecholamines as discussed elsewhere in this book and summarized in this chapter. Thus, it is conceivable that hormone-induced alterations of adrenergic receptors and/or their linked intramembrane or intracellular systems might play a role in the pathophysiology of human endocrine-metabolic diseases. For example, increments in beta-adrenergic receptor densities resulting from thyroid hormone excess could explain some of the clinical manifestations of hyperthyroidism. Alternatively, since catecholamines modify the secretion of most hormones, it is conceivable that altered catecholamine effector systems might play a role in the pathogenesis of human endocrine-metabolic disorders. For example, decreased dopaminergic inhibition of prolactin secretion could result in hyperprolactinemia.

II. CATECHOLAMINES

Catecholamines modify adrenergic receptors in vivo (1-71). We (1) found that infused isoproterenol produced a biphasic change—an initial increase and a late decrease—in mononuclear leukocyte (MNL) $beta_2$-adrenergic receptor density in humans. Infused epinephrine also produced an initial increase in MNL $beta_2$-adrenergic receptors;

the late effects of epinephrine were not studied. Although we did not
measure MNL sensitivity to agonists, isoproterenol induced increments
in MNL beta$_2$-adrenergic receptor density were paralleled by increments
in the cardiac chronotropic response to isoproterenol in vivo in normal
human subjects, a finding interpreted to indicate parallel, isoproterenol-
induced changes in MNL and myocardial beta-adrenergic receptors.
Further, beta-adrenergic receptor density on circulating cells is
decreased by the long-term administration of agonists which have
been reported to decrease adenylate cyclase responsiveness to agonists
in vitro and to decrease the bronchodilator response to agonists in
vivo (2). Fraser et al. (3) studied the effects of variations in sympathoadrenal activity produced by varying sodium intake in normal humans;
plasma catecholamine levels are increased by sodium restriction and
tend to be decreased by sodium loading. They found lymphocyte
beta-adrenergic receptor density to be inversely related to endogenous
plasma norepinephrine and epinephrine concentrations, and that
cardiac chronotropic sensitivity to isoproterenol, interpreted as an
index of myocardial beta-adrenergic receptor activity, was proportional
to lymphocyte beta-adrenergic receptor levels. Administration of the
nonselective beta-adrenergic antagonist propranolol to humans was
reported to increase lymphocyte beta-adrenergic receptor density (4).
Indeed, in rats infused with propranolol, Aarons and Molinoff (5)
found parallel changes in beta-adrenergic receptor densities in lymphocytes, heart, and lung. In contrast to the cited data suggesting
reciprocal regulation of beta-adrenergic receptors by endogenous
catecholamnes, however, we (6) found no relationship between MNL
beta$_2$-adrenergic receptor density and basal plasma epinephrine or
norepinephrine, a finding that has been confirmed by us (7) and
others (8). Further, Sowers et al. (8) found no relationship between
lymphocyte beta-adrenergic receptor density or affinity for antagonist
ligand and plasma norepinephrine during diurnal variation of the latter
in recumbent subjects, or during norepinephrine release stimulated
by upright posture. However, Feldman et al. (9) have reported that
upright activity (as well as norepinephrine infusion) is associated
with decreased MNL beta-adrenergic receptor affinity for agonists,
a finding associated with decreased maximal MNL adenylate cyclase
responsiveness to agonist in vitro. They found no alteration of receptor
density. These findings alone are somewhat difficult to reconcile.
Although decreased affinity for agonist alone would lead to decreased
sensitivity to agonist, it should not result in a decreased maximal
adenylate cyclase response. Thus, the latter finding implies a decrease
in receptor-cyclase coupling coordinate with the decrease in receptor
affinity.
     In general, catecholamine regulation of alpha-adrenergic receptors
has been difficult to demonstrate in vivo (10), although selective down-
regulation has been reported (10,11) as discussed below.

If physiological variations in sympathoadrenal activity modulate adrenergic receptors in vivo, pathophysiological variations—markedly increased catecholamines in pheochromocytoma or decreased catecholamines in degenerative diseases of the autonomic nervous system—should produce clear, reciprocal changes in adrenergic receptors.

Patients with hypoadrenergic postural hypotension due to degenerative disorders of the autonomic nervous system have been shown to have MNL beta-adrenergic receptor and platelet alpha-adrenergic receptor densities that are increased substantially (12-15). These are associated with increased cardiac chronotropic sensitivity to isoproterenol (12,13), a beta-adrenergic event, and increased pressor sensitivity to phenylephrine (12-15), an alpha-adrenergic event. Further, treatment with the alpha-adrenergic agonist ergotamine has been reported to reduce platelet alpha-adrenergic receptors to normal or subnormal levels in such patients (15).

Fat cells from patients with pheochromocytomas exhibit a decreased cyclic AMP response to norepinephrine in vitro (16). Although both polymorphonuclear leukocyte beta-adrenergic receptor and platelet alpha-adrenergic receptor densities were reported to be decreased in a patient with a pheochromocytoma and to increase to normal after surgical removal of the tumor (15), platelet $alpha_2$-adrenergic receptors did not differ from those of normal subjects in a series of patients with pheochromocytomas (10). Studies of rats bearing a transplanted pheochromocytoma have disclosed selective down-regulation of adrenergic receptors (10,11,17). These changes include: decreased $alpha_1$-adrenergic receptor densities in kidney and lung with no change in liver; no change in kidney or liver $alpha_2$-adrenergic receptors; decreased $beta_1$-adrenergic receptors in kidney, lung, and heart; and no change in kidney or lung $beta_2$-adrenergic receptors. Along with the unchanged platelet $alpha_2$-adrenergic receptors in patients with pheochromocytomas (10), these findings suggest a pattern: Catecholamine induced down-regulation of $alpha_1$- and $beta_1$-adrenergic receptors but not of $alpha_2$- and $beta_2$-adrenergic receptors when norepinephrine levels are elevated long term.

Thus, patients with chronic sympathetic hypofunction exhibit increased $beta_2$- and $alpha_2$-adrenergic receptor densities on circulating cells. Although studies of their responses to agonists in vivo can be interpreted to indicate increased $beta_1$- and $alpha_1$-adrenergic receptor function in cardiovascular tissues, these have not been documented directly and the data are subject to other interpretations. For example, increased vascular $beta_2$-adrenergic receptor function could result in an enhanced chronotropic response to isoproterenol and increased vascular $alpha_2$-adrenergic receptor function could result in an enhanced pressor response to phenylephrine. On the other hand, altered adrenergic receptors have not been established to occur in patients with pheochromocytomas, although the data are limited. Further, pheochromocytoma-bearing rats exhibit decreased $beta_1$- and $alpha_1$-adrenergic receptor densities in some tissues without demonstrable changes in those of $beta_2$- or $alpha_2$-adrenergic receptors.

III. GROWTH HORMONE AND PROLACTIN

Catecholamines modify pituitary hormone release indirectly by altering hypothalamic hormone release, and directly at the anterior pituitary level. Both alpha-adrenergic and dopaminergic stimulation normally increase growth hormone secretion in vivo; on the other hand, adrenergic and dopaminergic agonists have been reported to suppress growth hormone release from normal pituitary tissue in vitro (18). In contrast to its effect on growth hormone secretion, dopaminergic stimulation suppresses prolactin secretion in vivo, at least in part owing to a direct effect on pituitary lactotrophs. Dopaminergic agonists suppress prolactin release from normal human pituitary tissue in vitro (18). Indeed, it is thought that dopamine is the hypothalamic prolactin-inhibiting factor that tonically restrains pituitary prolactin secretion.

Adrenergic regulation of growth hormone secretion is qualitatively normal in patients with chronic hypersecretion of growth hormone resulting in clinical acromegaly (19). In contrast, the growth hormone response to L-dopa administration is qualitatively abnormal—suppressive rather than provocative—in most patients with acromegaly. This pattern is not altered by adrenergic blockade (20). Basal and posturally stimulated plasma norepinephrine and epinephrine concentrations are normal in acromegalic patients (21). Although plasma norepinephrine, epinephrine, and dopamine were reported to be less responsive to suppression by bromocriptine (22), others found no suppression in either patients or controls (23). When studied in vitro, pituitary somatotrophs from different patients with acromegaly exhibited in-creased, decreased, or unaltered growth hormone release in response to dopaminergic agonists (18). Thus, although there are altered growth hormone response patterns to catecholamines in acromegaly, it remains to be established that these play a role in the pathogenesis of chronic growth hormone hypersecretion.

Excessive prolactin secretion from pituitary prolactinomas is almost always responsive to suppression by dopaminergic agonists, including bromocriptine. Sensitivity to prolactin suppression by dopamine in vivo has been correlated with that to suppression of prolactin release from prolactinoma tissue from the same patient in vitro (18). Thus, it could be that defective dopamine receptors or postreceptor mechanisms might play a role in the pathogenesis of hyperprolactinemia in some patients. Clearly, this remains to be determined.

Rats bearing prolactin-secreting tumors have been found to have evidence of beta-adrenergic hyporesponsiveness (24). They exhibit attenuated isoproterenol-induced thirst and chronotropic and tail skin temperature responses to isoproterenol. The mechanism(s) of these changes is unknown.

IV. THYROID HORMONES

There is a large literature on the effects of thyroid hormones on adrenergic receptors and responsiveness to adrenergic agonists in animals. As it relates to adrenergic receptors studied by ligand binding, that literature has been reviewed in detail by Bilezikian and Loeb (25). I would draw several conclusions from the existing data: (1) Studies in animals indicate clearly that thyroid hormones can affect adrenergic receptors and responsiveness to catecholamines but these effects are not uniform from tissue to tissue. For example, in myocardium hyperthyroidism results in an increase in beta-adrenergic receptor density, maximum agonist-induced adenylate cyclase stimulation and cyclic adenosine monophosphate (AMP) accumulation as well as contractile sensitivity to agonist, whereas hypothyroidism results in decreased beta-adrenergic receptor density, adenylate cyclase activity, and inotropic and chronotropic responses to agonist (25). In contrast, in liver (26), the changes in beta-adrenergic receptors, adenylate cyclase activity, cyclic AMP accumulation, and physiological responses in hyperthyroidism are the opposite of those in the myocardium. (2) Thyroid hormone effects do not result in simple reciprocal changes in alpha- and beta-adrenergic receptors. For example, although reciprocal changes in myocardial alpha- and beta-adrenergic receptors occur in experimental hyperthyroidism, both alpha- and beta-adrenergic receptors are reported to be decreased in the myocardium of hypothyroid animals (25). (3) As emphasized by Bilezikian and Loeb (25), although thyroid hormone-related changes in adrenergic receptor numbers are commonly reflected in corresponding changes in sensitivity to adrenergic agonists, this is not invariably true. Thus, thyroid hormones affect catecholamine action at sites in addition to the receptor.

Many of the clinical manifestations of thyrotoxicosis, e.g., increased myocardial contractility, tachycardia, enhanced thermogenesis, diaphoresis, tremor, lid lag, and hyperkinesis, resemble those produced by catecholamines. Since catecholamine release is not increased in thyrotoxic humans (6, 27-31), increased sensitivity to catecholamines would plausibly explain these similarities. Although, in my judgment, the evidence that thyrotoxic humans have increased sensitivity to catecholamines is not compelling, there is considerably evidence of such hypersensitivity in some animal tissues (reviewed in Refs. 6, 25).

Data concerning the effect of thyroid hormone excess on mononuclear leukocyte (MNL) beta-adrenergic receptors are conflicting. Scarpace and Abrass (32) found that thyroid hormone doses sufficient

to increase myocardial beta-adrenergic receptor density had no effect
on MNL beta-adrenergic receptors in rats. No differences in MNL
beta-adrenergic receptors were found when MNL from thyrotoxic
patients were compared with those from euthyroid controls (33,34).
However, in view of marked intraindividual variation (6), that experimental design does not convincingly exclude an effect. In a double-
blind, placebo-controlled study of normal humans, Ginsberg et al.
(6) found that short-term (1 week) mild triiodothyronine-induced
thyrotoxicosis resulted in increased MNL beta-adrenergic receptor
densities. Mononuclear leukocyte sensitivity to agonist was not examined
in the latter study, but it was examined by Andersson et al. (35).
They found increased MNL beta-adrenergic receptor density and
adenylate cyclase sensitivity to agonist in thyrotoxic patients compared
to those in the same patients studied again after restoration of euthy-
roidism.

To my knowledge, there are no reported studies of platelet alpha-
adrenergic receptors or of tissue adrenergic receptors assayed with
radioligand binding techniques in patients with hyperthyroidism.

Thus, although there is not uniform agreement, the balance of
current evidence favors the view that thyroid hormone-induced increments in beta-adrenergic receptor density explain some of the clinical
manifestations of hyperthyroidism and, therefore, the improvement
in those manifestations during treatment with beta-adrenergic antagonists. It should be noted, however, that the latter is not in itself
compelling evidence of increased sensitivity to catecholamines, since
the drugs have other effects. For example, propranolol is known
to decrease the conversion of thyroxine to triiodothyronine.

If thyroid hormone excess results in increased beta-adrenergic
receptor density, it is conceivable that thyroid hormone deficit results
in decreased beta-adrenergic density, which might explain some of the
clinical manifestations in hypothyroidism. Fat cells from patients with
hypothyroidism have been found to exhibit a decreased lipolytic response to norepinephrine in vitro (36), an effect attributed to enhancement of alpha-adrenergic responsiveness (37). Further, hypothyroid
persons have been shown to have a decreased lipolytic response to
intravenous norepinephrine; thyroxine replacement led to an increased
response (38). Smith et al. (39) found no significant alteration in
MNL adenylate cyclase sensitivity to isoproterenol in samples from
hypothyroid patients. However, Cognini et al. (40) found that compared with preoperative values MNL beta-adrenergic receptor density
decreased significantly shortly after thyroidectomy. This reduction
was associated with a decrease in serum triiodothyronine in those
patients and did not appear to be a nonspecific result of surgery, since
there were no MNL beta-adrenergic receptor changes after cholecystec-
tomy.

V. GLUCOCORTICOIDS AND MINERALOCORTICOIDS

There is evidence that glucocorticoids modify adrenergic receptors in
a variety of tissues, including lung (41-45), heart (46), liver (47,48),
muscle (49), fat cells (50,51), and leukocytes (52-55) in vivo. Glucocorticoid administration results in increased $beta_1$-adrenergic receptor
density in rat lung (41,42) and fetal rabbit lung (43), whereas
receptor density is decreased by adrenalectomy (41), an effect
reversed by exogenous hydrocortisone (41). Similarly, glucocorticoids
increase beta-adrenergic receptor density in explants of fetal rabbit
lung (44), and increase the rate of synthesis of beta-adrenergic
receptors in cultured human lung cells (45). Scarpace and Abrass
(42) studied the effects of administration of the glucocorticoid methylprednisolone in rats whose pulmonary beta-adrenergic receptors were
down-regulated by prior agonist administration. They found that
the glucocorticoid increased beta-adrenergic receptor number and
isoproterenol-stimulated adenylate cyclase activity. Glucocorticoid
administration to adrenalectomized rats has been found to prevent a
decrease in myocardial beta-adrenergic receptors (46). Adrenalectomy
results in an increase in hepatic beta-adrenergic receptors, but not
alpha-adrenergic receptors, in rats (47,48). In contrast, glucocorticoids inhibited myogenesis and isoproterenol-stimulated adenylate
cyclase, and decreased beta-adrenergic receptor density in cultured
muscle cells (49). Adrenalectomy results in decreased fat cell beta-
adrenergic receptors and maximum adenylate cyclase response to
isoproterenol in rats (50). Interestingly, dexamethasone promotes
the loss of $beta_1$-adrenergic receptors and the appearance of $beta_2$-
adrenergic receptors in cultured 3T3-L1 adipocytes (51).
In humans, cortisone acetate administration has been reported
to induce a prompt rise in granulocyte beta-adrenergic receptors and
adenylate cyclase activity, and a fall in lymphocyte beta-adrenergic
receptor density (52). Cortisone acetate in vivo, and hydrocortisone
in vitro, resulted in an increase in the proportion of granulocyte beta-
adrenergic receptors in the high-affinity state (53). In other studies,
methylprednisolone alone had no effect on lymphocyte beta-adrenergic
receptors, but it reversed the down-regulation produced by administration of the agonist terbutaline (54). In in vitro studies of human
neutrophils, an isoproterenol-induced decrease in adenylate cyclase
responsiveness was found to be attenuated by glucocorticoids, an
effect attributed to attenuation of receptor-cyclase uncoupling by
isoproterenol rather than alteration of the isoproterenol induced decrease in beta-adrenergic receptor density (55).
Deoxycorticosterone, a mineralocorticoid hormone, has been
reported to cause a decrease in cardiac (56) and mesenteric arterial
(57) beta-adrenergic receptors.

I am not aware of data concerning the direct assessment of adrenergic receptors in humans with endogenous overproduction or underproduction of adrenocortical hormones.

VI. SEX STEROIDS

There is considerable evidence that sex steroids alter adrenergic receptors and responsiveness to catecholamines in a variety of animal tissues, including uterus (58-63), ovary (64-66), brain (67-69), lung (70), kidney (71), vasculature (72), and platelets (73-74) in vivo. Progesterone has been reported to decrease alpha-adrenergic receptor density in rabbit myometrium (58), although that finding is potentially attributable to estrogen withdrawal (62). Estrogens increase alpha-adrenergic receptor density in rabbit (61,62) and rat (63) myometrium; this is the result of an increase in $alpha_2$-adrenergic receptors (61,62). These estrogen-induced changes are paralleled by increased sensitivity to alpha-adrenergic stimulation of uterine contraction (62). In contrast, neither changes in beta-adrenergic receptors nor in the ratio of beta- to alpha-adrenergic receptors appear to explain adequately the uterine beta-adrenergic predominance that occurs when estrogen is followed by progesterone administration (62). Myometrial beta-adrenergic receptor density is increased during proestrus and estrus in rats (59); it is reduced by castration, an effect reversed by estradiol administration. The estrogen-induced increase appears to be in $beta_1$-adrenergic receptors (60). Ovarian beta-adrenergic receptor density varies during the estrus cycle in rats; it is highest during proestrus and decreases on the morning of estrus (64). Ovarian $beta_2$-adrenergic receptors mediate progesterone secretion (65,66) and begin to do so at puberty in that species. Estrogen administration to ovariectomized rats has been reported to decrease cyclic AMP accumulation in response to isoproterenol, and to decrease beta-adrenergic receptor density in the cerebral cortex of rats (67); these results are consistent with the finding of decreased adenylate cyclase responsiveness and beta-adrenergic receptor density in cortical membranes of female compared with male rats, differences eliminated by ovariectomy (68). Testosterone has also been reported to decrease rat cerebral cortical beta-adrenergic receptors as well as those in the hypothalamus and the pineal gland (69). Diethylstibestrol increases, and progesterone decreases, beta-adrenergic receptor density in rabbit lung (70), and testosterone increases beta-adrenergic receptors in mouse kidney (71). Estrogen treatment of male rats was found to lead to increased affinity of vascular $alpha_1$-adrenergic receptors for epinephrine (72). Lastly, estrogen administration to rabbits decreases platelet aggregation and platelet $alpha_2$-adrenergic receptors (73, 74). Notably, as pointed out earlier, estrogens produce the opposite change, an increase, in rabbit myometrial alpha$_2$-adrenergic receptors (61, 62).

Finally, in humans, platelet alpha$_2$-adrenergic receptors, measured in an unstated number of women through the menstrual cycle, appeared to decrease with increasing plasma estradiol levels (75). A small decrease in platelet alpha$_2$-adrenergic receptor density 7-10 days postpartum has also been reported (76).

We (7) have examined the effects of physiological variations of testosterone, estradiol, and progesterone on adrenergic receptors and their linked adenylate cyclase systems by sampling from normal men and women; in the latter, in both the follicular and luteal phases of their menstrual cycles. As expected, testosterone levels were 10-fold higher in men, estradiol levels were lowest in men and highest in luteal-phase women. Despite these physiologically disparate sex steroid levels, there were no differences in mononuclear leukocyte beta$_2$-adrenergic receptor density or affinity for the antagonist ligand or adenylate cyclase sensitivity to stimulation by isoproterenol. In addition, there were no differences in platelet alpha$_2$-adrenergic receptor density or affinity for the ligand or adenylate cyclase sensitivity to inhibition by epinephrine. Thus, physiological variations of sex steroids do not appear to have a generalized effect on human adrenergic receptors or their linked adenylate cyclase systems. To the extent that these measurements on circulating cells reflect adrenergic receptors and adenylate cyclase systems in extravascular target cells, the data provide no support for the concept that physiological variations in testosterone, estradiol, and progesterone modulate catecholamine action.

To my knowledge, there is no published information concerning the effects of pathological alterations of sex steroid production on adrenergic receptors in humans.

VII. INSULIN

The effect of diabetes, produced by streptozotocin administration to rats, on adrenergic receptors has been studied by several investigators. Streptozotocin diabetes has been reported to be associated with decreased myocardial beta-adrenergic receptor density (77, 78), although neither basal nor isoproterenol-stimulated adenylate cyclase activity was altered (78). Further, the use of rather high concentration of ligand and of propranolol to define nonspecific binding suggests that much of the measured specific binding was nonreceptor binding. Increased sensitivity of aorta from diabetic rats to contraction by norepinephrine has been reported (79).

Fat cells from streptozotocin diabetic rats have been shown to exhibit increased sensitivity to the lipolytic, adenylate cyclase-stimulating and protein kinase-stimulating effects of epinephrine (80). However, decreased beta-adrenergic receptor densities and basal and maximal isoproterenol-stimulated adenylate cyclase activities in fat cells from such animals have also been reported (81). These discrepancies are in part explicable on the basis of expression of the membrane data (adenylate cyclase and beta-adrenergic receptors) per cell in the latter study and per mg protein in the former study. Lastly, evidence that lipolysis is accelerated and that low K_m phosphodiesterase activity is partially inhibited in adipose tissue from human diabetics has been reported (82).

Patients with diabetes have been found to have increased pressor sensitivity to infused norepinephrine (and angiotensin) (83). The mechanisms of this observation are not known, although the possibilities that it is the result of excess body sodium, structural alterations in the vasculature, or both have been raised (83).

Patients with insulin-dependent diabetes mellitus (IDDM) have been found to have a heightened hyperglycemic response to epinephrine (84). However, their mononuclear leukocyte $beta_2$-adrenergic receptors and adenylate cyclase sensitivity to isoproterenol do not differ from those of nondiabetic controls (85). Thus, there is not a generalized alteration of beta-adrenergic receptors or adenylate cyclase sensitivity in IDDM. To the extent that mononuclear leukocyte beta-adrenergic receptors and adenylate cyclase activities reflect those of extravascular catecholamine target cells, these findings suggest that the heightened hyperglycemic response to epinephrine in vivo is not due to increased sensitivity of cellular $beta_2$-adrenergic receptor-effector systems per se, and is best attributed to the altered hormonal milieu of the insulin-deficient state.

VIII. PARATHYROID HORMONE

More than 40 years ago, Albright (86) recognized pseudohypoparathyroidism as being the first human disorder caused by resistance to the action of a hormone. Although the clinical manifestations are generally attributable to resistance to the action of parathyroid hormone (PTH), partial (and usually subclinical) resistance to the actions of many hormones—including thyrotropin, gonadotropins, glucagon (plasma cyclic AMP but not glucose responses), vasopressin, and thyrotropin-releasing hormone (prolactin secretion) as well as parathyroid hormone—is now known to occur in such patients (87). Thus, pseudohypoparathyroidism has been characterized as a ". . . genotypically diverse group of syndromes of primary resistance to hormones whose actions are mediated by cyclic AMP" (87).

Patients with pseudohypoparathyroidism, type I, have markedly reduced urinary cyclic AMP and phosphaturic responses to infused parathyroid hormone (88), but normal phosphaturic responses to a cyclic AMP analog (89), indicating defective parathyroid hormone receptor-adenylate cyclase coupling. Other patients exhibit a decreased phosphaturic response, but a normal urinary cyclic AMP response to parathyroid hormone, and are said to have pseudohypoparathyroidism, type II (90). Many patients with pseudohypoparathyroidism, type I, have reduced guanine nucleotide regulatory protein (N-protein) activity in erythrocytes, platelets, fibroblasts, and virus-transformed lymphoblasts (87) as well as renal membranes (91). These patients are classified as pseudohypoparathyroidism, type Ia, whereas those with normal N-protein activities are termed type Ib (87). It is the subtype Ia that correlates with the clinical phenotype termed Albright's hereditary osteodystrophy—short stature, obesity, and short metacarpals (87). Hypocalcemia is attributable to deficient parathyroid hormone stimulation of 1, 25-dihydroxyvitamin D formation. Resistance to the actions of other hormones is most often demonstrable in subtype Ia patients, but it has also been detected in subtype Ib patients (87). Platelets of patients with subtype Ia have reduced adenylate cyclase-stimulating N-protein activity but normal cyclase-inhibiting N-protein, evidence that the two N-proteins are distinct entities (92).

Carlson and Brickman (93) demonstrated blunted plasma cyclic AMP responses to isoproterenol both in patients with pseudohypoparathyroidism, type Ia, and in those with type Ib. These findings indicate that deficient N-protein activity is not the only mechanism of defective coupling of beta-adrenergic receptors to adenylate cyclase in pseudohypoparathyroidism. It should be noted, however, that the cardiovascular and plasma glucose responses to isoproterenol were normal in these patients. The blunted plasma cyclic AMP responses to isoproterenol in pseudohypoparathyroidism are not likely the result of agonist-induced desensitization, since endogenous plasma catecholamine levels have been found to be normal or low (94). Heterologous desensitization of beta-adrenergic receptors by PTH remains a theoretical possibility. To my knowledge, beta-adrenergic receptors per se have not been studied in patients with pseudohypoparathyroidism.

IX. CONCLUSIONS

As is apparent from the foregoing, available evidence concerning the physiological effects of hormones on human adrenergic receptors and responsiveness to catecholamines is fragmentary. A role of altered adrenergic mechanisms in the pathophysiology of human endocrine-metabolic diseases remains to be firmly established. Although

patients with sympathetic hypofunction (not an endocrine-metabolic
disorder) have increased $beta_2$- and $alpha_2$-adrenergic receptors
on circulating cells, patients with pheochromocytoma have not
been shown to have down-regulated adrenergic receptors (although
selective down-regulation occurs in rats with such tumors).
Altered catecholamine action has not been shown convincingly to
be involved in the pathogenesis of abnormal growth hormone or
prolactin secretion. There is evidence that thyroid hormone-
induced increments in beta-adrenergic receptor density explain
some of the clinical manifestations of hyperthyroidism; decreased
beta-adrenergic receptor function could explain some of those
of hypothyroidism. The possibility that steroid-induced alterations
of adrenergic receptors might occur has not been studied in patients
with abnormal adrenocortical or gonadal hormone secretion. Available
evidence does not support the presence of abnormal adrenergic receptors per se in patients with diabetes. Diminished plasma cyclic
AMP responses have been demonstrated in patients with pseudohypoparathyroidism, but the clinical relevance, if any, of this finding
remains to be established.

In my judgment, there are substantial limitations to assessment
of adrenergic status by measurement of adrenergic receptors on circulating cells, the common approach in humans. First, catecholamine-
receptor interaction represents only one step in a series of steps in
catecholamine action. Thus, receptor measurements should be coupled
with measurements of a later step, such as adenylate cyclase sensitivity
for those receptors linked to adenylate cyclase, and preferably with
a biological response of the cell under study. Second, circulating
cells permit assessment of $beta_2$- and $alpha_2$-adrenergic receptors only;
they do not permit study of $beta_1$- or $alpha_1$-adrenergic receptors.
Third, the assumptions that all regulated changes in beta- and alpha-
adrenergic receptors are coordinate and the same in all tissues, and
that those in extravascular catecholamine target cells are faithfully
reflected by those in circulating cells have not been established rigorously. Although one can cite a substantial body of circumstantial
evidence that these assumptions are generally valid, there are examples
of dissociation of changes on different tissues as cited earlier in this
chapter. Changes in adrenergic receptors need not be coordinate or
reflected in circulating cells: Thyroid hormone excess results in increased myocardial (25) but decreased hepatic beta-adrenergic receptor
densities in rats (26); catecholamine excess results in decreased $beta_1$-
and $alpha_1$-adrenergic receptor densities in several tissues but no
change in $beta_2$- or $alpha_2$-adrenergic receptors in tissues or platelets
in rats (10,11,17); estrogens decrease platelet alpha-adrenergic
receptor density but increase myometrial alpha-adrenergic receptor
density in rabbits (73,74).

Thus, creative approaches to the study of adrenergic receptors and sensitivity to catecholamines in humans are needed. The potential roles of alterations of catecholamine action in human endocrine-metabolic physiology and in the pathogenesis and pathophysiology of human endocrine-metabolic diseases may represent a fertile area of future study.

REFERENCES

1. Tohmeh, J. F., and Cryer, P. E.: J. Clin. Invest. 65:836-840, 1980.

2. Galant, S. P., Duriseti, L., Underwood, S., and Insel, P. A.: N. Engl. J. Med. 299:933-936, 1978.

3. Fraser, J., Nadeau, J., Robertson, D., and Wood, A. J. J.: J. Clin. Invest. 67:1777-1784, 1981.

4. Wood, A. J. J., Feldman, R., and Nadeau, J.: Clin. Exp. Hypertens. A4:807-817, 1982.

5. Aarons, R. D., and Molinoff, P. B.: J. Pharmacol. Exp. Ther. 221:439-443, 1982.

6. Ginsberg, A. M., Clutter, W. E., Shah, S. D., and Cryer, P. E.: J. Clin. Invest. 67:1785-1791, 1981.

7. Rosen, S. G., Berk, M. A., Popp, D. A., Serusclat, P., Smith, E. B., Shah, S. D., Ginsberg, A. M., Clutter, W. E., and Cryer, P. E.: J. Clin. Endocrinol. Metab. (in press).

8. Sowers, J. R., Connelly-Fittinghoff, M., Tuck, M. L., and Krall, J. F.: Cardiovasc. Res. 17:184-188, 1983.

9. Feldman, R. D., Limbird, L. E., Nadeau, J., FitzGerald, G. A., Robertson, D., and Wood, A. J. J.: J. Clin. Invest. 72:164-170, 1983.

10. Snavely, M. D., Motulsky, H. J., O'Connor, D. T., Ziegler, M. G., and Insel, P. A.: Clin. Exp. Hypertens. A4:829-848, 1982.

11. Snavely, M. D., Mahan, L. C., O'Connor, D. T., and Insel, P. A.: Clin. Res. 31:254A, 1983.

12. Hui, K. K. P., and Conolly, M. E.: N. Engl. J. Med. 304:1473-1476, 1981.

13. Davies, B., Sudera, D., Mathias, C., Bannister, R., and Sever, P.: N. Engl. J. Med. 305:1017-1019, 1981.

14. Davies, B., Sudera, D., Sagnella, G., Marchesi-Saviotti, E., Mathias, C., Bannister, R., and Sever, P.: J. Clin. Invest. 69: 779-784, 1982.

15. Chobanian, A. V., Tifft, C. P., Sackel, H., and Pitruzella, A.: Clin. Exp. Hypertens. A4:793-806, 1982.

16. Smith, U., Sjostrom, L., Stenstrom, G., Isaksson, O., and Jacobsson, B.: Eur. J. Clin. Invest. 7:355-361, 1977.

17. Snavely, M. D., Motulsky, H. J., Moustafa, E., Mahan, L. C., and Insel, P. A.: Circ. Res. 51:504-513, 1982.

18. Tallo, D., and Malarkey, W. B.: J. Clin. Endocrinol. Metab. 53: 1278, 1981.

19. Cryer, P. E., and Daughaday, W. H.: J. Clin. Endocrinol. Metab. 39:658, 1974.

20. Cryer, P. E., and Daughaday, W. H.: J. Clin. Endocrinol. Metab. 44:977, 1977.

21. Cryer, P. E., J. Clin. Endocrinol. Metab. 41:542, 1975.

22. Van Loon, G. R.: J. Clin. Endocrinol. Metab. 48:784, 1979.

23. Bybee, D. E., Wiesen, M., Aronin, N., Krieger, D. T., Frohman, L. A., and Kopin, I. J.: J. Clin. Endocrinol. Metab. 54:648, 1982.

24. Katovich, M. J.: Life Sci. 32:1213, 1983.

25. Bilezikian, J. P., and Loeb, J. N.: Endocr. Rev. 4:378, 1983.

26. Malbon, C. C., and Greenberg, M. L.: J. Clin. Invest. 69:414-426, 1982.

27. Bayliss, R. I. S., and Edwards, O. M.: J. Endocrinol. 49:167, 1971.

28. Christensen, N. J.: Clin. Sci. Molec. Med. 45:163, 1973.

29. Coulombe, P., Dussault, J. H., and Walker, P.: Clin. Endocrinol. Metab. 25:973, 1976.

30. Coulombe, P., Dussault, J. H., LeTarte, J., and Simard, S. J.: J. Clin. Endocrinol. Metab. 42:125, 1976.

31. Coulombe, P., Dussault, J. H., and Walker, P.: J. Clin. Endocrinol. Metab. 44:1185, 1977.

32. Scarpace, P. J., and Abrass, I. B.: Endocrinology 108:1007, 1981.

33. Williams, R. S., Guthrow, C. E., and Lefkowitz, R. J.: J. Clin. Endocrinol. Metab. 48:503, 1979.

34. Hui, K. K. P., Wolfe, R. N., and Conolly, M. E.: Clin. Pharmacol. Ther. 32:161, 1982.

35. Andersson, R. G. G., Nilsson, O. R., and Kuo, J. F.: J. Clin. Endocrinol. Metab. 56:42, 1983.

36. Rosenquist, U.: Acta Med. Scand. (Suppl.) 532:1, 1972.

37. Grill, V., and Rosenquist, U.: Acta Med. Scand. 194:129, 1973.

38. Rosenquist, U., and Hylander, B.: Life Sci. 30:641, 1982.

39. Smith, B. M., Silas, J. H., and Yates, R. O.: Clin. Pharmacol. Ther. 29:327, 1981.

40. Cognini, G., Piantanelli, L., Paolinelli, E., Orlandoni, P., Pellegrini, A., and Masera, N.: Acta Endocrinol. 103:40, 1983.

41. Mano, K., Akbarzadeh, A., and Townley, R. G.: Life Sci. 25: 1925-1930, 1978.

42. Scarpace, P. J., and Abrass, I. B.: J. Pharmacol. Exp. Ther. 223:327-331, 1982.

43. Cheng, J. B., Goldfien, A., Ballard, P. L., and Roberts, J. M.: Endocrinology 107:1646-1648, 1980.

44. Giannopoulos, G., and Smith, S. K. S.: Life Sci. 31:795-802, 1982.

45. Fraser, C. M., and Venter, J.: Biochem. Biophys. Res. Commun. 94:390-397, 1980.

46. Abrass, I. B., and Scarpace, P. J.: Endocrinology 108:977-980, 1981.

47. Guellaen, G., Yates-Aggerbeck, M., Vauquelin, G., Strosberg, D., and Hanoune, J.: J. Biol. Chem. 253:1114-1120, 1978.

48. Guellaen, G., Aggerbeck, M., Schmelck, P., Barouki, R., and Hanoune, J.: J. Cardiovasc. Pharmacol. 4:546-550, 1982.

49. Smith, T. J., Dana, R., Krichevsky, A., Bilezikian, J. P., and Schonberg, M.: Endocrinology 109:2110-2116, 1981.

50. Thotakura, N. R., de Mazancourt, P., and Guidicelli, Y.: Biochim. Biophys. Acta 717:32-40, 1982.

51. Lai, E., Rosen, O. M., and Rubin, C. S.: J. Biol. Chem. 257: 6691-6696, 1982.

52. Davies, A. O., and Lefkowitz, R. J.: J. Clin. Endocrinol. Metab. 51:599-605, 1980.

53. Davies, A. O., and Lefkowitz, R. J.: J. Clin. Endocrinol. Metab. 53:703-708, 1981.

54. Hui, K. K., Conolly, M. E., and Tashkin, D. P.: Clin. Res. 30: 26A, 1982.

55. Davies, A. O., and Lefkowitz, R. J.: J. Clin. Invest. 71:565-571, 1983.

56. Woodcock, E. A., Funder, J. W., and Johnston, C. I.: Circ. Res. 45:560-565, 1979.

57. Woodcock, E. A., Olsson, C. A., and Johnston, C. I.: Biochem. Pharmacol. 29:1465-1468, 1980.

58. Williams, L. T., and Lefkowitz, R. J.: J. Clin. Invest. 60:815-818, 1977.

59. Krall, J. F., Mori, H., Tuck, M. L., LeShon, S. L., and Korenman, S. G.: Life Sci. 23:1073-1082, 1978.

60. Johansson, S. R. M., Andersson, R. G. G., and Wikberg, J. E.: Pharmacol. Toxicol. 47:252-258, 1980.

61. Hoffman, B. B., Lavin, Y. N., Lefkowitz, R. J., and Ruffolo, R. R., Jr.: J. Pharmacol. Exp. Ther. 219:290-295, 1981.

62. Roberts, J. M., Insel, P. A., and Goldfein, A.: Mol. Pharmacol. 20:52-58, 1981.

63. Kano, T.: Jpn. J. Pharmacol. 32:535-549, 1982.

64. Jordan, A. W.: Biol. Reproduct. 24:245-248, 1981.

65. Adashi, E. Y., Hsueh, A. J. W.: Endocrinology 108:2170-2178, 1981.

66. Aguado, L. I., Petrovic, S. L., and Ojeda, S. R.: Endocrinology 110:1124-1132, 1982.

67. Wagner, H. R., Crutcher, K. A., and Davis, J. N.: Brain Res. 171:147-151, 1979.

68. Wagner, H. R., and Davis, J. N.: Brain Res. 201:235-239, 1980.

69. Vacas, M. I., Lowenstein, P. R., and Cardinali, D. P.: J. Neural Transm. 53:49-57, 1982.

70. Moawad, A. H., River, L. P., and Kilpatrick, S. J.: Am. J. Obstet. Gynecol. 144:608-613, 1982.

71. Petrovic, S. L., Stanic, M. A., Haugland, R. P., and Dowben, R. M.: Biochim. Biophys. Acta 676:329-337, 1981.

72. Colucci, W. S., Gimbrone, M. A., Jr., and Alexander, R. W.: Clin. Res. 28:162A, 1980.

73. Roberts, J. M., Goldfein, R. D., Tsuchiya, A. M., Goldfien, A., and Insel, P. A.: Endocrinology 104:722-728, 1979.

74. Elliott, J. M., Peters, J. R., and Grahame-Smith, D. G.: Eur. J. Pharmacol. 66:21-30, 1980.

75. Brodde, O. E.: J. Recept. Res. 3:151-162, 1983.

76. Metz, A., Cowen, P. J., Gelder, M. G., Stump, K., Elliott, J. M., and Grahame-Smith, D. G.: Lancet 1:495-498, 1983.

77. Savarese, J. J., and Berkowitz, B. A.: Life Sci. 25:2075, 1979.

78. Ingebretsen, C. G., Hawelu-Johnson, C., and Ingebretsen, Jr., W. R.: J. Cardiovasc. Pharmacol. 5:454, 1983.

79. MacLeod, K. M., and McNeill, J. H.: Proc. West. Pharmacol. Soc. 25:245, 1982.

80. Zapf, J., Waldvogel, M., Zumstein, P., and Froesch, E. R.: F.E.B.S. Lett. 94:43, 1978.

81. Lacasa, D., Agli, B., and Giudicelli, Y.: Eur. J. Biochem. 130: 457, 1983.

82. Engfeldt, P., Arner, P., Bolinder, J., and Ostman, J.: J. Clin. Endocrinol. Metab. 55:983, 1982.

83. Beretta-Piccoli, C., and Weidmann, P.: Am. J. Med. 71:829, 1981.

84. Shamoon, H., Hendler, R., and Sherwin, R. S.: Diabetes 29: 284, 1980.

85. Serusclat, P., Rosen, S. G., Smith, E. B., Shah, S. D., Clutter, W. E., and Cryer, P. E.: Diabetes 32:825, 1983.

86. Abright, F., Burnett, C. H., Smith, P. H., and Parsons, W.: Endocrinology 30:922, 1942.

87. van Dop, C., and Bourne, H. R.: Ann. Rev. Med. 34:259, 1983.

88. Chase, L. R., Melson, G. L., and Aurbach, G. D.: J. Clin. Invest. 48:1832, 1969.

89. Bell, N. H., Avery, S., Sinha, T., Clark, C. M., Jr., Allen, D. O., and Johnston, C., Jr.: J. Clin. Invest. 51:816, 1972.

90. Drezner, M., Neelon, F. A., and Lebovitz, H. E.: N. Engl. J. Med. 289:1056, 1973.

91. Downs, R. W., Jr., Levine, M. A., Drezner, M. K., Burch, W. M., Jr., and Spiegel, A. M.: J. Clin. Invest. 71:231, 1983.

92. Motulsky, H. J., Hughes, R. J., Brickman, A. S., Farfel, Z., Bourne, H. R., and Insel, P. A.: Proc. Natl. Acad. Sci. U.S.A. 79:4193, 1982.

93. Carlson, H. E., and Brickman, A. S.: J. Clin. Endocrinol. Metab. 56:1323, 1983.

94. Brickman, A., Sowers, J., Golub, M., Asp, N., Berg, G., and Pettis, R.: Clin. Res. 29:82A, 1981.

11

Activation of Human Platelets by Epinephrine

SANFORD J. SHATTIL   *Hospital of the University of Pennsylvania,
Philadelphia, Pennsylvania*

I. INTRODUCTION

Platelets normally circulate in the bloodstream in a nonactivated state
such that they fail to interact with each other or with the vessel wall.
In contrast, when the endothelial lining of blood vessels becomes damaged, platelets in the area of injury become activated. This results
in their adherence to the vessel wall, their aggregation to each other,
and their undergoing a secretory response (1,2). These responses
of adhesion, aggregation, and secretion culminate in the formation of
a platelet plug. On the one hand, this plug is a necessary component
of the body's hemostatic response to traumatic damage of the vessel
wall. On the other hand, platelet plug formation at endothelial surfaces
damaged by atherosclerosis or inflammation can result in both thrombo-
embolism and accelerated atherosclerosis (3,4). Thus, the activation
of platelets is of major physiological and pathophysiological importance.

A number of substances endogenous to man are capable of activating platelets in vitro. Some or all of these may act as agonists in vivo.
These agonists are chemically diverse and include proteins (thrombin,
collagen), biogenic amines (epinephrine, serotonin), a nucleotide
(adenosine diphosphate [ADP]), and a phospholipid (platelet-activating
factor) (1,2,5,6). The purpose of this chapter is to summarize our
current understanding of the mechanism whereby one of these agonists,
epinephrine, activates human platelets. I will discuss (1) the array
of functional responses of platelets to agonists; (2) the initial activation
event, the binding of epinephrine to its platelet membrane receptor;
(3) the biochemical reactions within the platelet that link epinephrine

binding to the final platelet responses of aggregation and secretion;
(4) the role of cyclic adenosine monophosphate [AMP] in platelet
function; (5) whether epinephrine is a physiologically relevant platelet
agonist; and (6) the clinical conditions associated with abnormal platelet
responses to epinephrine.

It should be emphasized that virtually nothing is known about
the interaction of catecholamines and platelets in vivo. Therefore,
the bulk of the information to be discussed is derived from in vitro
studies. Much of our understanding of platelet activation has been
obtained from the study of other agonists, in particular, thrombin,
collagen, and ADP. As a result, reference to these other agonists
will be frequent. However, there are qualitative and quantitative
differences among the agonists with respect to the platelet responses
they evoke and these differences will also be stressed.

II. PLATELET RESPONSES TO AGONISTS

Epinephrine is capable of stimulating certain platelet responses, such
as aggregation and secretion. However, it does not directly stimulate
or participate in other platelet responses, such as adhesion, shape
change, or the development of platelet procoagulant activity. All of
these platelet responses originate at the platelet surface membrane,
but they also involve to a variable degree intracellular membranes,
cytoplasmic proteins, and other organelles within the cell (Fig. 1).

The *adhesion* of platelets to subendothelial components of the
damaged vessel wall is dependent upon the presence of a multimeric
protein, von Willebrand protein, particularly at the high shear rates
present in capillaries (7,8). Von Willebrand protein is synthesized
in endothelial cells and megakaryocytes and is normally present in
plasma, platelet α granules, and in the vessel wall just abluminal to
the endothelium (8-11). Adhesion presumably requires the binding
of von Willebrand protein to a 150,000 mol. wt. glycoprotein on the
platelet surface membrane known as Ib (12). This binding does not
require metabolically active platelets, and it can be demonstrated in
vitro on the addition of an antibiotic, ristocetin, to formalin-fixed
platelets (13). Platelets from individuals with the Bernard-Soulier
syndrome lack glycoprotein Ib, do not bind von Willebrand protein
in response to ristocetin, and adhere poorly to the vessel wall (14,15).
The in vivo counterpart of ristocetin has not been identified. In
addition, the specific components of the vessel wall to which platelets
(and von Willebrand protein) adhere have not been completely characterized, although collagen is one of these (16,17). In response to
thrombin, ADP, or ADP plus epinephrine, von Willebrand protein also
has been shown to bind to a different platelet membrane glycoprotein

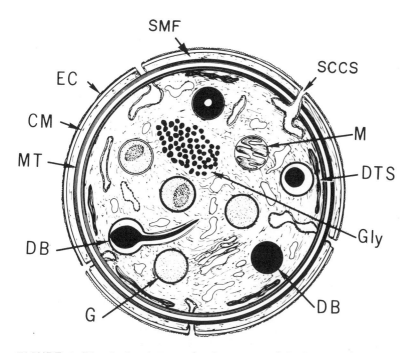

FIGURE 1 Blood platelet cut in the equatorial plane as it appears in thin sections by electron microscopy. EC, electron-dense exterior coat; CM, surface membrane; SCCS, surface-connected open canalicular system; DTS, dense tubular system; M, mitochondria; G, α granules; DB, dense granules; MT, circumferential band of microtubules; Gly, glycogen; SMF, submembrane microfilaments. (From Ref. 103, used with permission.)

complex, known as IIb-IIIa (18-20). Since this latter binding is inhibited by levels of fibrinogen normally found in plasma, its physiological role is unknown. There is no known role for epinephrine alone, either in the binding of von Willebrand protein to platelets or in platelet adhesion.

In contrast to their undefined role in platelet adhesion, agonists such as thrombin, collagen, ADP, and epinephrine are capable of stimulating platelet *aggregation* and *secretion*. These platelet responses require metabolic energy (21). However, there are several important qualitative and quantitative differences among the agonists with respect to the pattern of responses they evoke. First, both ADP and thrombin trigger an immediate change in the shape of platelets, from smooth discs to spheres with pseudopodial projections. This shape change

is believed to result from the polymerization of actin filaments and
their interaction with the cytoplasmic surface of the plasma membrane.
Local changes in cytoplasmic free Ca^{2+} and the phosphorylation of the
light chain of myosin may be involved in this process (22,23). Shape
change is not observed in response to epinephrine. Second, platelet
aggregation begins immediately on the addition of epinephrine, ADP,
or thrombin to mechanically agitated platelet suspensions. At low
concentrations, these agonists stimulate either aggregation alone
("primary aggregation") or primary aggregation followed by secretion
and more aggregation ("secondary aggregation") (Fig. 2). In contrast,
aggregation in response to collagen is delayed and begins coincident
with platelet secretion (24). Third, ADP, epinephrine, and low concentrations of thrombin or collagen stimulate secretion of the contents
of platelet δ granules (ADP, serotonin, calcium) and α granules
(fibrinogen, thrombospondin, fibronectin, von Willebrand protein,
coagulation factor V, platelet-derived growth factor, platelet factor 4,
β-thromboglobulin) (25,26). Large concentrations of thrombin or
collagen also cause secretion of hydrolytic enzymes from a third class
of platelet granules (λ granules). Platelets also are capable of taking
up, storing, and secreting catecholamines, but it is not known if
this pool of catecholamines plays a role in platelet function (27).
Fourth, the platelet secretory response to ADP and epinephrine is
completely dependent upon primary platelet aggregation and prostaglandin synthesis, whereas high concentrations of thrombin and
collagen are capable of stimulating platelet secretion independent of
aggregation and prostaglandin synthesis (24,28,29). These differences
in the aggregation and secretion responses to agonists suggest that
the mechanisms of stimulus-response coupling within the platelet are
not the same for all agonists. As a result, specific details of the
mechanism of platelet activation by one agonist do not necessarily
pertain to other agonists.

In addition to adhesion, aggregation, and secretion, platelets
contribute a *procoagulant surface* upon which coagulation proteins
are localized and thrombin generation is facilitated. The surface of
nonactivated platelets contains a specific binding site for coagulation
factor Va, and platelet-bound Va serves as a binding site for coagulation factor Xa (30). Platelet activation by thrombin facilitates the
procoagulant function of platelets in part because it releases coagulation factor V from platelet α granules and converts it to Va (31). The
anionic phospholipids of the platelet membrane are capable of providing
a procoagulant surface (32). Whether other components of the surface
membrane participate in procoagulant activity remains to be determined.
There is no known role for epinephrine in the procoagulant function
of platelets.

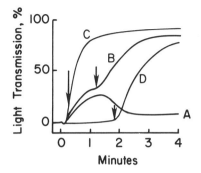

FIGURE 2 Typical spectrophotometric traces of stirred platelet-rich plasma at 37°C. The platelets were agitated with a magnetic stir bar, and each agonist was added at time "0". The vertical arrow above each trace indicates when the products of platelet secretion, such as ADP and serotonin, were first detected outside of the platelets. Upward deflection of the trace indicates an increase in light transmittance through the platelet-rich plasma. On addition of most agonists (but not epinephrine), there is an initial small decrease in light transmittance due to shape change. This is followed by a progressive increase in light transmittance as platelets aggregate. If the concentration of ADP, epinephrine, or thrombin is low, aggregation is partial and reversible and is not accompanied by platelet secretion (A). This is often referred to as "primary" aggregation. With intermediate concentrations of ADP or thrombin and with high concentrations of epinephrine, an additional and irreversible "secondary" wave of aggregation is seen (B). This is due to the formation of larger aggregates and correlates with prostaglandin synthesis and platelet secretion. Secondary aggregation may be caused directly by prostaglandin endoperoxides, by thromboxane A_2, or by secreted ADP. With high concentrations of thrombin or ADP, a single irreversible wave of aggregation is observed (C). In this case, prostaglandin synthesis and platelet secretion begin earlier than in curve B. In response to ADP, curves B and C and platelet secretion are not observed if prostaglandin synthesis is blocked with aspirin. The secretory response and secondary aggregation to low concentration of thrombin are also blocked by aspirin. However, high concentrations of thrombin can still cause secretion and produce aggregation curve C even in the presence of aspirin. Curve D is typical of collagen. An initial lag phase is followed by a single wave of platelet aggregation that is coincident with platelet secretion. Thus, except for the initial shape change, only curves A and B are typical of those observed with epinephrine.

III. THE PLATELET EPINEPHRINE RECEPTOR

Epinephrine and norepinephrine stimulate platelet aggregation and secretion at concentrations above 0.5 μM, and these responses are maximal at 5-10 μM (33). At lower concentrations (50-100 nM), these agonists are also capable of potentiating aggregation and secretion induced by subthreshold concentrations of unrelated platelet agonists, such as ADP, thrombin, collagen, vasopressin, and platelet-activating factor (33-36). The (-)-enantiomers of epinephrine and norepinephrine are more potent than their respective (+)-enantiomers. α_2-Adrenergic antagonists such as yohimbine are much more potent than α_1 or β antagonists such as prazosin or propranalol in preventing these platelet responses to epinephrine (37-39). Although most other α-adrenergic agonists fail to stimulate aggregation and secretion directly, they do potentiate the platelet responses to unrelated agonists. α_2-Agonists such as clonidine and guanabenz are more potent as potentiators than α_1 agonists (33,39,40). Furthermore, such α_2-"partial" agonists or "mixed agonist-antagonists" inhibit platelet responses to epinephrine (40).

As reviewed in detail in Chapters 6 and 7, epinephrine and norepinephrine are capable of inhibiting basal and hormone (prostaglandins [PGs] E_1, I_2)-stimulated adenylate cyclase activity in platelet lysates and cyclic AMP levels in intact platelets (41-45). These effects are mediated by α_2-adrenergic receptors (46,47). Occupancy of the α_2 receptor by epinephrine triggers an interaction of the receptor with a heterotrimeric, inhibitory guanine nucleotide regulatory protein of the adenylate cyclase complex (G_i) (48,49). Thus, the bulk of the pharmacological data support the proposal that epinephrine causes aggregation and secretion and inhibits adenylate cyclase of human platelets by binding to α_2-adrenergic receptors on the platelet surface. In contrast, the β-adrenergic agonist isoproterenol stimulates a small increase in both adenylate cyclase activity and cyclic AMP levels in the presence of a phosphodiesterase inhibitor (47,50). In addition, isoproterenol does not inhibit agonist-induced aggregation or secretion when added to platelet suspensions, but it does cause weak inhibition when added along with an inhibitor of phosphodiesterase (51,52).

Direct radioligand binding studies using intact platelets and crude platelet membranes also support the proposal that human platelets contain significant numbers of α_2-adrenergic receptors and a very small number of β receptors (53,54). For example, nonselective α-adrenergic antagonists such as [³H]dihydroergocryptine (55,56) and [³H]phentolamine (57) bind specifically and with high affinity (K_d = 1-10 nM) to a single class of binding sites on intact platelets (200-400 sites per platelet). Furthermore, the selective α_2-antagonist [³H]-yohimbine binds with high affinity (K_d = 2 nM) to about 150-300 sites per platelet (53,54,58), whereas nanomolar concentrations of the α_1 antagonist [³H]prazosin fail to bind (59). Competition by unlabeled adrenergic compounds for [³H]yohimbine binding sites demonstrates the stereoselectivity and rank order of potency expected for binding to an α_2-adrenergic receptor. Using a β antagonist, [^{125}I]hydroxy-benzylpindolol, Steer and Atlas detected a very small number of β receptors (24/platelet) on human platelets (60). These receptors are not typical of either the β_1 or β_2 subtype, but they presumably mediate the weak increase in adenylate cyclase activity induced by isoproterenol.

The nonselective α-agonists [^{3}H]epinephrine and [^{3}H]norepinephrine and the selective α_2-agonists [^{3}H]clonidine and [^{3}H]para-aminoclonidine bind to platelet membranes with high affinity (K_d = 1-30 nM). However, these radiolabeled agonists detect only 20-50% the number of binding sites per platelet or per milligram of membrane protein detected by radiolabeled antagonists (40,48,54,61,62). These data, taken together with data obtained from computer modeling of competition experiments with a labeled α antagonist and unlabeled α agonists suggest that the platelet α_2 receptor exists both in high-affinity and low-affinity states with respect to agonists (61,63,64). For example, of the approximately 200 α_2 receptors recognized by [^{3}H]yohimbine, epinephrine recognizes 71% with high affinity (K_d = 11 nM), and the remaining 29% with a lower affinity (K_d = 520 nM). Furthermore, the intrinsic activity of an α agonist in inhibiting adenylate cyclase is proportional to the ratio of the K_ds of the lower and higher affinity states of the receptor for that agonist (64).

Recently, the platelet α_2 receptor has been purified to homogeneity and identified as a 64,000-M_r protein containing an essential sulfhydryl residue (71). Peptide maps of this receptor reveal little, if any, homology between this receptor and the α_1 receptor from smooth muscle cells or the β receptor from lung.

According to the model proposed by Lefkowitz and co-workers, the binding of an α-adrenergic agonist to the high-affinity state of the α_2 receptor and conversion of that receptor to the lower affinity state may be linked to the cellular responses evoked by that agonist (63,64). As summarized in Chapters 6 and 7, this concept gains support from studies using platelet membranes and solubilized and purified α receptors from these membranes. The studies to date suggest that the binding of an α_2 agonist to its platelet receptor stimulates an association between that receptor and G_i, the guanine nucleotide regulatory component of the adenylate cyclase complex (48,49,65). This results in a release of the guanosine diphosphate (GDP) bound to G_i and an exchange of guanosine triphosphate (GTP) for GDP (66). The GTP-liganded G_i is then capable of interacting with and inhibiting the catalytic component of adenylate cyclase. This presumably results from a GTP-induced dissociation of G_i into its α_i, β, and γ subunits. The catalytic component of adenylate cyclase is then inhibited, either directly by α_i or indirectly by the binding of the β subunit to α_s, the α subunit of the stimulatory G protein, G_s (67). GTP also "destabilizes" the receptor, converting its ability to bind agonist from a higher to a lower affinity state. In addition to regulation of agonist binding by GTP, the binding of agonists to the platelet α_2 receptor is increased by Mg^{2+} and decreased by Na^+ (40, 68-70).

The above working model is consistent with most of the pharmacological and ligand binding data that pertain to the platelet α_2-adrenergic receptor. Many questions about the molecular interactions of this receptor with the components of the adenylate cyclase complex and with other constituents of the platelet plasma membrane remain to be answered. The recent availability of α_2-affinity labels and purification and reconstitution of the platelet α_2 receptor should accelerate progress in this area (71,72). Does the agonist-induced link between the receptor and inhibition of adenylate cyclase play any role in platelet aggregation or secretion by epinephrine? Why do epinephrine and norepinephrine directly cause aggregation and secretion, whereas other α agonists only potentiate the functional responses of the platelet to unrelated agonists such as ADP or thrombin?

IV. EPINEPHRINE-INDUCED PLATELET AGGREGATION

Initial or "primary" aggregation begins immediately upon the addition of epinephrine to platelets in plasma (Fig. 2). In order for aggregation to occur, the platelets must be mechanically agitated or stirred in the presence of extracellular fibrinogen and Ca^{2+} and be within the pH range 6.8-8.0 (74,75). The molecular basis for the fibrinogen and Ca^{2+} requirements in the aggregation process is partly understood.

A. The Role of Fibrinogen

Fibrinogen is a symmetric, dimeric glycoprotein with a molecular weight of 340,000 that is present in plasma at a concentration of 2-4 mg/ml ($\sim$6-12 μM). It is also stored within platelet α granules (26). Extracellular fibrinogen does not bind to the surface of unstimulated platelets. However, platelets stimulated with ADP, epinephrine, thrombin, collagen, or platelet-activating factor bind fibrinogen (76-78). Maximal concentrations of ADP stimulate the binding of approximately 40,000-50,000 fibrinogen molecules per platelet with a K_d of approximately 100 nM (76,77). Fibrinogen binding in response to epinephrine is quantitatively either similar to or slightly less than that observed with ADP (76,79,80). Saturation of these receptors is observed at a fibrinogen concentration of approximately 300 nM, about one-twentieth the normal plasma concentration. Fibrinogen binding is rapid and initially reversible, but an increasing proportion of the ligand becomes irreversibly bound to platelets with incubations longer than 10 min (81). The D-domain of fibrinogen, and more specifically the 12 amino acid sequence at the carboxy terminus of the gamma chain of the fibrinogen molecule,

is a major site of interaction of fibrinogen with its platelet receptor
(78,82).

Fibrinogen binds to stimulated platelets whether or not they are
stirred, but the platelets will not aggregate unless the stimulated
platelets are stirred (76). There is compelling evidence that platelet
aggregation requires the binding of fibrinogen. First, the extent of
fibrinogen binding generally correlates directly with the extent of
aggregation. Second, platelets from individuals with Glanzmann's
thrombasthenia fail to bind fibrinogen and do not aggregate after stimulation with agonists (76). Finally, platelets from individuals with hereditary
afibrinogenemia and barely detectable levels of plasma fibrinogen fail
to aggregate in response to ADP or epinephrine (83).

The process of fibrinogen binding to platelets involves two
separate processes: (1) an agonist-induced "exposure" or induction
of the receptor, and (2) fibrinogen binding to the exposed receptor.
Studies using a variety of techniques all suggest that the fibrinogen
receptor is localized to a surface membrane glycoprotein complex.
This heterodimer complex contains the integral membrane glycoproteins
IIb (unreduced molecular weight 136,000 daltons) and IIIa (unreduced
molecular weight 95,000 daltons). When studied after solubilization
in detergents, this complex is held together by Ca^{2+} (84-88). These
two glycoproteins may also exist as heterodimer complexes in unstimulated platelets because murine monoclonal antibodies specific for this
complex (or the Ca^{2+}-dependent conformation of the complex) bind to
40,000-50,000 sites per unstimulated platelet (89). Evidence for this
complex being the fibrinogen receptor includes: (1) Some murine
monoclonal antibodies specific for the complex inhibit both fibrinogen
binding and platelet aggregation induced by ADP, epinephrine, or
thrombin (89-91). (2) Glanzmann's thrombasthenia platelets, which
lack glycoproteins IIb and IIIa, are unable to bind fibrinogen (76,92).
(3) Fibrinogen complexed to a photoreactive, heterobifunctional cross-
linking reagent binds to glycoprotein IIIa on ADP-stimulated platelets
(93).

The events within the platelet plasma membrane that link the
binding of epinephrine at α_2-adrenergic receptors to the exposure of
the fibrinogen receptors are unknown. Moreover, little is known regarding the changes involving IIb/IIIa that "expose" these glycoproteins, thus enabling them to bind fibrinogen. Fibrinogen receptor
exposure requires platelet ATP as a source of metabolic energy (94).
Also, despite the fact that platelet prostaglandins may be synthesized
within the intracellular dense tubular membranes rather than the
plasma membrane of the platelet (95), products of prostaglandin synthesis (prostaglandin endoperoxides and thromboxane A_2) are partly
involved in the process of fibrinogen receptor exposure. Under appropriate experimental conditions, blockade of prostaglandin endoperoxide
synthesis with aspirin or indomethacin decreases the number of fibrinogen receptors on epinephrine-stimulated platelets by over 50% (96).

ADP secreted from platelets may itself induce fibrinogen receptor
exposure. However, experiments using ATP, an ADP receptor antagonist, suggest that fibrinogen binding in response PGH_2 is not solely
the result of secreted ADP (80, 96). The binding of PGH_2 or thromboxane A_2 to their receptors on the platelet can also stimulate fibrinogen
binding. Induction of the fibrinogen receptor may actually involve a con-
formational change in glycoproteins IIb/IIIa (78) or the lateral movement
and clustering of IIb/IIIa complexes within the plane of the membrane
(97). It is conceivable that some of the early changes in membrane
lipid fluidity observed when agonists are added to platelet suspensions
are reflective of changes in lipid-protein interactions during fibrinogen
receptor exposure (98).

B. The Role of Extracellular Ca^{2+}

At least 1 μM extracellular Ca^{2+} is required for induction of the
fibrinogen receptor (78). Also, the requirement for extracellular
Ca^{2+} during platelet aggregation reflects, in part, the requirement
for divalent cations in fibrinogen binding. Fibrinogen binding to
stimulated, gel-filtered platelets is maximal at 0.5 mM Ca^{2+} or 2.5 mM
Mg^{2+} and minimal at divalent cation concentrations below 0.01 mM or
above 5 mM (76). It is not clear whether the Ca^{2+} or Mg^{2+} required
for fibrinogen binding interacts with fibrinogen, with the platelet
receptor, or with both. Furthermore, it is not known if Ca^{2+} forms
a metal ion bridge between fibrinogen and the receptor or whether
it is required to maintain the proper conformation of fibrinogen or
the receptor. Extracellular Ca^{2+} does bind reversibly to fibrinogen
and to glycoproteins IIb/IIIa (99,100). There are approximately two
Ca^{2+} bound per IIb/IIIa complex with a K_d of 5 nM and six Ca^{2+} bound
per complex with a K_d of 400 nM (100). Ca^{2+} bound with high affinity
may participate in the formation of the IIb/IIIa complex within the
plasma membrane (103).

Although platelets stimulated with epinephrine or ADP bind
fibrinogen, they do not aggregate unless mechanically agitated.
Thus, fibrinogen binding is necessary but not sufficient for platelet
aggregation. Apparently the close cell contact resulting from agitation
is necessary for bound fibrinogen to mediate platelet to platelet interaction. Additional factors may be required to convert reversible
aggregation to irreversible aggregation. There is some evidence
that thrombospondin, released from α granules during secretion, is
capable of interacting with fibrinogen and platelets to promote irreversible aggregation (101). Alternatively, products of the lipoxygenase
pathway have been suggested as functioning to promote irreversible
aggregation (102).

V. EPINEPHRINE-INDUCED PLATELET SECRETION

Platelets metabolize glucose by glycolysis and oxidative phosphorylation. They contain subcellular organelles common to many secretory cells, such as storage granules, microfilaments, microtubules, and a dense tubular membrane system that is capable of sequestering Ca^{2+}. Epinephrine stimulates the secretion of substances stored in platelet α granules and δ granules, but it does not stimulate secretion from λ granules (25-27). Secretion is triggered by an increase in cytoplasmic free Ca^{2+} that initiates Ca^{2+}-dependent phosphorylation reactions, a contractile response involving actin and myosin, and the fusion of storage granule membranes to surface or surface-connected membranes. However, the source of the increased cytoplasmic Ca^{2+} and the precise sequence of biochemical events that link the binding of epinephrine to the final secretory response are unclear. Any model for platelet secretion in response to epinephrine must accommodate the following facts: (1) it occurs only if platelets are first stirred and allowed to aggregate, (2) it requires the presence of extracellular free Ca^{2+} (100,104,105), and (3) it requires the production of prostaglandin endoperoxides and/or thromboxane A_2 (28).

A. The Role of Cytoplasmic Free Ca^{2+}

A number of indirect methods have suggested that platelet agonists stimulate an increase in cytoplasmic free Ca^{2+} just prior to platelet secretion (106-109). Recent studies with the intracellular fluorescent Ca^+ indicator, Quin 2, have now demonstrated this directly (110,111). The cytoplasmic free Ca^{2+} concentration in unstimulated platelets is about 100 nM. In the presence of 1 mM external Ca^{2+}, the Ca^{2+} ionophore, ionomycin, causes Ca^{2+} influx and induces platelet shape change at a cytoplasmic free Ca^{2+} concentration of 500 nM. It induces secretion of serotonin at a cytoplasmic free Ca^{2+} concentration of 800 nM and platelet aggregation of 2 μM. In the presence of 1 mM extracellular Ca^{2+}, thrombin also stimulates a rapid rise in cytoplasmic free Ca^{2+} to 3 μM. However, in a Ca^{2+}-free medium, thrombin causes an increase in cytoplasmic free Ca^{2+} to only 300 nM. Despite this minimal rise in Ca^{2+}, thrombin induces both shape change and secretion (110). At 1 mM extracellular Ca^{2+}, epinephrine, collagen, and ADP increase the cytoplasmic free Ca^{2+} of platelets to 300 nM, 300 nM, and 800 nM, respectively. The epinephrine response is inhibited by aspirin (111). With all of these agonists, the increase in cytoplasmic free Ca^{2+} is sufficiently rapid to be causally involved in secretion. However, since the extent of secretion does not generally correlate with the extent of rise in cytoplasmic free Ca^{2+}, factors other than a global rise in Ca^{2+} must regulate the extent of the secretory response.

B. The Source of Cytoplasmic Ca^{2+} Required for Secretion

In the case of high concentrations of thrombin or collagen, where platelet secretion is observed in the virtual absence of extracellular Ca^{2+}, an increase in cytoplasmic Ca^{2+} must be derived from Ca^{2+} ordinarily sequestered within the platelet. The most likely location for the sequestered Ca^{2+} is the dense tubular membranes. Crude vesicles enriched in dense tubular membranes are capable of sequestering Ca^{2+} in the presence of ATP and releasing Ca^{2+} in response to prostaglandin endoperoxides and thromboxane A_2 (112,113). This Ca^{2+} release is inhibited by 13-azaprostanoic acid, a specific endoperoxide/thromboxane A_2 antagonist (114). It should be emphasized that the other membrane systems in platelets, e.g., the surface membrane, the surface-connected open-canalicular membrane, and the mitochondria, have not been rigorously excluded as sources of Ca^{2+} for platelet activation.

In the case of epinephrine and ADP, which require extracellular Ca^{2+} for secretion, the possibility that the increase in cytoplasmic Ca^{2+} results from a net influx of extracellular Ca^{2+} has been examined. Influx of extracellular Ca^{2+} has been linked to both α_1- and α_2-receptor responses in other cells (115). Using $^{45}Ca^{2+}$, Owen and colleagues have suggested that epinephrine, but not ADP, stimulates a rapid influx of $^{45}Ca^{2+}$ into platelets (116,117). However, more recently, a compartmental, kinetic analysis of platelet $^{45}Ca^{2+}$ has demonstrated that most of the $^{45}Ca^{2+}$ associated with platelets is rapidly exchangeable, dissociates rapidly from platelets upon the addition of ethylene glycol titraacetic acid (EGTA) or lanthanides, and is, therefore, presumably surface bound. A small, more slowly exchangeable fraction of platelet $^{45}Ca^{2+}$ is not removable by EGTA or lanthanides, and is presumably intracellular (118). Although both epinephrine and ADP increase the amount of surface-bound Ca^{2+} and the rate of Ca^{2+} exchange into and out of the intracellular Ca^{2+} pool, neither agonist increases the size of the intracellular exchangeable pool (118). These studies demonstrate that an increase in platelet-associated $^{45}Ca^{2+}$ after epinephrine can be accounted for in ways other than a net Ca^{2+} influx. Thus, the proposition that epinephrine stimulates a net flux of Ca^{2+} into platelets must be viewed as likely but unproven. It is unlikely that depolarization of the platelet surface membrane or a change in the conductance of ions other than Ca^{2+} is primarily involved in the mechanism of stimulus-secretion coupling in platelets (119).

Processes other than a net influx of Ca^{2+} could explain the dependence of epinephrine-induced platelet secretion on extracellular Ca^{2+}. First, extracellular Ca^{2+} in the μM range is required for induction of the fibrinogen receptor by epinephrine (78). Second, extracellular Ca^{2+} (or Mg^{2+}) in the 0.05-1.0-mM range is required for fibrinogen binding and platelet aggregation (76). Platelet secretion stimulated by epinephrine does not occur in the absence of this aggregation. Third, removal of Ca^{2+} from high-affinity binding sites ($K_d < \mu M$) on the platelet surface by either EGTA or by the lanthanide gadolinium prevents epinephrine and ADP-induced platelet secretion (100). Thus, association of Ca^{2+} with specific surface membrane components may be required for secretion to occur in response to epinephrine or ADP.

C. Physical and Biochemical Reactions Involved in Platelet Secretion

1. Platelet-Platelet Interaction

It is not known why secretion by epinephrine or ADP depends upon initial platelet aggregation. Platelet-platelet interaction induced by the cationic polymer polylysine and platelet agglutination by ristocetin also stimulate platelet secretion (120,121). Once the surface membrane is perturbed by an agonist, subsequent platelet-to-platelet contact may promote prostaglandin synthesis. In addition, it has been suggested that the initial platelet-platelet contact causes release of ADP from platelets in amounts small enough to go undetected by measurements of extracellular ADP, but large enough to potentiate the secretory response of the initial agonist (120).

2. Phospholipid Hydrolysis

Most of the studies of phospholipid metabolism in activated platelets have used thrombin as the stimulus. Thrombin induces the release of arachidonic acid from platelet phospholipids by stimulating (1) hydrolysis of phosphatidylcholine and phosphatidylethanolamine by membrane-bound phospholipase A_2, and (2) sequential hydrolysis of phosphatidylinositol by cytoplasmic phospholipase C and membrane-bound diglyceride and monoglyceride lipases (113,122,123) (Fig. 3). It is unsettled which of these two sources of arachidonic acid is the initial or major source of arachidonic acid for prostaglandin synthesis.

The product of phospholipase C action, a diacylglyceride, can also be converted by a diglyceride kinase to phosphatidic acid (Fig. 3) (124-126). The rapid but transient *elevation* of both diacylglyceride and phosphatidic acid after thrombin has aroused interest in these compounds as potential endogenous Ca^{2+} ionophores or as membrane fusogens (126,127). Phosphatidic acid can be converted either to lysophosphatidic acid by a phosphatidic acid-specific phospholipase A_2, or back to phosphatidylinositol, completing the "phosphatidylinositol cycle" (126,128) (Fig. 3). The hydrolysis of phosphotidylinositol has been linked to Ca^{2+} gating in some cells. However, in the platelet,

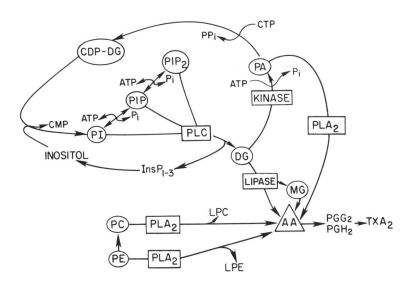

FIGURE 3 Phospholipid metabolism in stimulated platelets. Lipids are
indicated by circles. Enzymes are indicated by squares. See text
for discussion of each pathway. Abbreviations—Lipids: PI, phosphatidylinositol; PIP, phosphatidylinositol 4-phosphate; PIP_2, phosphatidylinositol 4,5-bisphosphate; DG, diacylglyceride; PA, phosphatidic
acid; CDP-DG, cytidine diphosphate diglyceride; MG, monoacylglycerol;
PC, phosphatidylcholine; PE, phosphatidyl ethanolamine; LPC, lyso-
phosphatidylcholine; LPE, lysophosphatidylethanolamine; AA, arachidonic acid. Enzymes: PLC, phospholipase C; PLA_2, phospholipase A_2.
Prostaglandin compounds: PGG_2, prostaglandin G_2; PGH_2, prostaglandin
H_2; TXA_2, thromboxane A_2. InsP, L-myoinositol 1-phosphate; $InsP_2$,
L-myoinositol 4,5-bisphosphate; IP_3, L-myoinositol 1,4,5-triphosphate.
Potential sources of arachidonic acid are: (1) hydrolysis of PI by PI-
specific PLC, (2) hydrolysis of PIP_2 by PI-specific PLC, (3) hydrolysis
of PA by PA-specific PLA_2, and (4) hydrolysis of PC and PE by PLA_2.
The classic "PI cycle" contains PI, DG, PA, and CDP-DG. PE can be
sequentially N-methylated to form PC via the "methylation pathway."

activation of this phosphatidylinositol cycle may be the result of
thrombin-induced changes in membrane or cytoplasmic Ca^{2+} rather
than the cause of these changes (129). In fact, phospholipase C,
phospholipase A_2, and diglyceride lipase activities are all Ca^{2+} dependent, but it is not known if changes in platelet Ca^{2+} actually regulate
these activities (122). Platelet phospholipase A_2 activity is also stimu-

lated by calmodulin plus Ca^{2+}, but not through a direct effect of
calmodulin on the enzyme (130).

A rapid but transient *decrease* in triphosphatidylinositol (PIP_2),
which represents only a small fraction of platelet phosphatidylinositol,
has been observed during thrombin stimulation of human and horse
platelets (131,132). This occurs through the action of a phospholipase C
(131,133). Since PIP_2 binds Ca^{2+} with a K_d of ~10 μM and is probably
localized to the plasma membrane, it has been suggested that the
hydrolysis of PIP_2 might serve as a source of cytoplasmic Ca^{2+} for
platelet secretion. A more likely source of the Ca^{2+} is the platelet-
dense tubular system, since IP_3, a product of PIP_2 hydrolysis, can
cause the release of Ca^{2+} from this membrane system (134). Decreases
in PIP_2 have also been observed during stimulation of human platelets
by platelet-activating factor (128).

Collagen also stimulates platelet phosphatidylinositol hydrolysis
resulting in transient elevations of diacylglyceride and phosphatidic
acid and synthesis of prostaglandins. However, in contrast to thrombin,
collagen-induced hydrolysis is significantly inhibited by inhibitors of
cyclo-oxygenase (135,136). This suggests a role for prostaglandin
endoperoxides and/or thromboxane A_2 in the process of PIP_2 hydrolysis.

A single study with epinephrine has demonstrated that a relatively
small amount of diacylglyceride is transiently formed in platelets when
epinephrine and ADP are used as costimuli (137). These minor changes
were prevented by inhibiting prostaglandin synthesis. Epinephrine
also stimulates de novo synthesis of phosphatidylinositol, but this
begins well after secretion is underway (138,139). It appears unlikely
that PIP_2 hydrolysis plays a primary role in epinephrine-induced
platelet secretion. Finally, activation of the phospholipid methylation
pathway, which has been suggested to be involved in stimulus-
secretion coupling in a number of cell types (140), plays no role
in epinephrine-induced platelet activation (141,142). Since platelet
secretion after epinephrine is totally dependent upon arachidonic
acid release from phospholipids and its conversion to prostaglandins,
additional studies of phospholipid metabolism during epinephrine-
induced platelet activation are warranted.

3. Prostaglandin Synthesis

Compared to thrombin-stimulated platelets, platelets stimulated
with epinephrine produce relatively small amounts of prostaglandin
endoperoxides and thromboxane A_2 (143). Nonetheless, prostaglandin
synthesis is necessary for epinephrine-induced platelet secretion (28).
Epinephrine may activate prostaglandin synthesis even in the absence
of stirring, but stirring and initial platelet aggregation appear to
amplify the process (76). In addition, the presence of extracellular

Na^+ appears to be required (144). The prostaglandin endoperoxides G_2 and H_2 and thromboxane A_2 can directly initiate platelet aggregation and secretion, and specific receptors for these substances exist on the platelet surface (145, 146). It is possible that the binding of these compounds to their surface membrane receptors stimulates or regulates phospholipid hydrolysis (135, 136). Alternatively, thromboxane A_2 may act as Ca^{2+} ionophore, and thereby secondarily stimulate the enzymes of phospholipid hydrolysis that are Ca^{2+} dependent (112). Indeed, stimulation of prostaglandin synthesis in platelets by the addition of arachidonic acid or prostaglandin endoperoxides results in phospholipase C-dependent phosphatidylinositol hydrolysis (126, 135, 136).

In the case of epinephrine, the precise sequence of events between α_2-receptor occupancy and prostaglandin synthesis has not been determined. Since arachidonic acid availability is the rate-limiting step in this synthetic pathway, epinephrine-induced phospholipid hydrolysis must be involved early in the stimulus-response process.

4. Protein Phosphorylation

Stimulation of platelets with a number of agonists results in an increased phosphorylation of at least two proteins (147-149). One of these is the 20,000-dalton light chain of myosin. This light chain is phosphorylated by a myosin light chain kinase that is itself activated by Ca^{2+} and calmodulin-dependent protein kinase (150). Platelets are rich in calmodulin, and inhibitors of this ubiquitous regulatory protein inhibit phosphorylation of the myosin light chain in stimulated platelets (113, 151, 152). Since the interaction of myosin with platelet actin is regulated by this phosphorylation step, this phosphorylation reaction is believed to play a regulatory role in platelet secretion. It may also be involved in platelet shape change (153). A second protein phosphorylated during platelet stimulation is a 40,000-47,000-dalton protein whose identity and function in the secretory process are unknown. This protein is phosphorylated by protein kinase C (113, 149, 154, 155). C-Kinase is a soluble and membrane-bound enzyme that is reversibly activated by Ca^{2+}, diacylglycerol, and phospholipid. Partially purified protein kinase C from rat brain is irreversibly activated on limited proteolysis by a Ca^{2+}-dependent neutral protease (156). A similar protease activity is present in stimulated platelets (see Sect. V.C.5). Phosphorylation of the 40,000-dalton platelet protein by protein kinase C is closely linked to agonist-induced platelet secretion (157). It is believed that phosphorylation of this protein and an increase in cytoplasmic Ca^{2+} must both occur simultaneously to cause platelet secretion (158). Phosphorylation of the 40,000-dalton protein during the secretory process is a particularly attractive reaction from a mechanistic viewpoint because it links receptor-mediated phosphatidylinositol hydrolysis (diacylglyceride formation) to a phosphorylation reaction

involved in secretion. At this time, however, the functional significance of phosphorylation of platelet proteins in the 40,000-47,000 range
remains speculative. Platelets also contain a calmodulin-dependent,
but Ca^{2+}-independent, protein kinase activity that phosphorylates
large molecular weight basic proteins (159). The functional significance
of this kinase is unclear.

Phosphorylation of platelet proteins has not been adequately
studied during epinephrine-induced platelet activation. In a single
study, epinephrine (1 μM) or ADP (5 μM) caused little or no increase
in platelet diacylglyceride, phosphorylation of the 20,000- and 40,000-
dalton proteins, or secretion of serotonin (137). The addition of the
epinephrine and ADP simultaneously did cause an increase in diacylglyceride, protein phosphorylation, and secretion. All of these reactions were inhibited by inhibition of cyclo-oxygenase. The amount
of diacylglyceride produced by epinephrine plus ADP was only 20%
of the amount produced by thrombin, whereas the extent of protein
phosphorylation was similar to that produced by thrombin.

5. Activation of Ca^{2+}-Dependent Proteases

Platelets contain at least two Ca^{2+}-activated neutral proteases,
one with a $K_{Ca^{2+}}$ of 25 μM, the other with a $K_{Ca^{2+}}$ of 500 μM (160-163).
The latter protease hydrolyzes a number of platelet substrates, including actin-binding protein and a 235,000-dalton protein (P-235),
both of which interact with and may regulate the cytoskeleton (160-163).
This protease also can hydrolyze and inactivate membrane glycoprotein
Ib and coagulation protein Va (164,165). Since the proteases are
activated by Ca^{2+} and hydrolyze proteins potentially involved in
platelet secretion, evidence for their activity in stimulated platelets
has been sought. In one study, the Ca^{2+} ionophore A23,187 stimulated
the proteolysis of actin-binding protein and several other platelet
proteins, but maximum concentrations of thrombin, ADP, and epinephrine did not (166). However, in another study, Ca^{2+}-dependent
proteolysis of several intracellular proteins was observed during
platelet aggregation by thrombin or collagen (164). Thrombin-
stimulated platelets may also exhibit Ca^{2+}-dependent protease activity
on their surface (167). In conclusion, there may be a role for Ca^{2+}-
dependent proteolysis in agonist-induced platelet activation. On the
other hand, protease activity during epinephrine-induced platelet
activation has yet to be demonstrated.

VI. ROLE OF CYCLIC AMP IN PLATELET FUNCTION

Substances such as PGE_1, PGI_2, PGD_2, and adenosine can stimulate
platelet adenylate cyclase. An increase in adenylate cyclase activity
raises the level of cyclic AMP in platelets, and this causes a dose-

dependent inhibition of shape change, aggregation, and secretion
(168). Platelets contain numerous cyclic AMP binding proteins,
including both soluble and membrane-bound cyclic AMP-dependent
protein kinases (169,170). Both the type I and type II regulatory
subunits of these protein kinases are present in human platelets (169).
The inhibitory effect of cyclic AMP results from the phosphorylation
of platelet proteins by these protein kinases. There are several
specific areas of the platelet activation sequence inhibited by cyclic
AMP: (1) Increases in cyclic AMP decrease the level of cytoplasmic
free Ca^{2+} (171). This reduction of Ca^{2+} may be mediated by the
phosphorylation of a 22,000-dalton protein that regulates a $Ca^{2+} + Mg^{2+}$-
ATPase pump in the dense tubular membranes (172). (2) Cyclic AMP-
dependent phosphorylation of the myosin light chain kinase of platelets
results in a decrease in both the V_{max} and K_d of the enzyme for
calmodulin (173). (3) Cyclic AMP prevents or limits the exposure of
platelet fibrinogen receptors (174). (4) Cyclic AMP may also modulate
cyclo-oxygenase activity by decreasing the affinity of this enzyme
for its substrate, arachidonic acid (175).

Inasmuch as epinephrine and other agonists decrease basal and
PGI_2-stimulated adenylate cyclase activity, a role for this decrease in
cyclic AMP levels during platelet activation has been considered. Inhibition of adenylate cyclase with a synthetic inhibitor of the enzyme
does not by itself stimulate platelet shape change, aggregation, or
secretion (176). Thus, a primary role for adenylate cyclase inhibition
as a cause of aggregation and secretion by epinephrine remains to be
demonstrated.

VII. PHYSIOLOGICAL SIGNIFICANCE OF EPINEPHRINE-
INDUCED PLATELET ACTIVATION

In resting man, the blood level of epinephrine is normally about 0.1-
0.2 nM and that of norepinephrine about 1-3 nM (177). With physical
or emotional stress or in the presence of a pheochromocytoma, the
blood levels of these catecholamines may rise more than 10-fold (177,
178). Since platelet aggregation and secretion requires at least 0.5
μM epinephrine or norepinephrine, it is unlikely that the catecholamines
by themselves stimulate platelet plug formation in vivo. However,
lower concentrations of epinephrine or norepinephrine (50-100 nM)
are capable of potentiating the stimulatory effects of unrelated agonists,
such as ADP, thrombin, collagen, vasopressin, and platelet-activating
factor (33-36). Thus, the catecholamines may work synergistically
with one or more unrelated agonists to stimulate platelet aggregation
in vivo.

Several studies have examined the mechanism of synergy between
epinephrine and unrelated agonists. Synergy with respect to platelet

aggregation and secretion has been observed either when a subthreshold concentration of epinephrine is added before, simultaneously with, or after an unrelated agonist. In each case, the pattern of the resultant aggregation tracing and the time course of secretion are typical for that member of the agonist pair added in higher concentration relative to its threshold concentration (179). Potentiation of the second agonist by the first may require the simultaneous occupancy by both agonists of their specific receptors on the platelet surface (180). Synergistic aggregation responses can be observed even in the absence of agonist-induced secretion. Thus, addition of a subthreshold concentration of epinephrine before the addition of a low concentration of ADP to aspirin-treated platelets converts a partial aggregation response into a complete and irreversible one (181). A synergistic aggregation response has also been observed with epinephrine and platelet-activating factor (36,182,183).

The molecular mechanism(s) responsible for synergy between agonist pairs remains to be elucidated. In theory, such potentiation of the platelet's final responses could be occurring at one or more levels of the stimulation-response pathway: at the receptor for the second agonist, at the fibrinogen receptor, and at biochemical reactions involved in secretion, including those responsible for control of cytoplasmic free Ca^{2+}. However, potentiation by ADP of the platelet response to epinephrine does not occur at the level of the α_2-adrenergic receptor or at the level of the guanine nucleotide regulatory protein, G_i, to which the receptor binds (184). On the other hand, there is evidence for synergy at the level of the fibrinogen receptor. The number of fibrinogen receptors exposed in response to low concentrations of ADP plus epinephrine is much greater than the sum of receptors exposed by either agonist alone (78,80,185). It is quite possible that synergy also occurs at the level of control of cytoplasmic free Ca^{2+} or at the level of Ca^{2+} sensitivity of the many platelet proteins regulated by Ca^{2+} (186).

VIII. ABNORMAL PLATELET RESPONSES TO EPINEPHRINE IN MAN

There are many variables to consider when interpreting platelet aggregation responses to agonists. Both age and sex may influence the sensitivity of the response of the donor's platelets (187). Methodological variables such as platelet concentration, temperature, pH, and duration of platelet storage also influence platelet responsiveness (75). Platelet responsiveness to epinephrine in particular is quite labile (188). In an individual subject, therefore, abnormalities of platelet function should be considered significant only if documented on several occasions.

There are many clinical conditions associated with diminished
aggregation and secretory responses to one or another agonist (2).
Decreased platelet responsiveness to epinephrine is the dominant or
sole abnormality in only a few clinical situations.

A. Essential Thrombocythemia

Platelets from most individuals with the myeloproliferative disease
essential thrombocythemia exhibit markedly reduced or absent aggregation and secretion in response to epinephrine. This abnormality
is usually not corrected by increasing the concentration of epinephrine.
It may be associated with decreased platelet responses to other agonists,
particularly ADP (189). Decreased aggregation and secretion in response to epinephrine or ADP is so characteristic of the disease that
its presence is useful in differentiating essential thrombocythemia
from reactive causes of thrombocytosis (190). The platelet functional
abnormality has been associated with a greater than 50% reduction in
the number of platelet α-adrenergic receptors (Fig. 4) (191). The
affinity of the remaining receptors for both adrenergic agonists and
antagonists appears to be normal. Direct binding studies with labeled
α_2-adrenergic agonists or computer-modeling of agonist competition
data in order to directly assess the high-affinity state of the epinephrine receptor have not been carried out. The deficiency of α-
adrenergic receptors is one of only several plasma membrane abnormalities observed in thrombocythemic platelets (2). This may reflect
generalized membrane protein and lipid abnormalities of platelets that
are derived from a single clone of malignant megakaryocytes (192).

B. Familial Platelet Hyporesponsiveness
   to Epinephrine

Platelets from as many as 5-10% of "normal" donors exhibit diminished
aggregation and secretory responses to epinephrine. Scrutton and
colleagues have studied five such unrelated donors and found that
this abnormality was present in the blood relatives of four of the
donors (193). Epinephrine caused the expected decrease in PGE_1-
stimulated cyclic AMP levels in these platelets, suggesting that the
α_2-adrenergic receptors themselves and the adenylate cyclase complex
were normal. Moreover, epinephrine normally potentiated subthreshold
concentrations of ADP. The diminished responsiveness of these platelets
to epinephrine could be restored by merely incubating the platelet-rich
plasma at room temperature or by adding a subthreshold concentration
of A23,187. These authors (193) concluded that some step in stimulus-
response coupling distal to the α-adrenergic receptor and adenylate
cyclase is responsible for the platelet defect in these individuals.

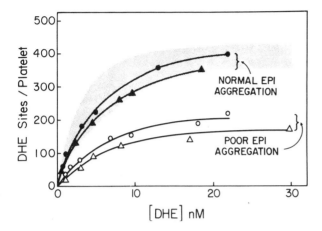

FIGURE 4 Specific binding of an α-adrenergic antagonist, [³H]di-
hydroergocryptine (DHE), to intact platelets in essential thrombocythemia as a function of [³H]dihydroergocryptine concentration. The
shaded area represents the range of [³H]dihydroergocryptine binding
to platelets from nine normal donors. Binding is expressed as DHE
sites per platelet. Open symbols represent experiments with platelets
from two patients whose platelets failed to aggregate in response to
100 μM epinephrine. Closed symbols represent two patients with
essential thrombocythemia whose platelets aggregated normally to
5 μM epinephrine. (From Ref. 191, used with permission.)

C. Bartter's Syndrome

Bartter's syndrome is a rare disorder that is associated with renal
juxtoglomerular hyperplasia, hypokalemia, hypertension, and increased
blood levels of renin, angiotensin II, and aldosterone. Afflicted
individuals excrete increased amounts of prostaglandins in their urine
and a fundamental defect in renal prostaglandin metabolism has been
proposed (194).

Several reports have demonstrated prolonged bleeding times and
decreased platelet aggregation and secretion in response to epinephrine
and ADP in this disease. These abnormalities are caused by a plasma
factor, with the platelets themselves being intrinsically normal (195,
196). It has been suggested that the plasma factor is a prostaglandin(s)
inhibitory to platelets, and a shortening of the prolonged bleeding
times by aspirin is consistent with this suggestion. However, systemic
prostaglandin production is normal in Bartter's syndrome (197).
Thus, a renal prostaglandin that gains access to the systemic circulation may be the platelet inhibitor. However, further studies are needed

in this disease, including measurement of platelet cyclic AMP levels
and formal identification of the plasma inhibitor.

D. Administration of Penicillin Antibiotics

Large doses of several β-lactam antibiotics can cause a bleeding diathesis resulting from platelet dysfunction (198). These antibiotics also
inhibit platelet function in vitro, but severalfold higher drug concentrations are required for this effect (199). Carbenicillin and penicillin
G inhibit platelet aggregation and secretion induced by ADP and epinephrine (200). This inhibition can be overcome by increasing the
concentrations of the agonists. In vitro, at antibiotic concentrations
that inhibit platelet function by more than 80%, there is a decreased
affinity of the α_2-adrenergic receptor for both α-adrenergic agonists
and antagonists (200). Thus, the affinity of epinephrine, [^{3}H]di-
hydroergocryptine, and [^{3}H]yohimbine are reduced twofold by carbenicillin and sixfold by penicillin G (Fig. 5). Moreover, these antibiotics
inhibit the incorporation of an ADP-affinity label into the putative ADP
receptor on the platelet surface. These antibiotics also inhibit the
interaction of von Willebrand protein with its platelet receptor, membrane glycoprotein Ib (200). Thus, these antibiotics cause a global
impairment of the interaction of agonists with the platelet surface
membrane. Were this mechanism operative in vivo, it could account
for the prolonged bleeding times frequently observed and the bleeding
diathesis occasionally observed in individuals receiving large doses
of β-lactam antibiotics (198, 201- 204).

E. Short-Term Exposure to Endogenous
and Exogenous Catecholamines

Hollister and co-workers studied platelet α_2-adrenergic receptors with
[^{3}H]yohimbine during short-term physiological and pharmacological
elevations of plasma catecholamine levels (177). Two hours of upright
posture and exercise caused a severalfold increase in plasma levels
of norepinephrine and epinephrine, and this was associated with a
3.4-fold decrease in the affinity of platelet α_2 receptors for (-)-
epinephrine. Platelets were similarly affected by infusions of norepinephrine or epinephrine that raised the level of these catecholamines by
severalfold. There was no change in the number of platelet α_2 receptors, or in their affinity for antagonists. In vitro exposure of intact
platelets to 10^{-6}-10^{-10} M epinephrine for 2 hr led to a concentration-
dependent decrease in α_2-receptor agonist affinity, and incubation
with 10^{-6}-10^{-8} M epinephrine led to a dose- and time-dependent loss
of platelet aggregatory response to epinephrine. These studies demonstrate that physiological elevations in plasma catecholamines can acutely

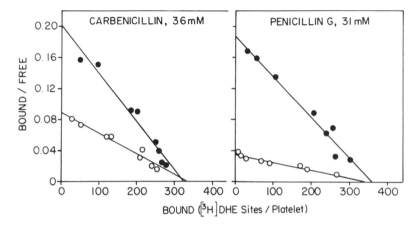

FIGURE 5 Scatchard analysis of equilibrium binding of [³H]dihydro-ergocryptine to intact platelets: effect of carbenicillin and penicillin G. Closed circles depict control platelets and open circles depict platelets incubated with 36 mM carbenicillin or 31 mM penicillin G. The lines were calculated from the data by the method of least squares. The K_d of [³H]dihydroergocryptine for control platelets was 4.0 nM, for carbenicillin-treated platelets 9.5 nM, and for penicillin-treated platelets 20.2 nM. Similar results were obtained using [³H]yohimbine as the α_2-adrenergic ligand. (From Ref. 200, used with permission.)

modulate the affinity of platelet α_2 receptors for adrenergic agonists and possibly alter the sensitivity of platelets to these agonists. The clinical significance of these observations remains to be defined.

IX. CONCLUSIONS

Epinephrine initiates platelet activation by binding to high-affinity α_2-adrenergic receptors on the platelet surface. The resulting agonist-receptor complex interacts with a guanine nucleotide regulatory protein that inhibits platelet membrane adenylate cyclase. This inhibition of adenylate cyclase and the resultant decrease in cyclic AMP may be causally related to platelet aggregation and secretion, but current evidence does not support such a relationship. In some as yet un-determined way, the epinephrine-α_2 receptor interaction stimulates the induction of fibrinogen receptors on the platelet surface. In the presence of extracellular Ca^{2+}, fibrinogen binds to these receptors. If cell-cell contact is facilitated by stirring, platelet aggregation ensues.

Epinephrine stimulates a secretory response involving the release of substances from platelet α and δ granules. Both platelet aggregation and prostaglandin synthesis are required to initiate this secretory response. Aggregation and secretion are facilitated by release of ADP from the granules. How epinephrine initiates phospholipid hydrolysis to provide the initial arachidonic acid needed for prostaglandin synthesis is not known. Nonetheless, once prostaglandin endoperoxides and thromboxane A_2 are formed, they stimulate further phospholipid hydrolysis and an increase in the concentration of cytoplasmic free Ca^{2+}. Platelet secretion is dependent upon (1) this increase in cytoplasmic free Ca^{2+}, and (2) phosphorylation of platelet proteins. The stimulus-induced increase in cytoplasmic Ca^{2+} appears to be derived from intra-platelet sources. Epinephrine may stimulate a net influx of extracellular Ca^{2+}, but this remains to be demonstrated. The 20,000-dalton myosin light chain is phosphorylated by a Ca^{2+}- and calmodulin-dependent protein kinase, and a 40,000-dalton protein is phosphorylated by a Ca^{2+}-, diacylglyceride-, and phospholipid-dependent protein kinase C. In contrast, an increase in platelet cyclic AMP prevents or reverses agonist-induced platelet activation by stimulating additional sites of phosphorylation. This results in a decrease in cytoplasmic free Ca^{2+}

The physiological role of epinephrine and norepinephrine in platelet function is uncertain. Involvement of either of these catecholamines in platelet plug formation would have to take place at nanomolar concentrations by means of their potentiating the aggregation and secretion induced by unrelated agonists, such as ADP, thrombin, and collagen, or platelet-activating factor.

Platelets respond poorly to epinephrine in several clinical conditions. In the case of essential thrombocythemia and the administration of large doses of penicillins, this is accounted for by abnormalities at the level of the platelet $α_2$-adrenergic receptor. In the case of Bartter's syndrome, a plasma factor that increases platelet cyclic AMP may be responsible for the abnormalities of platelet function. In the case of a familial decrease in platelet responsiveness to epinephrine, the locus of the underlying defect has not been determined but is apparently distal to the $α_2$ receptor.

REFERENCES

1. Weiss, H. J.: *Platelets: Pathophysiology and Antiplatelet Drug Therapy*, New York, Liss, 1982.

2. Shattil, S. J., and Bennett, J. S.: Ann. Int. Med. 94:108, 1980.

3. Stemerman, M. B.: Cardio. Med. 3:17, 1978.

4. French, J. E.: Semin. Hematol. 8:84, 1971.

5. McManus, L. M., Hanahan, D. J., and Pinkard, R. N.: J. Clin. Invest. 67:903, 1981.

6. Marcus, A. J., Safier, L. B., Ullman, H. L., Wong, K. T. H., Broekman, M. J., Weksler, B. B., and Kaplan, K. L.: Blood 58: 1027, 1981.

7. Weiss, H. J., Turrito, V. T., and Baumgartner, H. R.: J. Lab. Clin. Med. 92:750, 1978.

8. Zimmerman, T. S., and Ruggeri, Z. M.: Prog. Hemost. Thromb. 6:203, 1982.

9. Jaffe, E. A., Hoyer, L., and Nachman, R. L.: J. Clin. Invest. 52:2757, 1973.

10. Wagner, D. D., and Marder, V. J.: J. Biol. Chem. 258:2065, 1983.

11. Rand, J. H., Gordon, R. E., Sussman, I. I., Cher, S. V., and Solomon, V.: Blood 60:627, 1982.

12. Ruan, C., Tobelem, G., McMichael, A. J., Drovet, L., Legrand, Y., Degos, L., Kieffer, N., Lee, H., and Caen, J. P.: Br. J. Haematol. 49:511, 1981.

13. Kao, K.-J., Pizzo, S. V., and McKee, P. A.: J. Clin. Invest. 63:656, 1979.

14. Weiss, H. J., Tschopp, T. B., Baumgartner, H. R., Sussman, I. I., Johnson, M. M., and Egan, J. J.: Am. J. Med. 57:920, 1974.

15. Clemetson, K., McGregor, J. L., James, E., Dechewanne, M., and Luscher, E. F.: J. Clin. Invest. 70:304, 1982.

16. Sakariassen, K. S., Bolhuis, P. A., and Sixma, J. J.: Nature 279:636, 1979.

17. Santoro, S. A.: Thromb. Res. 21:689, 1981.

18. Fujimoto, T., and Hawiger, J.: Nature 297:154, 1982.

19. Ruggeri, Z. M., DeMarco, L., Gatti, L., Bader, R., Montgomery, R. R.: J. Clin. Invest. 72:1, 1983.

20. Gralnick, H., and Coller, B.: Clin. Res. 31:482A, 1983.

21. Holmsen, H., Kaplan, K. L., and Dangelmaier, C. A.: Biochem. J. 208:9, 1982.

22. Lind, S. E., and Stossel, T. P.: Prog. Hemost. Thromb. 6:63, 1982.

23. Daniel, J. L, Molish, H. R., and Holmsen, H.: J. Biol. Chem. 256:7510, 1981.

24. Charo, I. F., Feinman, R. D., and Detwiler, T. C.: J. Clin. Invest. 60:866, 1977.

25. Witte, L. D., Kaplan, K. L., Nossel, H. L., Lages, B. A., Weiss, H. J., and Goodman, D. S.: Circ. Res. 42:402, 1978.

26. Kaplan, K. L., Broekman, M. J., Chernoff, A., Lesznik, G. R., and Drillings, M.: Blood 53:604, 1979.

27. Zweifler, A. J., and Julius, S.: N. Engl. J. Med. 306:890, 1982.

28. Zucker, M. B., and Peterson, J.: Proc. Soc. Exp. Biol. Med. 127: 547, 1968.

29. Kinlough-Rathbone, R. L.: Packham, M. A., Reimers, H.-J., Cazenave, J.-P., and Mustard, J. F.: J. Lab. Clin. Med. 90: 707, 1979.

30. Kane, W. H., Lindhout, M. J., Jackson, C. M., and Majerus, P. W.: J. Biol. Chem. 255:1170, 1980.

31. Miletich, J. P., Jackson, C. M., and Majerus, P. W.: J. Biol. Chem. 253:6908, 1978.

32. Zwaal, R. F. A., and Hemker, H. C.: Haemostasis 11:12, 1982.

33. Scrutton, M. C., and Wallis, R. B.: In: *Platelets in Biology and Pathology—2*. J. Gordon (ed.). Amsterdam, Elsevier/North-Holland, 1981, p. 179.

34. Packham, M. A., Guccione, M. A., Chang, P. Y., and Mustard, J. F.: Am. J. Physiol. 225:38, 1973.

35. Grant, J. A., and Scrutton, M. C.: Br. J. Haematol. 44:109, 1980.

36. Vargaftig, B. B., Fouque, F., Benveniste, J., Odiot, J.: Thrombos. Res. 28:557, 1982.

37. Glusa, E., Markwardt, F., and Barthel, W.: Pharmacology 19: 196, 1979.

38. Grant, J. A., and Scrutton, M. C.: Nature 227:659, 1979.

39. Grant, J. A., and Scrutton, M. C.: Br. J. Pharmacol. 71:121, 1980.

40. Shattil, S. J., McDonough, M., Turnbull, J., and Insel, P. A.: Molec. Pharmacol. 19:179, 1981.

41. Kafka, M. S., Tallman, J. F., and Smith, C. C.: Life Sci. 21: 1429, 1977.

42. Salzman, E. W., and Neri, L. L.: Nature 224:609, 1969.

43. Jacobs, K. H., Sauer, W., and Schultz, G.: F.E.B.S. Lett. 85: 167, 1978.

44. Steer, M. L., and Wood, A.: J. Biol. Chem. 254:10791, 1979.

45. Jacobs, K. H., and Schultz, G.: Nauyn-Schmiedeberg's Arch. Pharmacol. 310:113 (1979).

46. Lasch, P., and Jacobs, K. H.: Naunyn Schmiedebergs Arch. Pharmacol. 306:119, 1979.

47. Jacobs, K. H., Saur, W., and Schultz, G.: Naunyn Schmiedebergs Arch. Pharmacol. 302:285, 1978.

48. Smith, S. K., and Limbird, L. E.: Proc. Natl. Acad. Sci. U.S.A. 78:4026, 1981.

49. Michel, T., Hoffman, B. B., Lefkowitz, R. J., and Caron, M. G.: Biochem. Biophys. Res. Commun. 100:1131, 1981.

50. Mills, D. C. B., and Smith, J. B.: Biochem. J. 121:185, 1971.

51. Yu, S. K., and Latour, J. G.: Thrombs. Haemost. 37:413, 1977.

52. Mills, D. C. B., and Smith, J. B.: Ann. N.Y. Acad. Sci. 201: 391, 1972.

53. Motulsky, H. J., and Insel, P. A.: N. Engl. J. Med. 307:18, 1982.

54. Bylund, D. B., and U'Prichard, D. C.: Int. Rev. Neurobiol. 24: 343, 1983.

55. Alexander, R. W., Cooper, B., and Handin, R. I.: J. Clin. Invest. 61:1136, 1978.

56. Newman, K. D., Williams, L. T., Bishopric, N. H., and Lefkowitz, R. J.: J. Clin. Invest. 61:395, 1978.

57. Steer, M L., Khorana, J., and Galgoci, B.: Molec. Pharmacol. 16:719, 1979.

58. Motulsky, H. J., Shattil, S. J., and Insel, P. A.: Biochem. Biophys. Res. Commun. 97:1562, 1980.

59. Daiguji, M., Meltzer, H. Y., and U'Prichard, D. C.: Life Sci. 28: 2705, 1981.

60. Steer, M. L., and Atlas, D.: Biochim. Biophys. Acta 686:240, 1982.

61. Lynch, C. J., and Steer, M. L.: J. Biol. Chem. 256:3298, 1981.

62. Garcia-Sevilla, J. A., Hollingsworth, P. J., and Smith, C. B.: Eur. J. Pharmacol. 74:329, 1981.

63. Hoffman, B. B., Mullikin-Kilpatrick, D., and Lefkowitz, R. J.: J. Biol. Chem. 255:4645, 1980.

64. Hoffman, B. B., Michel, T., Brenneman, T. B., and Lefkowitz, R. J.: Endocrinol. 110:926, 1982.

65. Smith, S. K., and Limbird, L. E.: J. Biol. Chem. 257:10471, 1982.

66. Michel, T., and Lefkowitz, R. J.: J. Biol. Chem. 257:13557, 1982.

67. Gilman, A. G.: Cell 36:577, 1984.

68. Limbird, L. E., Speck, J. L., and Smith, S. K.: Mol. Pharmacol. 21:609, 1982.

69. Connolly, T. M., and Limbird, L. E.: J. Biol. Chem. 258:3907, 1983.

70. Motulsky, H. J., and Insel, P. A.: J. Biol. Chem. 258:3913, 1983.

71. Regan, J. W., Nakata, H., DeMarinis, R. M., Caron, M. G., and Lefkowitz, R. J.: J. Biol. Chem. 261:3894, 1986.

72. Cerione, R. A., Regan, J. W., Nakata, H., Codina, J., Benovic, J. L., Gierschik, P., Somers, R. L., Spiegel, A. M., Birnbaumer, L., Lefkowitz, R. J., and Caron, M. G.: J. Biol. Chem. 261:3901, 1986.

73. Lomasney, J. W., Leeb-Lundberg, M. F., Cotecchia, S., Regan, J. W., DeBernardis, J. F., Caron, M. G., and Lefkowitz, R. J.: J. Biol. Chem. 261:7710, 1986.

74. Mustard, J. F., Perry, D. W., Ardlie, N. G., and Packham, M. A.: Br. J. Haematol. 22:193, 1972.

75. Packham, M. A., Kinlough-Rathbone, R. L., and Mustard, J. F.: In: DHEW Publication 78-1087. H. J. Day, H. Holmsen, and M. B. Zucker (eds.). United States Department of Health, Education and Welfare, Washington, D.C., 1978, p. 66.

76. Bennett, J. S., and Vilaire, G.: J. Clin. Invest. 64:1393, 1979.

77. Marguerie, G., Plow, E. F., and Edgington, T. S.: J. Biol. Chem. 254:5357, 1979.

78. Shattil, S. J., Hoxie, J. A., Cunningham, M. C., and Brass, L. F.: J. Biol. Chem. 260:11107, 1985.

79. Plow, E. F., and Marguerie, G. A.: J. Biol. Chem. 255:10971, 1980.

80. Peerschke, E. I.: Blood 60:71, 1982.

81. Marguerie, G. A., Edgington, T. S., and Plow, E. F.: J. Biol. Chem. 255:154, 1980.

82. Kloczewiak, M., Timmons, S., and Hawiger, J.: Thromb. Res. 29:249, 1983.

83. Weiss, H. J., and Rogers, J.: N. Engl. J. Med. 285:369, 1971.

84. Kunicki, T. J., Pidard, D., Rosa, J.-P., and Nurden, A. T.: Blood 58:268, 1981.

85. Jennings, L. K., and Phillips, D. R.: J. Biol. Chem. 257:10458, 1982.

86. Hagen, I., Bjerrum, O. J., Gogstad, G., Korsmo, R., and Solum, N. O.: Biochim. Biophys. Acta 701:1, 1982.

87. Howard, L., Shulman, S., Sadanandan, S., and Karpatkin, S.: J. Biol. Chem. 257:8331, 1982.

88. Pidard, D., Rosa, J.-P., Kunicki, T. J., and Nurden, A. T.: Blood 60:894, 1982.

89. Bennett, J. S., Hoxie, J. A., Leitman, S. F., Vilaire, G., and Cines, D. B.: Proc. Natl. Acad. Sci. U.S.A. 80:2417, 1983.

90. Coller, B. S., Peerschke, E. J., Scudder, L. E., and Sullivan, C. A.: J. Clin. Invest. 72:325, 1983.

91. McEver, R. P., Bennett, E. M., and Martin, M. N.: J. Biol. Chem. 258:5269, 1983.

92. Nurden, A. T., and Caen, J. P.: Br. J. Haematol. 28:253, 1974.

93. Bennett, J. S., Vilaire, G., and Cines, D. B.: J. Biol. Chem. 257:8049, 1982.

94. Peerschke, E. I., and Zucker, M. B.: Blood 57:663, 1981.

95. Gerrard, J. M., and White, J. G.: Prog. Hemost. Thromb. 4:87, 1978.

96. Bennett, J. S., Vilaire, G., and Burch, J. W.: J. Clin. Invest. 68:981, 1981.

97. Polley, M. J., Leung, L. L. K., Clark, F. Y., and Nachman, R. L.: J. Exp. Med. 154:1058, 1981.

98. Nathan, I., Fleischer, G., Levine, A., Dvilansky, A., and Parola, A. H.: J. Biol. Chem. 254:9822, 1979.

99. Nieuwenhuizen, W., van Ruijven-Vermeer, I. A. M., Nooijen, W. J., Vermond, A., Haverkate, F., and Hermans, J.: Thromb. Res. 22:653, 1981.

100. Brass, L. F., and Shattil, S. J.: J. Clin. Invest. 257:14,000, 1984.

101. Leung, L. L. K., and Nachman, R. L.: J. Clin. Invest. 70:542, 1982.

102. Dutilh, C. E., Haddeman, E., Don, J. A., and ten Hoor, F.: Prostaglandins Med. 6:111, 1981.

103. Shattil, S. J., Brass, L. F., Bennett, J. S., and Pandhi, P.: Blood 66:92, 1985.

104. Scrutton, M. C., and Egan, C. M.: Thromb. Res. 14:713, 1979.

105. Lages, B., and Weiss, H. J.: Thromb. Haemost. 45:173, 1981.

106. LeBreton, G. C., Dinerstein, R. J., Roth, L. J., and Feinberg, H.: Biochem. Biophys. Res. Commun. 71:362, 1976.

107. Charo, I. F., Feinman, R. D., Detwiler, T. C.: Biochem. Biophys. Res. Commun. 72:1462, 1976.

108. Feinstein, M. B.: Biochem. Biophys. Res. Commun. 93:593, 1980.

109. Knight, D. E., and Scrutton, M. C.: Thromb. Res. 20:437, 1980.

110. Rink, T. J., Smith, S. W., and Tsien, R. Y.: F.E.B.S. Lett. 148:21, 1982.

111. Rink, T. J.: Thromb. Haemost. 50:92, 1983.

112. Gerrard, J. M., Peterson, D. A., and White, J. G.: In: *Platelets in Biology and Pathology—2.* J. Gordon (ed.). Amsterdam, Elsevier/North-Holland, 1981, p. 407.

113. Feinstein, M. B.: Prog. Hemos. Thromb. 6:25, 1982.

114. Rybicki, J. P., Venton, D. L., LeBreton, G. C.: Biochim. Biophys. Acta 751:66, 1983.

115. Exton, J. H.: Mol. Cell. Endocrinol. 23:233, 1981.

116. Owen, N. E., Feinberg, H., and LeBreton, G. C.: Am. J. Physiol. 239:483, 1980.

117. Owen, N. E., LeBreton, G. C.: Thromb. Res. 17:855, 1980.

118. Brass, L. F., and Shattil, S. J.: J. Biol. Chem. 257:14000, 1982.

119. MacIntyre, D. E., and Rink, T. J.: Thromb. Haemost. 47:22, 1982.

120. Massini, P., and Luscher, E. F.: Thromb. Diath. Haemorrh. 27:121, 1972.

121. Jenkins, C. S. P., Meyer, D., Dreyfus, M., and Larrieu, M. J.: Br. J. Haematol. 28:561, 1974.

122. Rittenhouse-Simmons, S., and Deykin, D.: In: *Platelets in Biology and Pathology—2.* J. Gordon (ed.). Amsterdam, Elsevier/North-Holland, 1981, p. 349.

123. Prescott, S. M., and Majerus, P. W.: J. Biol. Chem. 258:764, 1983.

124. Rittenhouse-Simmons, S.: J. Clin. Invest. 63:580, 1979.

125. Broekman, M. J., Ward, J. W., and Marcus, A. J.: J. Clin. Invest. 60:275, 1980.

126. Lapetina, E. G., and Seiss, W.: Life Sci. 33:1011, 1983.

127. Serhan, C., Anderson, P., Goodman, E., Dunham, P., and Weissman, G.: J. Biol. Chem. 256:2736, 1981.

128. Lapetina, E. G.: Life Sci. 32:2069, 1983.

129. Billah, M. M., and Lapetina, E. G.: J. Biol. Chem. 257:11856, 1982.

130. Ballou, L. R., and Cheung, W. Y.: Proc. Natl. Acad. Sci. U.S.A. 80:5203, 1983.

131. Agranoff, B. W., Murthy, P., and Seguin, E. G.: J. Biol. Chem. 258:2076, 1983.

132. Billah, M. M., and Lapetina, E. G.: J. Biol. Chem. 257:12705, 1982.

133. Rittenhouse, S. E.: Proc. Natl. Acad. Sci. U.S.A. 80:5417, 1983.

134. Brass, L. F., and Joseph, S. K.: J. Biol. Chem. 260:15172, 1985.

135. Rittenhouse, S. E., and Allen, C. L.: J. Clin. Invest. 70:1216, 1982.

136. Seiss, W., Siegel, F. L., and Lapetina, E. G.: J. Biol. Chem. 258:11236, 1983.

137. Kawahara, Y., Yamanishi, J., Tsunemitsu, M., and Fukuzaki, H.: Thromb. Res. 30:477, 1983.

138. Deykin, D., and Snyder, D.: J. Lab. Clin. Med. 82:554, 1973.

139. Wallace, M. A., Agarwal, K. C., Garcia-Sainz, J. A., and Fain, J. N.: J. Cell. Biochem. 18:213, 1982.

140. Hirata, F., and Axelrod, J.: Science 209:1082, 1980.

141. Shattil, S. J., McDonough, M., and Burch, J. W.: Blood 57: 537, 1981.

142. Shattil, S. J., Montgomery, J. A., and Chiang, P. K.: Blood 59:906, 1982.

143. Best, L. C., Jones, P. B. B., McGuire, M. B., and Russell, R. G. G.: Adv. Prostaglandin Thromboxane Leukotriene Res. 6:297, 1980.

144. Connolly, T. M., and Limbird, L. E.: Proc. Natl. Acad. Sci. U.S.A. 80:5320, 1983.

145. MacIntyre, D. E., Salzman, E. H., and Gordon, J. L.: Biochem. J. 175:921, 1978.

146. LeBreton, G. C., Venton, D. L., Enke, S. E., and Halushka, P. V.: Proc. Natl. Acad. Sci. U.S.A. 76:4097, 1979.

147. Lyons, R. M., Stanford, N., and Majerus, P. W.: J. Clin. Invest. 56:924, 1975.

148. Haslam, R. J., Lynham, J. A., and Fox, J. E. B.: Biochem. J. 178:397, 1979.

149. Takai, Y., Kishimoto, A., Kikkawa, U., Mori, T., Nishizuka, Y., Tamura, A., and Fujii, T.: Biochem. Biophys. Res. Commun. 91:1218, 1979.

150. Hathaway, D. R., and Adelstein, R. S.: Proc. Natl. Acad. Sci. U.S.A. 76:1653, 1979.

151. Nishikawa, M., Tanaka, H., and Hidaka, H.: Nature 287:863, 1980.

152. Feinstein, M. B., and Hadjian, R. A.: Mol. Pharmacol. 21:422, 1982.

153. Daniel, J. L., and Molish, I. R.: Thromb. Haemost. 50:123, 1983.

154. Takai, Y., Kishimoto, A., Iwasa, Y., Kawahara, Y., Mori, T., and Nishizuka, Y.: J. Biol. Chem. 254:3692, 1979.

155. Takai, Y., Kishimoto, A., Kikkawa, U., Mori, T., and Nishizuka, Y.: J. Biochem. (Tokyo) 85:575, 1979.

156. Kishimoto, A., Kajikawa, N., Shiota, M., and Nishizuka, Y.: J. Biol. Chem. 258:1156, 1983.

157. Sano, K., Takai, Y., Yamanishi, J., and Wishizuka, Y.: J. Biol. Chem. 258:2010, 1983.

158. Kaibuchi, K., Takai, Y., Sawamura, M., Hoshijima, M., Fujikura, T., and Nishizuka, Y.: J. Biol. Chem. 258:6701, 1983.

159. Yamaki, T., Nishikawa, M., and Hidaka, H.: Biochem. Biophys. Acta 714:257, 1982.

160. Phillips, D. R., and Jakabova, M.: J. Biol. Chem. 252:5602, 1977.

161. Truglia, J. A., and Stracher, A.: Biochem. Biophys. Res. Commun. 100:814, 1981.

162. Tsujinaka, T., Shiba, E., Kambayashi, J., and Kosaki, G.: Biochem. Int. 6:71, 1983.

163. Yoshida, N., Weksler, B., and Nachman, R.: J. Biol. Chem. 258:7168, 1983.

164. Fox, J. E. B., Reynolds, C. C., and Phillips, D. R.: J. Biol. Chem. 258:9973, 1983.

165. Kane, W. H., Mruk, J. S., and Majerus, P. W.: J. Clin. Invest. 70:1090, 1982.

166. White, G. C., II: Biochem. Biophys. Acta 631:130, 1980.

167. Lucas, R. C., Lawrence, J. J., and Stracher, A.: J. Cell. Biol. 83:77a, 1979.

168. Haslam, R. J., Davidson, M. D., Fox, J. E., and Lynham, J. A.: Thromb. Haemost. 40:232, 1978.

169. Salama, S. E., and Haslam, R. J.: F.E.B.S. Lett. 130:230, 1981.

170. Chambers, D. A., Nachman, R. L., Evarts, J., and Kinoshita, T.: Biochim. Biophys. Acta 719:208, 1982.

171. Feinstein, M. B., Egan, J. J., Shaafi, R. I., and White, J.: Biochem. Biophys. Res. Commun. 113:598, 1983.

172. Kaser-Glanzmann, R., Gerber, E., Luscher, E. F.: Biochim. Biophys. Acta 558:344, 1979.

173. Hathaway, D. R., Eaton, C. R., and Adelstein, R. S.: Nature 291:252, 1981.

174. Graber, S. E., and Hawiger, J.: J. Biol. Chem. 257:14606, 1982.

175. Lindgren, J. A., Classon, H. E., Kindahl, H., and Hammarstrom, S.: In: *Advances in Prostaglandin and Thromboxane Research*, Vol. 6. B. Samuelsson, P. Ramwell, and R. Paoletti (eds.). New York, Raven, 1980, p. 275.

176. Salzman, E. W., MacIntyre, D. E., Steer, M. L., and Gordon, J. L.: Thromb. Res. 13:1089, 1978.

177. Hollister, A. S., Fitzgerald, G. A., Nadean, J. H. J., and Robertson, D.: J. Clin. Invest. 72:1498, 1983.

178. Simpson, M. T., Olewine, D. A., Jenkins, C. D., Ramsey, F. H., Zyzanski, S. J., Thomas, G., and Haines, C. G.: Psychosom. Med. 36:476, 1974.

179. Huang, E. M., and Detwiler, T. C.: Blood 57:685, 1981.

180. Hallam, T. J., Scrutton, M. C., and Wallis, R. B.: Thromb. Res. 27:435, 1982.

181. Rao, G. H. R., Johnson, G. J., and White, J. G.: Prostaglandins Med. 5:45, 1980.

182. Rao, G. H. R., and White, J. G.: Prostaglandins Med. 9:459, 1982.

183. Cameron, H. A., and Ardlie, N. G.: Prostaglandins Med. 9:117, 1982.

184. Motulsky, H. J., and Insel, P. A.: Fed. Proc. 42:1850, 1983.

185. DiMinno, G., Capitanio, A. M., Thiagarajan, P., Martinez, J., and Murphy, S.: Blood 61:1054, 1983.

186. Meyers, K. M., Huston, L. Y., and Clemmons, R. M.: Am. J. Physiol. 245:R100, 1983.

187. Johnson, M., Ramey, E., and Ramwell, P. W.: Nature 253:355, 1975.

188. Rossi, E. C., and Louis, G.: J. Lab. Clin. Med. 85:300, 1975.

189. Spaet, T. H., Lejnieks, I., Gaynor, E., and Goldstein, M. L.: Arch. Int. Med. 124:135, 1969.

190. Ginsberg, A. D.: Ann. Int. Med. 82:605, 1975.

191. Kaywin, P., McDonough, M., Insel, P. A., and Shattil, S. J.: N. Engl. J. Med. 299:505, 1978.

192. Fialkow, P. J., Faguet, G. B., Jacobson, R. J., Vardya, K., and Murphy, S.: Blood 58:926, 1981.

193. Scrutton, M. C., Clare, K. A., Hutton, R. A., and Bruckdorfer, K. R.: Br. J. Haematol. 49:303, 1981.

194. Bartter, F. C., Gill, J. R., and Frolich, J. C.: Adv. Nephrol. 7:191, 1977.

195. Regan, S. O., Rivard, G. E., Mongeau, J.-G., and Robtaille, P. O.: Pediatrics 64:939, 1979.

196. O'Regan, S., Rivard, G. E., Cole, C., and Robtaille, P. O.: Prostaglandins Med. 2:321, 1979.

197. Watson, M. J., Branch, R. A., Gill, J. R., Oates, J. A., Brash, A. R.: Lancet 2:368, 1983.

198. Haburchak, D. R., Head, D. R., and Everett, E. D.: Am. J. Surg. 134:630, 1977.

199. Cazenave, J. P., Packham, M. A., Guccione, M. A., and Mustard, J. F.: Proc. Soc. Exp. Biol. Med. 142:159, 1973.

200. Shattil, S. J., Bennett, J. S., McDonough, M., and Turnbull, J.: J. Clin. Invest. 65:329, 1980.

201. Andrassy, K., Ritz, E., Hasper, B., Scherz, M., Walter, E., Storck, H., and Vomel, W.: Lancet 2:1039, 1976.

202. Brown, C. H., Natelson, E. A., Bradshaw, W., Williams, T. W., and Alfrey, C. P.: N. Engl. J. Med. 291:265, 1974.

203. Zacombe, M. J., Veret, B., Godeau, P., Herremany, G., and Cenac, A.: Nouv. Presse Med. 3:1435, 1974.

204. Weitekamp, M. R., and Aber, R. C.: J.A.M.A. 249:69, 1983.

12

Adrenergic Receptors in Neurological Disorders

DAVID ROBERTSON and ALAN S. HOLLISTER   *Vanderbilt University Medical Center, Nashville, Tennessee*

I. INTRODUCTION

Many neuropathological processes can result in abnormalities in autonomic nervous system function. Unfortunately, until very recent
times, the character and completeness of such abnormalities could not
be adequately quantitated. In the case of Parkinson's disease, the
autonomic involvement has been most dramatically underscored by the
high incidence of orthostatic hypotension in patients receiving therapy
with levodopa, and the frequent recognition that this orthostatic hypotension does not entirely disappear even when levodopa is withdrawn.
     Several laboratories have recently begun to employ radioligand
binding methodology to the problems of neurological and psychiatric
disorders in human subjects and animal models of human disease (1-4),
but our knowledge in this area is unfortunately still extremely limited.
Our understanding of adrenoreceptors in neurological disease is therefore best illustrated in the set of disorders which have been most
carefully investigated from this point of view. These are the disorders
generally referred to as autonomic dysfunctional states.

II. AUTONOMIC DYSFUNCTION

Human autonomic dysfunction includes many disease processes, having
as their final common pathway the interruption of the efficient functioning of the autonomic nervous system (5). It occurs in approximately
1:5000 population in Tennessee. In most cases, such patients seek

medical attention because of symptoms of orthostatic hypotension
(Table 1). The three principal symptoms of orthostatic hypotension
are (1) lightheadedness, (2) discomfort in the neck or head, and
(3) dimming of vision. While these complaints bring the patient to
medical attention, further questioning often reveals impotence and
reduced sweating as associated symptoms. Pupillary abnormalities
are common. Usually there is also nasal stuffiness, especially on
assumption of the supine posture at the end of the day. Gastrointestinal complaints, such as diarrhea or constipation, occur in a minority
of patients (6,7).

Clearly, the hemodynamic abnormality is the major therapeutic
challenge in giving the severely affected patient an improved quality
of life; the impotence has thus far been resistant to treatment in all
cases in our experience. Therapy of orthostatic hypotension, however,
has been complicated by a number of unusual features of autonomic
failure.

First, the majority of patients with autonomic failure and severe
orthostatic hypotension have supine hypertension (6). This greatly
limits the strategy of using long-acting pressor agents to raise blood
pressure, since unacceptably high levels of supine hypertension are
frequently encountered. The nature of the supine hypertension in
such patients is still uncertain. It is known that supine hypertension
is worse at the end of a day's activity than on awakening, and that
it is associated with increased levels of arteriolar resistance (arteriolar
resistance is also raised in the upright posture in such patients).
Since urine output during the day is quite small in such patients,
whereas a brisk nocturnal diuresis often occurs, it is likely that
volume factors may be involved in the phenomenon.

Second, postural change is not the only major stress faced by
the patient with autonomic dysfunction. The ingestion of food is a
potent vasodepressor stimulus (8). In health, vasodilatory factors
related to food ingestion are adequately compensated by the normally
functioning autonomic nervous system. When the buffering capacity
of this system is compromised, however, large shifts in blood pressure
can occur. In autonomic dysfunction, it is not unusual for systolic
blood pressure in the supine position to fall by more than 50 mm Hg
within minutes of ingesting a meal of moderate size.

Third, the cardiovascular responses to a great number of drugs
and maneuvers are altered in the presence of a dysfunctional autonomic
nervous system (9). This makes it very difficult to manage even
mundane illness, such as upper respiratory infections, without inad-
vertently eliciting major unwanted cardiovascular effects. Many of
these abnormal responses relate directly to alterations in adrenergic
receptor number, affinity, and function in patients with autonomic
dysfunction, and these will be discussed in greater detail later in this
chapter.

TABLE 1 Symptoms of Autonomic Failure

1. Orthostatic hypotension
   a. Dimming of vision
   b. Dizziness
   c. Neck pain
   d. Syncope

2. Supine hypertension

3. Nasal stuffiness

4. Nocturia

5. Impotence

6. Anhidrosis

7. Abnormalities of bowel control

8. Abnormalities of bladder control

9. Night blindness

10. Angina pectoris

III. TAXONOMY OF AUTONOMIC DYSFUNCTION

Many pathological processes may result in autonomic dysfunction, and the character of the clinical abnormality may sometimes suggest the nature of the etiology. Some of the disorders that result in secondary autonomic dysfunction are listed in Table 2. The two most common secondary causes of autonomic dysfunction, diabetes mellitus and amyloidosis, together account for almost half of the cases of severe orthostatic hypotension due to autonomic pathology seen in hospital practice.

In certain individuals, the autonomic dysfunction does not seem to be related to any other clearly defined pathological process. These patients may have autonomic failure alone, or the autonomic abnormality may occur in concert with dysfunction in some other neurological system (10,11). Patients with an acquired generalized autonomic dysfunction but no other neurological complaints are said to have progressive autonomic failure (Bradbury-Eggleston syndrome or idiopathic orthostatic hypotension), whereas patients with acquired abnormalities in two or more neurological systems, including the autonomic nervous system, are said to have multiple system atrophy (Shy-Drager syndrome). There is a somewhat more favorable prognosis in the Bradbury-Eggleston syndrome than in the Shy-Drager syndrome, and the lesions

TABLE 2 Causes of Autonomic Failure

---

1. Primary autonomic failure
   a. Bradbury-Eggleston syndrome (idiopathic orthostatic hypotension
   b. Shy-Drager syndrome (multiple system atrophy)
   c. Riley-Day syndrome (familial dysautonomia)
   d. Selective noradrenergic dysfunction
   e. Baroreceptor dysfunction

2. Secondary autonomic failure
   a. Diabetes millitus
   b. Amyloidosis
   c. Paraneoplastic
   d. Porphyria
   e. Tabes dorsalis
   f. Guillain-Barré syndrome
   g. Wernicke-Korsakoff syndrome (thiamine deficiency)
   h. Pernicious anemia (B_{12} deficiency)
   i. Central nervous system tumors
   j. Cord lesions above T6
   k. Drug toxicity

---

in the former tend to be most severe in the periphery, whereas there is marked central nervous system involvement in the latter (12).

Some individuals manifest a congenital abnormality in autonomic function. The Riley-Day syndrome (familial dysautonomia) is an inherited disorder of Ashkenazic Jews in which autonomic abnormalities are combined with absence of tears, insensitivity to pain, emotional lability, and frequent pulmonary infections due to aspiration.

Another congenital disorder has recently been described (13). Profound orthostatic hypotension and ptosis of the eyelids are the clinical hallmarks of this disorder; it is noteworthy that ptosis is not pronounced in either the Riley-Day syndrome or in the acquired primary or secondary autonomic dysfunction syndromes. In this newly described syndrome, we have observed a complete sparing of the parasympathetic nervous system and also of the sympathetic cholinergic fibers that mediate sweating. Circulating norepinephrine is barely detectable in plasma; no more than 10% of normal levels. In marked contrast, circulating levels of dopamine are quite elevated (200 pg/ml). Although the precise lesion is still unknown, the high dopamine levels and the low norepinephrine levels, together with evidence that clonidine therapy reduces circulating dopamine levels, whereas upright posture raises them, suggest that an abnormality in peripheral conversion of dopamine to norepinephrine exists in such patients.

IV. ADRENERGIC RECEPTORS IN AUTONOMIC DYSFUNCTION

One of the most striking features of autonomic dysfunction is the marked hypersensitivity that develops to various pressor and depressor stimuli (14-17). The loss of the buffering capacity of the baroreceptor and cardiopulmonary receptors undoubtedly contributes to the hypersensitivity, but there is evidence that changes in the function of adrenergic receptors are also involved. Many studies have been carried out in patients with varying degrees of autonomic failure and various etiologies of neurological defect. In our own studies, we have found it useful to restrict our investigations to those individuals with as nearly complete loss of sympathetic and parasympathetic function as are encountered clinically. For practical purposes, this has entailed the study of severely affected patients with the Bradbury-Eggleston syndrome.

In 12 well-characterized patients with the Bradbury-Eggleston syndrome, we assessed adrenergic receptor sensitivity using standard tests of autonomic responsiveness (17). All these patients had plasma norepinephrine levels under 50 pg/ml (patients: 28 ± 3 pg/ml vs. controls: 368 ± 24 pg/ml). On assumption of upright posture, all the patients had a fall in systolic blood pressure greater than 60 mm Hg. None of these patients was able to remain standing longer than 60 sec.

Using boluses of phenylephrine, we determined the dose of this agent which would result in an increase in systolic blood pressure of 25 mm Hg (18). Whereas 204 ± 28 µg phenylephrine was required to raise blood pressure by this amount in age-matched control subjects, the same effect was achieved with only 42 ± 10 µg phenylephrine in the patients with the Bradbury-Eggleston syndrome. This suggested a fivefold increase in alpha$_1$-adrenergic receptor sensitivity (17).

We also used boluses of isoproterenol to assess beta-adrenergic receptor function (17,18). Since the chronotropic effect of isoproterenol is primarily mediated through cardiac beta$_1$-adrenergic receptors, whereas the vasodilatory effect of the drug derives from its effect on beta$_2$-adrenergic receptors, we used the response of heart rate as an estimate of beta$_1$ sensitivity and the vasodepressor response as an estimate of beta$_2$ sensitivity.

We calculated that 5.4 ± 2.1 µg isoproterenol was necessary to increase heart rate by 25 beats/min in control subjects, whereas only 0.9 ± 0.2 µg was required in the Bradbury-Eggleston patients. This indicates a sixfold enhancement in beta$_1$ sensitivity. However, when we determined the dosage of isoproterenol necessary to reduce blood pressure by 25 mm Hg, an even more striking finding emerged. There was a 17-fold increase in beta$_2$ responsiveness such that only 0.3 µg isoproterenol was adequate to reduce blood pressure by 25 mm Hg in the Bradbury-Eggleston patients, whereas 5.0 µg was required in age-matched controls.

V. BETA-ADRENERGIC RECEPTORS IN
AUTONOMIC FAILURE

Because satisfactory radioligands for the study of beta-adrenergic
receptors were developed before those for the alpha-adrenergic
receptors, our knowledge of the former is presently more advanced
(19, 20). In 1981, Bannister and co-workers used tritiated dihydroalprenolol [^{3}H]DHA to study beta-adrenergic receptors on lymphocyte
membranes of patients with multiple system atrophy in whom supersensitivity to isoproterenol had been observed (21). Those investigators
observed a sixfold increase in the number of [^{3}H]DHA sites with no
change in [^{3}H]DHA affinity when incubation was carried out at 37°C,
but this increase was less at lower incubation temperatures (21).

Similar observations have been made in other patients. One was
a 48-year-old woman who probably had a paracarcinomatous autonomic
neuropathy (22). She had an approximately threefold hypersensitivity
to the chronotropic effect of isoproterenol and a doubling in the number
of her lymphocyte beta-adrenergic receptor sites as detected using
[^{125}I]iodohydroxybenzylpindolol. In addition, Chobanian and co-
workers found a 60% increase in [^{3}H]DHA binding site number in the
polymorphonuclear leukocytes of patients with mild to moderate
autonomic failure whose plasma norepinephrine levels were approximately half normal (23). Thus, there is marked beta-adrenergic
hypersensitivity in autonomic failure that is accompanied by an increase
in the number of beta-adrenergic receptors on circulating lymphocytes.

Jennings and co-workers extended these observations by examining
the increase of cyclic adenosine monophosphate (AMP) in response to
isoproterenol during the incubation of lymphocytes from three patients
with multiple system atrophy, three patients with the Bradbury-
Eggleston syndrome, and seven normal subjects (24). The patients
with multiple system atrophy had a greater than normal generation of
cyclic AMP at each concentration of isoproterenol tested. Surprisingly,
this was not the case in the lymphocytes of the patients with the
Bradbury-Eggleston syndrome. It is unclear why these individuals
should not manifest the same evidence of supersensitivity as the
patients with multiple system atrophy, although it might be noted
that the patients discussed by Jennings and co-worders (24) had
mean circulating norepinephrine levels of 160 ± 4 pg/ml, suggesting
that they were in the mildly affected subgroup of Bradbury-Eggleston
patients.

VI. ALPHA-ADRENERGIC RECEPTORS IN
AUTONOMIC FAILURE

Comparing eight subjects with multiple system atrophy to five normal
male volunteers, Davies and co-workers used [^{3}H]dihydroergocryptine

([³H]DHE) and found a sevenfold greater number of [³H]DHE sites per platelet in the patients as in the normal subjects (25), with no difference in [³H]DHE affinity between the two groups. Using the same radioligand, Chobanian et al. found significant increases in platelet receptor numbers in six patients with orthostatic hypotension as compared with 22 control subjects (23). It is noteworthy that the absolute differences in numbers of [³H]DHE sites found by Chobanian et al. were much more modest than those observed by Davies and co-workers (25); there was slightly less than a doubling of receptor number in the patient group. At least one of the patients studied by Chobanian et al. (23) had hypotension associated with the anephric state rather than autonomic failure, and this may have contributed to the differences in results obtained. Chobanian and co-workers reported similar values for platelet alpha-adrenergic receptors in four patients in whom ergotamine was studied as a therapeutic agent (26); there may be overlap in the patients in this and in the previous study.

More recently, Kafka et al., also using [³H]DHE, reported that there was a mean 50% increase in platelet alpha-adrenergic receptor number in patients with idiopathic orthostatic hypotension and multiple system atrophy (27). However, the interindividual variation was quite large, and 80% of patients had [³H]DHE binding density below 350 fmol/mg protein, a level found in some normal subjects.

Unfortunately, radioligand binding studies utilizing [³H]DHE have been demonstrated to be unreliable for estimation of platelet alpha$_2$-adrenergic receptor number when compared with [³H]yohimbine (28, 29). We have found and Brodde et al. has recently reported that there is an approximate doubling in platelet alpha$_2$-adrenergic receptor number as determined using [³H]yohimbine binding in a patient with diabetic polyneuropathy (30). In view of this observation and because the patient had manifested a 10-fold hypersensitivity to phenylephrine, Brodde reasoned that yohimbine might be helpful in the treatment of his patient (30). In fact, his patient reportedly did well for 6 months on a regimen of 12.5 mg oral yohimbine daily.

At about this same time, we had become interested in the pharmacology of yohimbine and clonidine in normal subjects and patients with autonomic failure (31-33). Because animal studies had suggested that postjunctional alpha$_2$-adrenergic receptors might be especially numerous in the venous capacitance bed, while alpha$_1$-adrenergic receptors might predominate in the arteriolar bed (34), we examined the effect of clonidine as a potential therapeutic agent. We reasoned that reduced venous capacitance was an important goal of therapy, but that increased arteriolar tone, by increasing the already elevated systemic vascular resistance and impeding blood flow, was undesirable. Although clonidine is commonly employed as an antihypertensive agent, we predicted that no depressor response would be seen in individuals who lacked significant sympathetic tone.

In the most severely affected patients with either selective
noradrenergic dysfunction or the Bradbury-Eggleston syndrome,
clonidine even at usual "antihypertensive" doses profoundly raised
blood pressure, while heart rate was unchanged. We feel confident
that this pressor effect was due to stimulation of postjunctional
alpha$_2$-adrenergic receptors, since it was not attenuated by prazosin
pretreatment. In these individuals, no depressor response to clonidine
was observed over the dosage range 0.1-0.8 mg orally. Individuals
who responded in this way generally had plasma norepinephrine levels
near 25 pg/ml and no heart rate response to either atropine or propranolol. It should be noted, however, that in selective noradrenergic
dysfunction (13), atropine raises heart rate normally and bradycradia
in response to clonidine is intact, presumably because this response
is mediated through the parasympathetic nervous system.

In contrast, more mildly affected individuals with autonomic
dysfunction sometimes did have a depressor response to clonidine at
the lower end of the dosage range, although it was usually less than
seen in hypertensive patients. In addition, we observed a profound
hypotensive response when clonidine was administered to patients with
baroreceptor dysfunction. In one patient, a 70-mm Hg decrement in
systolic blood pressure occurred within 45 min of oral administration
of 0.2 mg clonidine. This latter response was much greater than that
seen when the drug is given to hypertensive subjects, and under-
scores the powerful sympathetic activation that can accompany baroreceptor dysfunction and the marked sensitivity to central alpha$_2$
stimulation that results. If clonidine therapy of autonomic failure is
contemplated, baroreceptor dysfunction must therefore be ruled out
carefully.

As more mildly affected patients with the Bradbury-Eggleston
syndrome were found to have a depressor response to clonidine, we
reasoned that sympathetic outflow, however diminished, must be
exerting some control over the vasculature, aided perhaps by a
"partial" denervation hypersensitivity. As indicated above, we find
about a sixfold hypersensitivity to alpha$_1$ and beta$_1$ stimuli in such
patients. Thus, we reasoned further that if some sympathetic outflow
were present, it would be advantageous to enhance it, perhaps thereby
increasing patient functional capacity.

Yohimbine, an alpha$_2$ antagonist, is known to enhance sympathetic
outflow in normal man, thereby nearly doubling plasma levels of norepinephrine (14,33). We therefore administered yohimbine to mildly
affected Bradbury-Eggleston patients. Blood pressure was dramatically
elevated. While the increment in circulating norepinephrine was slight,
it appears that in the face of hypersensitivity the increase in neurotransmitter was able to exert considerable pressor effect.

As we have extended therapy with yohimbine to more severely affected individuals, it has become clear that at least a modest pressor response can be found in everyone with the Bradbury-Eggleston syndrone, even those who have had a pressor response to clonidine. We interpret this to mean that all of these patients still have at least some preservation of sympathetic function, however small. Denervation hypersensitivity to the pressor effect of clonidine, however, is so great that a slight clonidine-induced reduction in the already minimal basal sympathetic outflow is masked. In spite of the pressor response to yohimbine in the Bradbury-Eggleston patients, there has not been a pressor response in selective noradrenergic dysfunction (13).

Long-term therapeutic studies with clonidine and yohimbine are currently underway. The two major limitations observed with clonidine have been a "narcotic" or "drunken" feeling and the development of partial tolerance. An additional practical problem is the profound hypotension seen with baroreceptor dysfunction, which in certain circumstances could be quite dangerous. Although the latter disorder can be adequately distinguished on clinical grounds, the possibility of misdiagnosis is always present. With yohimbine, nervousness, diuresis, and sleep disturbance can sometimes be seen; it is not yet known whether tolerance will occur with this agent.

VII. CONCLUSIONS

Many neurological diseases can result in abnormalities in autonomic nervous system function. Autonomic failure occurs in approximately one in 5000 people. The syndrome is characterized by orthostatic hypotension, impotence, anhidrosis, nasal stuffiness, and nocturia. Patients with autonomic failure exhibit marked hypersensitivity to both pressor and depressor stimuli.

Greatly increased numbers of beta adrenoreceptors have been identified in the lymphocytes of these patients and modestly elevated numbers of $alpha_2$-adrenergic receptors on platelets have also been found. Severely affected patients have a greater than fivefold hypersensitivity to the vasoconstrictor and cardioacceleratory effects of agonists and a 17-fold increased sensitivity to $beta_2$-mediated vasodepression.

The hypersensitivity to pressor and depressor stimuli has permitted novel treatment programs for selected patients. Thus, the partial $alpha_2$ agonist clonidine has been a useful pressor agent in severely affected patients with autonomic failure, and the $alpha_2$ adrenergic receptor blocking agent yohimbine has been useful in the management of more mildly affected patients with autonomic failure.

ACKNOWLEDGMENTS

Supported by grants from the Parkinson Disease Foundation, the
Dysautonomia Foundation, and GM 15431 (National Institutes of Health).
Dr. Robertson is a Burroughs Wellcome Scholar in Clinical Pharmacology,
and Dr. Hollister is Clinical Associate Physician in the Elliot V. Newman Clinical Research Center.

REFERENCES

1. Kato, G., and Ban, T.: Central nervous system receptors in
   neuropsychiatric disorders. Prog. Neuropsychopharmacol. Biol.
   Psychiatry 6:207-222, 1982.

2. Unnerstall, J. R., Kopajtic, T. A., and Kuhar, M. J.: Distribution of alpha-2 binding sites in the rat and human central nervous
   system: Analysis of some functional correlates of the pharmacologic
   effects of clonidine and related adrenergic agents. Brain Res.
   Rev. 7:60-101, 1984.

3. Kunos, G. (ed.): *Adrenoreceptors and Catecholamine Action*,
   Part B. New York, Wiley, 1983, pp. 1-327.

4. Segawa, T., Yamamura, H. I., and Kurriyama, K. (eds.):
   Molecular Pharmacology of Neurotransmitter Receptors. New
   York, Raven, pp. 1-304.

5. Bannister, R.: *Autonomic Failure. A Textbook of Clinical Disorders of the Autonomic Nervous System*. Oxford, England,
   Oxford University Press, 1983, pp. 1-666.

6. Hines, S., Houston, M., and Robertson, D.: The clinical spectrum
   of autonomic dysfunction. Am. J. Med. 70:1090-1096, 1981.

7. Robertson, D., and Ziegler, M. G.: Orthostatic hypotension.
   In: *Current Therapy in Cardiovascular Disease*. N. J. Fortuin
   (ed.). Philadelphia, Decker, 1984.

8. Robertson, D., Wade, D., and Robertson, R. M.: Postprandial
   alterations in cardiovascular hemodynamics in autonomic dysfunctional states. Am. J. Cardiol. 48:1048-1052, 1981.

9. Robertson, D.: Idiopathic orthostatic hypotension: Contraindication
   to the use of ocular phenylephrine. Am. J. Opthalmol. 87:819-
   822, 0000.

10. Bradbury, S., and Eggleston, C.: Postural hypotension: A report
    of three cases. Am. Heart J. 1:73-86, 1925.

11. Shy, G. M., and Drager, G. A.: A neurological syndrome associated with orthostatic hypotension. Arch. Neurol. 2:511-527, 1960.

12. Ziegler, M. G., Lake, C. R., and Kopin, I. J.: The sympathetic-nervous-system defect in primary orthostatic hypotension. N. Engl. J. Med. 296:293-297, 1977.

13. Robertson, D., Goldberg, M. R., Onrot, J., Hollister, A. S., Thompson, J. G., Wiley, R., Robertson, R. M.: Isolated failure of autonomic noradrenergic neurotransmission: Evidence for impaired β-hydroxylation of dopamine. N. Engl. J. Med. 314:1494-1497, 1986.

14. Demanet, J. C.: Usefulness of noradrenaline and tyramine infusion tests in the diagnosis of orthostatic hypotension. Cardiology (Suppl. I) 61:213-224, 1976.

15. Davies, I. B., Bannister, R., Hensby, and Sever, P. S.: The pressor actions of noradrenaline and angiotensin II in chronic autonomic failure treated with indomethacin. Br. J. Clin. Pharmacol. 10:223-229, 1980.

16. Mohring, J., Glanzer, K., Marciel, J. A., Jr., Dusing, R., Kramer, H. J., Arbogast, R., and Koch-Weser, J.: Greatly enhanced pressor response to antidiuretic hormone in patients with impaired cardiovascular reflexes due to idiopathic orthostatic hypotension. J. Cardiovasc. Pharmacol. 2:367-376, 1980.

17. Robertson, D., Hollister, A. S., Carey, E. L., Tung, C. S., Goldberg, M. R., and Robertson, R. M.: Vascular beta-2-adrenergic hypersensitivity in autonomic dysfunction. J. Am. Coll. Cardiol. 3:850-856, 1984.

18. Robertson, D.: Clinical pharmacology: Assessment of autonomic function. In: *Clinical Diagnostic Manual for the House Officer.* K. L. Baughman and B. M. Greene (eds.). Baltimore, Williams & Wilkins, pp. 86-101.

19. Fraser, J. A., Nadeau, J. H. J., Robertson, D., and Wood, A. J. J.: Down regulation of leukocyte beta-adrenoreceptor density by circulating plasma levels of catecholamines in man. J. Clin. Invest. 67:1777-1784, 1981.

20. Feldman, R. D., Limbird, L. E., Nadeau, J., Robertson, D., and Wood, A. J. J.: Leukocyte beta-receptor alterations in hypertensive subjects. J. Clin. Invest. 72:164-170, 1983.

21. Bannister, R., Boylston, A. W., Davies, I. B., Mathias, C. J., Sever, P. S., and Sudera, D.: Beta-receptor numbers and thermodynamics in denervation supersensitivity. J. Physiol. (Lond.) 319:369-377, 1981.

22. Hui, K. K. P., and Conolly, M. E.: Increased numbers of beta receptors in orthostatic hypotension due to autonomic dysfunction. N. Engl. J. Med. 304:1473-1476, 1981.

23. Chobanian, A. V., Tifft, C. P., Sackel, H., and Pitruzella, A.: Alpha and beta adrenergic receptor activity in circulating blood cells of patients with idiopathic orthostatic hypotension and pheochromocytoma. Clin. Exp. Hypertens. A4:793-806, 1982.

24. Jennings, G., Bobik, A., and Esler, M.: Beta-receptors in orthostatic hypotension. N. Engl. J. Med. 305:1019, 1981 (letter).

25. Davies, B., Sudera, D., Sagnella, G., Marchesi-Saviotti, E., Mathias, C., Bannister, R., and Sever, P.: Increased numbers of alpha receptors in sympathetic denervation supersensitivity in man. J. Clin. Invest. 69:779-784, 1982.

26. Chobanian, A. V., Tifft, C. P., Fayon, D. P., Creager, M. A., and Sackel, H.: Treatment of chronic orthostatic hypotension with ergotamine. Circulation 67:602-609, 1983.

27. Kafka, M. S., Polinsky, R. J., Williams, A., Kopin, I. J., Lake, C. R., Eyert, M. H., Tokola, N. A.: Alpha-adrenergic receptors in orthostatic hypotension syndromes. Neurology 34:1121-1125, 1984.

28. Macfarlane, D. E., Wright, B. L., Stump, D. C.: Use of (methyl-^{3}H)-yohimbine as a radioligand for alpha-2 adrenoreceptors on intact platelets. Comparison with dihydroergocryptine. Thromb. Res. 24:31-43, 1981.

29. Boon, N. A., Elliott, J. M., Grahame-Smith, D. G., St. John-Green, T., and Stump, K.: A comparison of α_2-adrenoreceptor binding characteristics of intact human platelets identified by (^{3}H)-yohimbine and (^{3}H)-dihydroergocryptine. J. Auton. Pharmacol. 3:89-95, 1983.

30. Brodde, O.-E., Anlauf, M., Arroyo, J., Wagner, R., Weber, F., and Buck, K. D.: Hypersensitivity of adrenergic receptors and blood-pressure response to oral yohimbine in orthostatic hypotension. N. Engl. J. Med. 308:1033-1034, 1983 (letter).

31. Robertson, D. Goldberg, M. R., Hollister, A. S., Wade, D., and Robertson, R. M.: Clonidine raises blood pressure in severe idiopathic orthostatic hypotension. Am. J. Med. 74:193-200, 1983.

32. Goldberg, M. R., Hollister, A. S., and Robertson, D.: Influence of yohimbine on blood pressure, autonomic reflexes and plasma catecholamines. Hypertension 5:772-778, 1983.

33. Goldberg, M. R., and Robertson, D.: Yohimbine: A pharmacological probe for study of the alpha-2-adrenergic receptor. Pharmacol. Rev. 35:143-180, 1983.

34. VanZwieten, P. A., and Timmermans, P.B.M.W.M.: Cardiovascular α_2-receptors. J. Mol. Cell. Cardiol. 15:717-733, 1983.

13

Adrenergic Receptors in Psychiatric Disease

THOMAS R. INSEL and ROBERT M. COHEN   *National Institute of Mental Health, National Institutes of Health, Bethesda, Maryland*

I. INTRODUCTION

The study of adrenergic receptors in psychiatric patients has evolved from earlier investigations of brain catecholamines and their relationship to behavior. Since Vogt's (1) original demonstration in 1954 that norepinephrine (NE) is distributed unevenly in the brain, the pathways of noradrenergic neurons have been extensively mapped. It is now well known that noradrenergic neurons in mammalian brain originate from one of two major systems of nuclei in the pons and medulla. The most widely studied nucleus of origin is a pontine cluster of about 1500 neurons called the locus ceruleus. The second system arises from the lateral tegmentum of the medulla and includes several smaller nuclei such as the dorsal motor nucleus of the vagus and the nucleus tractus solitarii. Axons from both systems project rostrally, particularly to the cortex and to the limbic system, including the structures believed to be vital to the mediation of mood and cognition in man. Norepinephrine is the predominant catecholamine in the nucleus accumbens, ventral septum, central amygdaloid nucleus, nucleus of the stria terminalis, the paramedian thalamic nuclei, and the mamillary bodies (2). Accordingly, brain NE has been linked to functions such as arousal, rage, reward, locomotion, memory, regulation of sleep-wake cycles, neuroendocrine control, and autonomic reactivity (3). Each of these functions may be disturbed in psychiatric illness.

The study of adrenergic receptors in psychiatric patients has provided one approach to assessing the noradrenergic system. Within

the past decade, alpha$_1$, alpha$_2$, beta$_1$, and beta$_2$ receptors have all
been identified by radioligand binding studies in rodent brain. There
is some suggestion that the pharmacological selectivity and possibly
the regional distribution of these receptors vary between species (4).
Remarkably, postmortem study of human brain has not yet clearly
defined which receptor subtype predominates in those brain regions
most relevant to psychopathology in man. Nevertheless, it is of some
reassurance that at least in rodent brain, these receptor subtypes
which have been characterized entirely on a pharmacological basis do
indeed show different regional distributions (see Table 1). For
instance, although both subtypes of beta receptors can be identified,
only beta$_1$ receptors appear to be functionally important in the forebrain, whereas beta$_2$ receptors appear to predominate in the cerebellum
(5-7). Following NE depletion (by either 6-hydroxydopamine [6-OHDA]
injection or dorsal noradrenergic bundle lesion), beta$_1$-receptor binding
increases in the frontal cortex, thalamus, and throughout the limbic
system except for the hypothalamus (5). Beta$_2$ receptors, by contrast,
do not change in the forebrain following NE depletion, but instead
respond reciprocally to NE changes in the cerebellum (5). Some binding to beta$_2$ receptors has been demonstrated throughout the brain,
but outside of the cerebellum this binding is homogeneous and, thus,
may be associated with either glial or vascular elements (6).

Both alpha$_1$ and alpha$_2$ receptors have been identified in the forebrain with in vitro binding techniques (8) as well as with autoradiography (9,10). When [^{3}H]WB4101 has been used as a ligand, the highest
densities of alpha$_1$ receptors have been noted in the external plexiform
layer of the olfactory bulb. Although the density of alpha$_1$ receptors
throughout the brain is generally less than that of alpha$_2$ receptors,
greater alpha$_1$ binding is evident in certain regions, such as the dentate
gyrus of the hippocampus, the central gray area of the midbrain, and
along the floor of the fourth ventricle. It is widely recognized that
while beta receptors are linked to adenylate cyclase, alpha$_1$-receptor
effects in peripheral tissues are mediated by elevations in intracellular
calcium ion concentrations. In the brain, however, the second messenger for alpha$_1$ receptors is not clear. Although there is some evidence
that calcium ions may be involved in central alpha$_1$-receptor effects
(11), Daly and co-workers (7) have demonstrated a rise in cyclic
adenosine-3'5'-monophosphate (cAMP) in selective brain regions following administration of the alpha$_1$-receptor agonist 6-fluoronorepinephrine.
Depletion of brain NE (with 6-OHDA or reserpine administration) increases alpha$_1$-receptor number in several forebrain areas, although
the changes are of a lesser magnitude than those observed with beta
receptors (12). Curiously, alpha$_1$ sites in the hippocampus do not
change with a 6-OHDA lesion. Some evidence for the relevance of
alpha$_1$ receptors to psychiatry emerges from psychotropic drug effects:

TABLE 1 Brain Adrenergic Receptors

Receptor	Location (5, 9, 10)	Effects of NE lesion (5)	Hypothesized effector (12)	Putative functional role
Alpha$_1$	Forebrain	↑ Frontal cortex No change hypothalamus, hippocampus	Ca ↑ cAMP (?)	Postsynaptic NE effects
Alpha$_2$	Cortex, limbic system, locus ceruleus, nucleus tractus solitarii	↓ Amygdala and septum ↑ Frontal cortex	↓ cAMP (?)	Inhibit NE release Growth hormone release
Beta$_1$	Forebrain	↑ Except in hypothalamus	↑ cAMP	Postsynaptic NE effects
Beta$_2$	Cerebellum (homogeneous probably nonneural distribution in forebrain) also in anterior pituitary	↑ Cerebellum	↑ cAMP	Postsynaptic NE effects; adrenocorticotropic hormone release

There is a strong correlation between the affinities of both tricyclic antidepressants and neuroleptic (i.e., antipsychotic) drugs for binding to alpha$_1$ sites in rat brain and for their relative sedative potency (13, 14).

The study of brain alpha$_2$ receptors has also suggested a heterogeneous distribution. Autoradiographic studies using the selective alpha$_2$ ligand [^{3}H]para-aminoclonidine show high densities of binding in the noradrenergic nuclei (the locus ceruleus and nucleus of the tractus solitarii) as well as in several sites within the limbic system (e.g., amygdala, olfactory bulb, lateral septum, pyriform cortex) (9,10). The location of these receptors suggests a role in regulating noradrenergic output. Functionally, there is now little question that alpha$_2$ stimulation inhibits release of NE (17) and decreases the firing of the locus ceruleus (18). Although many have assumed that these receptors are presynaptic "autoreceptors," the data from binding studies suggest that most are postsynaptic (8,19). Possibly, some postsynaptic receptors inhibit NE release indirectly through interneurons, although this has yet to be demonstrated. The molecular mechanism by which central alpha$_2$ receptors work is poorly defined. Alpha$_2$ receptors in the periphery inhibit adenylate cyclase (15). Brain alpha$_2$-receptor effects may also be mediated by an inhibition of adenylate cyclase; however, this has yet to be conclusively demonstrated, partly because these receptors may be poorly coupled to adenylate cyclase in cell-free preparations (16).

The most compelling suggestion that adrenergic receptors are involved in psychiatric illness derives from studies of receptor changes following administration of antidepressant drugs. Decreases in beta-receptor binding and in NE-stimulated cAMP production have been demonstrated in rat brain following chronic (but not acute) administration of tricyclic drugs, monoamine oxidase inhibitors (MAOIs), and electroconvulsant therapy (20-25). Decreases in alpha$_2$-receptor binding and in functional activity have also been observed in rat brain following chronic administration of tricyclic antidepressants or MAOIs (26-30). In addition, functional alpha$_1$ supersensitivity has been reported following antidepressant administration (31), although a decrease in alpha$_1$-receptor density has also been observed (29). All of these changes are of clinical interest as they mimic the time course of clinical antidepressant response (2-4 weeks). Moreover, similar antidepressant-induced changes in adrenergic receptor function have been reported in patients (32,33). But logically these drug-induced changes cannot serve as evidence of pretreatment adrenergic receptor dysfunction.

Unfortunately, clinical studies of receptor function in psychiatry are considerably more difficult than in other specialties of medicine. Problems range from the obvious inability to directly study central

receptor function to the more subtle difficulties of diagnostic ambiguity.
Still, strategies have been developed for indirectly assessing adrenergic
receptor function. Prior to reviewing the results of these studies, we
will briefly summarize these strategies and their shortcomings.

II. STRATEGIES FOR THE STUDY OF RECEPTORS IN PSYCHIATRIC DISORDERS

The most straightforward and yet least pursued approach is to study
the number and affinity of receptors from various brain regions in
postmortem specimens. As many psychiatric disorders are associated
with suicide, the use of postmortem specimens should be feasible.
There are, however, several methodological problems with this approach.
From a clinical perspective, diagnosis and drug use may be difficult
to ascertain retrospectively. More importantly, the method of suicide—
be it exsanguination or drug overdose—may have acute effects on
receptor regulation which would be impossible to assess. For instance,
peripheral beta-receptor number may change on almost a moment to
moment basis depending on posture, blood pressure, and plasma catecholamine level (34). It seems unlikely that one could control for the
circumstances of death in comparing receptors from the brains of
psychiatric and nonpsychiatric patients. Finally, although receptors
are probably more stable than neurotransmitters, changes in the
interval between death and binding studies remain a confound which
can never be entirely controlled.

To avoid some of these complications, several investigators have
studied peripheral receptors in psychiatric patients. This strategy
has the advantage of a prospective approach with possible control of
variation due to diagnosis, medications, or sampling technique. In
addition, the same subject can be tested serially, permitting an assessment of receptor function with recovery or relapse. The major problem
with this approach resides in the interpretation—to extrapolate from
peripheral receptor to a central receptor is clearly an exercise in
inference; to extrapolate from a peripheral receptor to a particular
subset of receptors in the subcortical limbic pool is a giant leap indeed.
Nevertheless, some basis for this strategy arises from comparisons of
blood platelet receptors to brain receptors. Platelets with their
amine uptake sites have previously proved useful as peripheral models
of neuronal synapses (35). The finding that these "cells" also contain
$alpha_2$ receptors (36) and that this receptor is regulated in a manner
similar to the brain $alpha_2$ receptor (12) lends support to this approach.
One must remember, however, that receptors on platelets are exposed
to peripheral and not central catecholamines and, thus, are regulated
in an entirely different environment than receptors in brain. Platelets

themselves may be abnormal in psychiatric groups, as studies of oxidizing enzymes and amine uptake systems have previously demonstrated (37). Receptors on platelets could be abnormal owing to local membrane effects from abnormal catecholamine metabolism or owing to a less specific effect, such as platelet age, and thus not reflect the number or function of brain alpha$_2$ receptors.

There are no studies of peripheral alpha$_1$ receptors in psychiatric patients, primarily because these receptors have not yet been found on circulating blood elements. The study of beta receptors on granulocytes and lymphocytes has afforded another peripheral index of adrenergic receptors; again subject to the same difficulties of interpretation as the platelet studies.

A third approach to central adrenergic receptor function has been the pharmacological challenge strategy. This clinical method involves administration of a selective adrenergic receptor agonist or antagonist. Following drug administration, assessment of behavioral, physiological, endocrine, or biochemical variables permits comparison of patients to controls. For instance, clonidine, an alpha$_2$-adrenergic receptor agonist, increases growth hormone. If growth hormone increases less in psychiatric patients than in controls following administration of clonidine, then one might conclude that the patients have a subsensitive alpha$_2$ receptor. Unfortunately, growth hormone as well as other indicators for adrenergic receptor responsiveness are not solely regulated by adrenergic receptors. Neuroendocrine variables are controlled by a complex balance of neurotransmitter and neuropeptide regulators. Furthermore, this strategy treats the brain as a black box, assuming that receptor sensitivity in one region, which may mediate blood pressure or growth hormone, will correlate with receptor sensitivity in another region, which mediates mood or cognition. Nevertheless, the challenge strategy does provide a "physiological window" into central receptor sensitivity, and thus, is currently the most promising approach for the study of receptor function in psychiatric patients.

III. AFFECTIVE DISORDERS

It has long been known that drugs such as reserpine and α-methylparatyrosine, which deplete brain norepinephrine, may induce a state of "depression" in normal volunteers. Conversely, amphetamine and cocaine, both of which increase noradrenergic transmission, lead to behavioral activation and, in susceptible individuals, to some of the symptoms of mania. On the basis of such pharmacological observations, Schildkraut (38), over 20 years ago, proposed the catecholamine hypothesis of affective illness. This hypothesis, in its simplest form, suggested that depression resulted from a deficit of brain catecholamines, particularly NE, and mania was associated with a relative excess of brain catecholamines. A generation of research has failed to find consistent abnormalities in the levels of NE or its major metabolite, (in brain) 3-methoxy-4-hydroxyphenylglycol (MHPG), in the cerebrospinal fluid, plasma, or urine of patients with depression or mania (39). As a result, the notion of a simple deficit or excess of brain catecholamines has been revised to include a more complicated model that presumes some functional abnormality of catecholamines that might be manifested as dysregulation of central noradrenergic transmission (40). A key aspect of this dysregulation hypothesis is that disorders such as depression and mania might be associated with abnormal adrenergic receptor sensitivity even in the presence of normal levels of brain catecholamines.

Although the disorders of mood—depression and mania—have been studied for adrenergic receptor regulation, there are as yet no published studies of adrenergic receptors from postmortem specimens, even though recent studies have reported on imipramine binding (41) and other aspects of amine function in the brains of suicide victims.

IV. PERIPHERAL STUDIES

Three sets of investigators have focused on the platelet alpha$_2$ receptor (Table 2). Garcia-Sevilla et al. (42) found a 29% higher number of binding sites (B$_{max}$) and an equivalent dissociation constant (K$_d$) for [^{3}H]clonidine binding in the platelets of depressed subjects (drug-free for 2 weeks) compared to normal controls. A similar finding from Kafka et al. (43) was reported using [^{3}H]dihydroergocryptine ([^{3}H]DHE) as the ligand. Kafka et al. found 41% more binding in affective disorder patients (half of whom were in remission, i.e., euthymic) compared to normal controls. As apparent affinity for [^{3}H]DHE tested over a range of concentrations appeared normal in a subset of these patients, binding at the K$_d$ concentration was used as a reflection of B$_{max}$. Prostaglandin (PGE$_1$)-stimulated cAMP production was reduced in the affective disorder patients, although this reduction did not quite reach statistical significance. Subsequent studies by this same group in patients with unipolar depression revealed significantly reduced cAMP production following PGE$_1$ stimulation (43a; see Table 2). By contrast, studies by U'Prichard and co-workers (44) failed to replicate the findings of Garcia-Sevilla and Kafka. Using [^{3}H]yohimbine, U'Prichard et al. found no difference between depressed subjects and normals, although four schizoaffective patients showed increased binding (B$_{max}$ = 306 ± 59 compared to normal B$_{max}$ = 231 ± 19 [mol/mg protein]). The discrepancies in these three

TABLE 2 Ligand Binding to Platelet Membranes of Psychiatric Patients and Controls (Mean ± SEM Indicated)

Reference	Diagnosis	N	Relative binding (fmol/mg prot)	K_d (nm)	Ligand
43	Affective disorders (includes remitted BP)	23	212 ± 21[a]	—	[³H]DHE
	Schizophrenics (includes chronic)	83	204 ± 11[a]	—	
	Control	51	150 ± 8	—	
42	Major depressive disorder (low-affinity site)	17	45 ± 3[a]	5 ± 0.7	[³H]clonidine
		(8)	102 ± 14	19 ± 3.4	
	Controls (low-affinity site)	21	34.2 ± 2.4	5.0 ± 2.5	
		(11)	76 ± 5	17 ± 2.5	
44	Depressed (UP 8, BP 3)	11	204 ± 20	0.97 ± 0.06	[³H]yohimbine
	Manic	3	168 ± 47	0.81 ± 0.11	
	Schizophrenic	7	192 ± 29	0.95 ± 0.12	
	Schizoaffective	4	306 ± 59[a]	1.02 ± 0.06	
	Controls	10	231 ± 19	0.95 ± 0.07	
72	Schizophrenia	29 M	205 ± 18	—	[³H]DHE
		20 F	155 ± 19	—	
	Controls	26 M	170 ± 13	—	
		25 F	128 ± 7	—	
43a	Depressed (UP	24	247 ± 22[a]	—	[³H]DHE
	Depressed (never treated)	8	236 ± 42[a]	—	
	Controls	51	150 ± 8	—	
84	Panic disorder	11	32 ± 2	—	[³H]clonidine
	Melancholia	29	46 ± 2	—	
	Controls		32 ± 2	—	
	Panic disorder	11	120 ± 10	—	[³H]yohimbine
	Melancholia	29	183 ± 9	—	
	Controls		165 ±	—	

[a] Significant difference from corresponding control subjects.
BP, bipolar affective disorder.
UP, unipolar affective disorder.

studies partly reflect the different radioligands used. [^{3}H]yohimbine, an alpha$_2$ antagonist, binds with higher affinity and lower nonspecific binding than [^{3}H]DHE (45,46). If [^{3}H]yohimbine is labeling "too few" sites, it is possible that a real difference between depressives and normals would be overlooked. On the other hand, the findings with [^{3}H]DHE and [^{3}H]clonidine may reflect differences in nonspecific binding that bear no relationship to adrenergic function. The question may be more than academic. If, for instance, one ligand were identifying a subset of receptors that was more relevant to brain alpha$_2$ receptors, then even subtle changes in receptor number or function would be of interest. More information is needed about the relevance of platelet receptors to adrenergic receptors in brain. One untested strategy for providing that information in man would be to study the responsiveness of platelet receptors and central responses to pharmacological challenge in the same patients.

There have been fewer studies of beta-receptor function in peripheral cells. Pandey et al. (47) reported that leukocytes from depressed patients produced significantly lower levels of cAMP following stimulation by isoproterenol or NE. Similar findings were obtained by Extein et al. (48) using lymphocytes from patients with affective illness. In this latter report, lymphocytes from both depressed and manic patients not only produced less cAMP following isoproterenol stimulation, but also showed decreased [^{3}H]dihydroalprenolol (DHA) binding. In both studies, cAMP production following PGE$_1$ stimulation was within normal limits. As PGE$_1$ is linked to adenylate cyclase by a separate receptor, the decreased cAMP generated in response to isoproterenol very likely reflects an alteration in beta receptors and not in adenylate cyclase per se. What is less clear is whether these changes in receptors on blood cells are relevant to brain beta receptors or whether beta receptors are merely down-regulated because of elevated peripheral catecholamines, as might be expected with stress and as has been reported in affective illness (49). In support of this explanation, euthymic (i.e., remitted) patients do not show decreases in number or function of lymphocyte beta receptors (48).

V. CHALLENGE STUDIES

Pharmacological challenges in depressed patients originated with amphetamine (50,51), a drug with effects on a variety of neurotransmitter systems. Clonidine has been used more recently in an attempt to focus on alpha$_2$ responsiveness (Table 3). A series of studies (52-55) has demonstrated a blunted increase of growth hormone following clonidine administration to depressed subjects. The growth hormone response to clonidine does not normalize following remission, suggesting that

TABLE 3 Response to Clonidine in Psychiatric Patient Groups

Diagnosis	MHPG ↓	Plasma growth hormone ↑	Hypotension	Sedation	Reference
Depression	Blunted	Blunted (unipolar)	Normal	Normal	52
	Normal	Blunted	Normal	—	53
	—	Blunted	Normal	Normal	54
	—	Blunted	—	—	55
Schizophrenia	Blunted	—	Normal	—	73
		Normal			55
Obsessive-compulsive	—	Blunted	—	—	86
Probable site of action	Locus ceruleus	Hypothalamus	Nucleus tractus solitarii	Unknown	

this abnormality may be more of a trait than state marker (52). Presumably, these patients have a subsensitive alpha$_2$ receptor which may make them vulnerable to episodes of affective illness.

As the growth hormone response appears to result from clonidine's stimulation of hypothalamic alpha$_2$ receptors, this response may not be as relevant to alpha$_2$ receptors which are inhibitory at the locus ceruleus. In this regard, one study has shown a blunted decrease of plasma methoxyhydroxyphenylglycol (MHPG) in the first hour following intravenous clonidine (2 µg/kg) in depressed patients (52). This effect was not observed, however, in an earlier study using oral clonidine (5 µg/kg) with blood sampling for MHPG at 3.5, 4.0, and 5.0 hr (53). With the longer term sampling, a difference between patients and controls in the diurnal rhythm of MHPG may confound the results (56), and, indeed, the placebo day in this oral clonidine study does show significant differences in MHPG changes from baseline between patients and controls. Correcting for this difference, one finds that depressed patients and controls show equivalent MHPG decreases following clonidine. Three studies (52-54) have failed to find a difference between patients and controls in the hypotensive response to clonidine. In the one study describing heart rate responses to clonidine, normal controls had a significant decrease, but depressed patients (who had a higher baseline heart rate) did not change (52).

One interpretation of the blunted responses to clonidine in depressed subjects is that the alpha$_2$ receptor in brain shows defective coupling to adenylate cyclase. Such a defect might induce a vulnerability to depression which would be expressed with stress because of inappropriate regulation of adrenergic receptor responsiveness. Specifically, one might expect that the presynaptic release of norepinephrine would be increased with "stress." The normal feedback modulation of amine release might be inadequate because of defective alpha$_2$-receptor responsiveness. This would suggest a high-output presynaptic state that could be compensated by a low response postsynaptic adrenergic system in the depressed subjects. Indeed, a subset of depressives with high MHPG excretion and high cerebrospinal fluid (CSF) NE have been described (52)--although it is not clear that the clonidine-subsensitive patients would be among them. Furthermore, it is difficult to integrate the observation of subsensitive receptors in untreated patients with the consistent finding that many antidepressants lead to a subsensitivity of alpha$_2$ receptors. Finally, the importance of the subsensitivity of the clonidine response in the pathogenesis of depression seems further diluted by finding a similar abnormality in other psychiatric disturbances as well as in postmenopausal women (52), although many of these states may be associated with vulnerability to depression.

VI. SCHIZOPHRENIA

Just as studies of affective illness have been organized around the catecholamine hypothesis, the pharmacological study of schizophrenia for the past decade has been largely guided by the dopamine hypothesis. The dopamine hypothesis, in brief, states that in the schizophrenic brain, some dopamine systems are supersensitive or hyperactive. In support of this hypothesis, drugs that reduce schizophrenic symptoms share the property of dopamine receptor antagonism and their potency for receptor blockade correlates well with their antipsychotic potency (57). Furthermore, some investigators have reported increased numbers of dopamine receptors in the basal ganglia of schizophrenic brains (58). Recently, however, these observations have been subject to growing criticism. The correlation of antipsychotic potency and dopamine receptor blockade has been questioned because of different time courses for these effects. Dopamine receptor blockade is an acute effect, but antipsychotic changes usually become manifest over several days. The increase in dopamine receptors likewise may be an epiphenomenon of drug treatment with little relevance to psychosis, particularly as receptor changes in the basal ganglia would more likely correspond to disorders of movement than disorders of cognition. Citing these inconsistencies of the dopamine hypothesis, Hornykiewicz (59) has suggested a role for norepinephrine in the pathophysiology of schizophrenia.

Indeed, a series of studies have reported increased levels of either norepinephrine or MHPG in the limbic forebrain, brainstem, or CSF of schizophrenics (60-65). Increases in norepinephrine seem particularly prominent within the paranoid subgroup of schizophrenics. Also implicating the noradrenergic system, there have been recent reports of the successful treatment of some schizophrenic symptoms with clonidine (66), and precipitation of psychotic symptoms following withdrawal of clonidine (67). As adrenergic activity is intricately involved with dopamine receptor sensitivity (68,69), it seems reasonable to study adrenergic receptor function in schizophrenia.

Studies comparing postmortem brains of schizophrenics to normals have found no differences for binding of the beta receptor ligand [^{3}H]DHA to frontal cortex (70) or caudate (61). Similarly, the alpha receptor ligand [^{3}H]WB4101 has shown equivalent binding in schizophrenics and controls (71). Unfortunately, none of these studies used Scatchard analysis of binding and none looked at subcortical limbic structures, the very place where one would expect a "receptor lesion" if one were present. Thus, these "negative" studies must be interpreted cautiously. As norepinephrine (but not dopamine) has been reported to be increased in several nonstriatal sites (including hypothalamus), the continued search for changes in adrenergic receptor function in other parts of the brain seems warranted.

Studies of platelet alpha receptors in schizophrenia are also limited in number. Kafka (43) originally reported a 36% increase in [³H]DHE binding in platelets from schizophrenics; however, in a later report (72) (with fewer chronic patients) the amount of binding was not statistically different from normals. In both reports, Kafka found decreased PGE₁-stimulated cAMP production, suggesting abnormal coupling of receptors to adenylate cyclase. U'Prichard et al. (44), the only other group to study the platelets of schizophrenics, have failed to find abnormalities of [³H]yohimbine binding in a small group (n = 7).

Schizophrenics appear to resemble normals more than depressives in their plasma growth hormone response to clonidine (52). However Sternberg et al. (73) have reported a blunted decrease in plasma MHPG in schizophrenic patients at 3.5, 4.0, and 5.0 hr following oral clonidine administration. This difference in plasma MHPG level is not evident if patients are compared to normals on just the response to clonidine, but becomes significant when changes following placebo administration are subtracted from the values obtained with clonidine (only the controls show a significant increase in plasma MHPG following placebo). The report of a blunted decrease in MHPG following clonidine administration might thus be more due to a blunted increase of plasma MHPG following placebo rather than to alpha receptor subsensitivity.

Blood pressure response to clonidine has not been abnormal in schizophrenic patients (73), nor are these patients noticeably resistant to the sedative effects of the drug (74).

In summary, the evidence for adrenergic receptor dysfunction in schizophrenia is very meager. Although Kafka's report of defective responsiveness in platelet alpha-receptor is intriguing, this group has reported similar findings in depressives, suggesting that the deficit, if present, is not limited to a single syndrome and not relevant to a single treatment response. More studies are needed to probe for beta-receptor function in schizophrenia and to look more closely at the regional distribution of catecholamines and their receptors in postmortem brain.

VII. ANXIETY SYNDROMES

Anxiety, as a symptom of stress, has long been associated with cate-cholamine release. It is not surprising then that, like affective dis-orders and schizophrenia, anxiety disorders have been the subject of a norepinephrine hypothesis. Specifically, the locus ceruleus has

been studied as the source of an "alarm" system in primate brain.
Redmond (75) has shown that increases in locus ceruleus activity
(e.g., following piperoxane or yohimbine administration) lead to
"anxietylike" behaviors in rhesus monkeys, whereas decreases in
firing (e.g., following clonidine or diazepam administration) lead to
a decrease in anxietylike behaviors. Consistent with this model,
Charney and co-workers (76) have reported that yohimbine administration to normal volunteers leads to increased self-ratings of anxiety
and to corresponding increases in plasma MHPG. Several groups have
reported that clonidine lowers anxiety ratings in both psychiatric
patients and normals (77-79), although this effect may be in part due
to sedation. Clonidine has been demonstrated to decrease panic
attacks (79). These are episodes of severe and spontaneous anxiety
which may progress to agoraphobia (e.g., fear of leaving home) and
chronic disability. Unfortunately, the "antipanic" effect of clonidine
wears off within a few weeks. Propranolol has also been used to
treat anxiety with considerable success (80-83). In some studies
(80,81), however, propranolol has appeared to reduce the somatic
symptoms of anxiety (e.g., palpitations, tremor) more than the
psychic symptoms (e.g., worries and fears). One other link between
the anxiety syndromes and adrenergic receptors is the recent finding
that several forms of anxiety—expecially panic disorder with agoraphobia
and obsessive compulsive disorder—have been responsive to tricyclic
antidepressants and MAO inhibitors, drugs which have potent effects
on adrenergic receptor sensitivity following chronic administration.

Direct studies of adrenergic receptors in anxious patients are
extremely limited. A report by Cameron and co-workers (84) described
decreased binding of [^{3}H]yohimbine to platelet membranes of panic
disorder patients. Binding of [^{3}H]clonidine in these same individuals
was not different from controls, although it was significantly lower
than depressives. Binding affinities with both ligands were unchanged
across the three diagnostic groups. Cameron and co-workers (84)
suggest that this decrease in receptor number measured by [^{3}H]yohim-
bine represents down-regulation in response to elevated levels of
plasma catecholamines; however, these levels were not reported. To
our knowledge, no other studies of peripheral adrenergic receptors
have been reported in anxiety disorder patients.

A preliminary report (85) of yohimbine administration to panic
disorder patients (n = 11) and controls found the patients to be more
sensitive to the anxiogenic effects of yohimbine, but not different in
their MHPG response to the drug.

Obsessive compulsive disorder patients are the only anxiety disorder group that has been studied with clonidine administration.
These patients (n = 9) resemble depressives in that their growth
hormone response to clonidine is blunted, although this effect does
not predict responsiveness to tricyclic antidepressants (86).

VIII. CONCLUSIONS

While there is by now an abundant literature demonstrating the effects of psychotropic drugs on adrenergic receptors, there remains no consistent evidence for receptor abnormalities in patients with psychiatric disorders. Most suggestive is the finding of blunted growth hormone increases following clonidine administration in patients with unipolar depression (Table 3). However, such endocrine responses are determined by multiple factors, making this finding, although consistent, difficult to interpret in terms of adrenergic receptor function.

Basic to the study of receptors in psychiatric patients is whether the hypothetical disturbance is primary (i.e., etiological) or a secondary manifestation (i.e., descriptive) of an underlying disturbance in adrenergic function. If a receptor protein itself were abnormal, one might expect a genetic basis and then search for similar defects in peripheral receptors. The report by Nadi and co-workers (87) that cell cultures from skin fibroblasts of depressed patients show reduced muscarinic cholinergic receptors suggests a new avenue for those in search of the genetic receptor defect. This technique, if it could be developed for a tissue with adrenergic receptor subtypes, would provide a strategy that circumvents drug effects and is noninvasive enough to allow extensive studies of families at risk.

Most likely, however, the receptor is at best a reflection of a more general disturbance in central adrenergic function. According to this assumption, the task is to develop better strategies for assessing central rather than peripheral receptor responsivity. For instance, the pharmacological challenge strategy might be modified to monitor electroencephalographic variables (such as the average evoked response) rather than such peripheral measures as plasma growth hormone or blood pressure. Even the monitoring of cerebral blood flow or behavioral measures, such as pain sensitivity, would provide a closer look at central receptor function.

The most promising new approach may be positron emission tomography (PET) scanning (88,89). If positron emitting ligands that are selective for adrenergic receptors can be developed, it should become possible for the first time to measure directly the functional status of brain receptors and to monitor their changes in various mental states. It may, for instance, be only a few years before patients are sent for an $alpha_2$-receptor brain scan to support a diagnosis of unipolar depression or to aid in the assessment of psychopharmacological treatment. In the interim, there is much to be learned about the role of different adrenergic receptor subtypes in the brain, their ontogeny, and their regulation. In addition, the complex interactions between the noradrenergic system and other monoamine and neuropeptide system is just beginning to be unraveled. As less than 1% of brain synapses are noradrenergic, the role of this system ultimately needs to be understood in the context of a vast network of interdependent transmitters and modulators.

REFERENCES

1. Vogt, M. The concentration of sympathin in different parts of the central nervous system under normal conditions and after the administration of drugs. J. Physiol. (Lond.) 123:451-481, 1954.

2. Moore, R. Y., and Bloom, F. E.: Central catecholamine neuron systems: Anatomy and physiology of the noradrenergic and adrenergic systems. Ann. Rev. Neurosci. 2:113-205, 1979.

3. Murphy, D. L., and Redmond, D. E.: The catecholamines: Possible role in affect, mood, and emotional behavior in man and animals. In: Catecholamines and Behavior, Vol. II. A. J. Friedhoff (ed.). New York, Plenum, 1975, pp. 73-117.

4. Summers, R. J., Barnett, D. B., and Nahorski, S. R.: The characteristics of adrenoreceptors in homogenates of human cerebral cortex labelled by (^{3}H)-rauwolscine. Life Sci. 33:1105-1112, 1983.

5. U'Prichard, D. C., Yamamura, H. I., and Reisine, T. D.: Characterization and differential in vivo regulation of brain adrenergic receptor subtypes. In: Receptors for Neurotransmitters and Peptide Hormones. G. Pepeu, M. J. Kuhar, and S. J. Enna (eds.). New York, Raven, 1980, pp. 212-220.

6. Minneman, K. P., Pittman, R. N., and Molinoff, P. B.: β Adrenergic receptor subtypes: Properties, distribution, and regulation. Ann. Rev. Neurosci. 4:419-461, 1981.

7. Daly, J. W., Padgett, W., Creveling, C. R., Cantacuzene, D., and Kirk, K. L.: Cyclic AMP-generating systems: Regional differences in activation by adrenergic receptors in rat brain. J. Neurosci. 1:49-59, 1981.

8. U'Prichard, D. C., Bechtel, W. D., Rouot, B. M., and Snyder, S. H.: Multiple apparent alpha-noradrenergic receptor binding sites in rat brain. Effects of 6-hydroxydopamine. Mol. Pharmacol. 16:47-60, 1979.

9. Young, W. S., and Kuhar, M. J.: Noradrenergic α_1 and α_2 receptors: Autoradiographic visualization. Eur. J. Pharmacol. 59:317-319, 1979.

10. Young, W. S., and Kuhar, M. J.: Noradrenergic α_1 and α_2 receptors: Light microscopic autoradiographic localization. Proc. Natl. Acad. Sci. U.S.A. 77:1696-1700, 1980.

11. Fain, J. N., and Garcia-Sainz, J. A.: Role of phosphatidylinositol turnover in alpha$_1$ and of adenylate cyclase inhibition in alpha$_2$ effects of catecholamines. Life. Sci. 26:1183-1194, 1980.

12. Bylund, D. B., and U'Prichard, D. C.: Characterization of alpha-1 and alpha-2 adrenergic receptors. Int. Rev. Neurobiol. 24:344-433, 1983.

13. U'Prichard, D. C., Greenberg, D. A., Sheehan, P. P., and Snyder, S. H.: Tricyclic antidepressants: Therapeutic properties and affinity for α-noradrenergic receptor binding sites in brain. Science 199: 197-198, 1978.

14. Peroutka, S. J., U'Prichard, D. C., Greenberg, D. A., and Snyder, S. H.: Neuroleptic drug interactions with norepinephrine alpha receptor binding sites in rat brain. Neuropharmacology 16: 549-566, 1977.

15. Hoffman, B. B., and Lefkowitz, R. J.: Alpha adrenergic receptor subtypes. N. Engl. J. Med. 302:1390-1396, 1980.

16. Maguire, M. E., Ross, E. M., and Gilman, A. G.: β-Adrenergic receptor: Ligand binding properties and the interaction with adenyl cyclase. Adv. Cyclic Nucl. Res. 8:1-83, 1977.

17. Langer, S. Z.: Presynaptic regulation of catecholamine release. Biochem. Pharmacol. 23:1793-1800, 1974.

18. Cedarbaum, J. M., and Aghajanian, G. K.: Noradrenergic neurons of the locus coeruleus: Inhibition by epinephrine and activation by the alpha-antagonist piperoxane. Brain Res. 112:413-419, 1976.

19. U'Prichard, D. C., Greenberg, D. A., and Snyder, S. H.: Binding characteristics of a radiolabeled agonist and antagonist at central nervous system alpha noradrenergic receptors. Mol. Pharmacol. 13:454-473, 1977.

20. Vetulani, J., and Sulser, F.: Action of various antidepressant treatments reduces reactivity of noradrenergic cyclic AMP generating system in the limbic forebrain of the rat. Nature 257:495-496, 1975.

21. Banerjee, S. P., Kung, L. S., Riggi, S. J., and Chanda, S. K.: Development of beta adrenergic receptor subsensitivity by antidepressants. Nature 268:455-456, 1977.

22. Pandey, G. N., Heinze, W. J., Brown, B. D., and Davis, J. M.: Electroconvulsive shock treatment decreases beta-adrenergic receptor sensitivity in rat brain. Nature 280:234-235, 1979.

23. Bergstrom, D. A., and Kellar, K. J.: Adrenergic and serotonergic receptor binding in rat brain after chronic desipramine treatment. J. Pharmacol. Exp. Ther. 209:256-261, 1979.

24. Sellinger, M., Frazer, A., and Mendels, J.: Control beta-adrenergic receptor subsensitivity develops after repeated administration of antidepressant drugs. In: *Catecholamines: Basic and Clinical Frontier.* E. Usdin, I. J. Kopin, and J. Barchas (eds.). New York, Pergamon, 1979.

25. Rosenblatt, J. E., Pert, C. B., Tallman, J. F., Pert, A., and Bunney, W. E., Jr.: The effect of imipramine and lithium on alpha and beta-receptor binding in rat brain. Brain Res. 160:186-191, 1979.

26. Crews, F. T., and Smith, C. B.: Presynaptic alpha-receptor subsensitivity after long term antidepressant treatment. Science 202: 322-324, 1978.

27. Svensson, T. H., and Usdin, T.: Feedback inhibition of brain noradrenaline neurons by tricyclic antidepressants: Alpha-receptor mediation. Science 202:1089-1091, 1978.

28. Spyraki, C., and Fibiger, H. C.: Functional evidence for subsensitivity of noradrenergic α_2 receptors after chronic desipramine treatment. Life Sci. 27:1863-1867, 1980.

29. Cohen, R. M., Campbell, I. C., Dauphin, M., Tallman, J. F., and Murphy, D. L.: Changes in α and β receptor densities in rat brain as a result of treatment with monoamine oxidase inhibiting antidepressants. Neuropharmacology 21:293-298, 1982.

30. Cohen, R. M., Campbell, I. C., Cohen, M. R., Torda, T., Pickar, D., Siever, L. J., and Murphy, D. L.: Presynaptic noradrenergic regulation during depression and antidepressant drug treatment. Psychiatry Res. 3:93-105, 1980.

31. Menkes, D. B., Kehne, J. H., Gallager, D. W., Aghajanian, G. K., and Davis, M.: Functional supersensitivity of CNS α_1-adrenoreceptors following chronic antidepressant treatment. Life Sci. 33:181-188, 1983.

32. Glass, I. B., Checkley, S. A., Shur, E., and Dawling, S.: The effect of desipramine upon central adrenergic function in depressed patients. Br. J. Psychiatry 141:372-376, 1982.

33. Siever, L. J., Uhde, T. W., and Murphy, D. L.: Possible subsensitization of alpha$_2$ adrenergic receptors by chronic monoamine oxidase inhibitor treatment in psychiatric patients. Psychiatry Res. 6:293-302, 1982.

34. Fitzgerald, G. A., Robertson, D., Fedy, J., and Wood, A. J. J.: Beta-2-adrenoreceptors are down regulated by upright posture and dynamic exercise in man. Clin. Res. 29:564A, 1981 (abstract).

35. Stahl, S. M., and Meltzer, H. Y.: A kinetic and pharmacologic analysis of 5-hydroxytryptamine transport by human platelets and platelet storage granules: Comparison with central serotonergic neurons. J. Pharmacol. Exp. Ther. 205:118-132, 1978.

36. Kafka, M. S., Tallman, J. F., Smith, C. C., and Costa, J. L.: Alpha-adrenergic receptors on human platelets. Life Sci. 21: 1429-1438, 1977.

37. Murphy, D. L., and Weiss, R.: Reduced monoamine oxidase activity in blood platelets from bipolar depressed patients. Am. J. Psychiatry 128:1351-1357, 1972.

38. Schildkraut, J. J.: The catecholamine hypothesis of affective disorders: A review of supporting evidence. Am. J. Psychiatry 122:509-520, 1965.

39. Koslow, S. N., Maas, J. W., Bowden, C. L., Davis, J. M., Hanin, I., and Javaid, J.: CSF and urinary biogenic amines and metabolites in depression and mania. Arch. Gen. Psychiatry 40: 9999-1010, 1983.

40. Cohen, R. M., and Campbell, I. C.: Receptor adaptation in animal models of mood disorders: A state change approach to psychiatric illness. In: *Neurobiology of the Mood Disorders*. R. M. Post and J. C. Ballenger (eds.). Baltimore, Williams & Wilkins, 572-586, 1984.

41. Stanley, M., Virgilio, J., and Gershon, S.: Tritiated imipramine binding sites are decreased in the frontal cortex of suicides. Science 216:1337-1339, 1982.

42. Garcia-Sevilla, J. A., Zis, A. P., Hollingsworth, P. J., Greden, J. F., and Smith, C. B.: Platelet α_2 adrenergic receptors in major depressive disorder. Arch. Gen. Psychiatry 38:1327-1333, 1981.

43. Kafka, M. S., van Kammen, D. P., Kleinman, J. E., Nurnberger, J. I., Siever, L. J., Uhde, T. W., and Polinsky, R. J.: Alpha adrenergic receptor function in schizophrenia, affective disorders and some neurologic diseases. Comm. Psychopharmacol. 4:477-486, 1980.

43a. Siever, L. J., Kafka, M. S., Targum, S., and Lake, C. R.: Platelet alpha-adrenergic binding and biochemical responsiveness in depressed patients and controls. Psychiatr. Res. 11:287-302, 1984.

44. U'Prichard, D. C., Daiguji, M., Tong, C., Mitrius, J. C., and Meltzer, H. Y.: α_2 Adrenergic receptors: Comparative biochemistry of neural and nonneural receptors, and *in vitro* analysis in psy-

chiatric patients. In: *Biological Markers in Psychiatry and Neurology*. I. Hanin and E. Usdin (eds.). New York, Pergamon, 1981.

45. Motulsky, H. J., and Insel, P. A.: Adrenergic receptors in man. N. Engl. J. Med. 307:18-30, 1982.

46. Daiguji, M., Meltzer, H. Y., Tong, C., U'Prichard, D. C., Young, M., and Kravitz, H.: α_2 Adrenergic receptors in platelet membranes of depressed patients: No change in number or ^{3}H-yohimbine affinity. Life Sci. 29:2059-2064, 1981.

47. Pandey, G. N., Dysken, M. W., Garver, D. L., and Davis, J. M.: Beta-adrenergic receptor function in affective illness. Am. J. Psychiatry 136:675-678, 1979.

48. Extein, I. G., Tallman, J., Smith, C. C., and Goodwin, F. K.: Changes in lymphocyte β-adrenergic receptors in depression and mania. Psychiatry Res. 1:191-197, 1979.

49. Siever, L. J., Pickar, D., Lake, C. R., Cohen, R. M., Uhde, T. W., and Murphy, D. L.: Extreme elevations in plasma norepinephrine associated with decreased α adrenergic responsivity in major depressive disorder: Two case reports. J. Clin. Psychopharmacol. 3:39-41, 1983.

50. Fawcett, J., and Siomopoulos, V.: Dextroamphetamine response as a predictor of improvement with tricyclic therapy in depression. Arch. Gen. Psychiatry 25:247-255, 1971.

51. Langer, G., Heinz, G., Reim, B., and Natussek, N.: Reduced growth hormone responses to amphetamine in "endogenous" depressive patients. Arch. Gen. Psychiatry 33:1471-1475, 1976.

52. Siever, L. J., and Uhde, T. W.: New studies and perspectives on the noradrenergic receptor system in depression. Biol. Psychiatry 19:131-156, 1984.

53. Charney, D. S., Heninger, G. R., Sternberg, D. E., Hafstad, K. M., Giddings, S., and Landis, D. H.: Adrenergic receptor sensitivity in depression. Arch. Gen. Psychiatry 39:290-294, 1982.

54. Checkley, S. A., Slade, A. P., and Shur, E.: Growth hormone and other responses to clonidine in patients with endogenous depression. Br. J. Psychiatry 138:51-55, 1981.

55. Matussek, N., Ackenheil, M., Hippius, H., Mueller, F., Schroeder, H.-Th., Shultex, H., and Wasilewski, B.: Effect of clonidine on growth hormone release in psychiatric patients and controls. Psychiatry Res. 2:25-36, 1980.

56. Wehr, T. A., Muscettola, G., and Goodwin, F. K.: Urinary 3-methoxy-4-hydroxyphenylglycol circadian rhythm. Arch. Gen. Psychiatry 37:257-265, 1980.

57. Snyder, S. H.: The dopamine hypothesis of schizophrenia: Focus on the dopamine receptor. Am. J. Psychiatry 133:197-202, 1976.

58. Owen, F., Crow, T. J., Poulter, M., Cross, J., Longden, A., and Riley, C. J.: Increased dopamine receptor sensitivity in schizophrenia. Lancet 2:223-226, 1978.

59. Hornykiewicz, O.: Brain catecholamines in schizophrenia—A good case for noradrenaline. Nature 299:484-486, 1982.

60. Farley, I. J., Price, K. S., McCullough, E., Deck, J. H. N., Hordynski, W., and Hornykiewicz, O.: Norepinephrine in chronic paranoid schizophrenia: Above normal levels in limbic forebrain. Science 200:456-458, 1978.

61. Kleinman, J. E., Karoum, F., Rosenblatt, J. E., Gillin, J. C., Hong, J., Bridge, T. P., Zaleman, S., Storch, F., del Carmen, R., and Wyatt, R. J.: Post-mortem neurochemical studies in chronic schizophrenia. In: *Markers in Psychiatry and Neurology*. E. Usdin and I. Hanin (eds.). New York, Pergamon, 1982, pp. 67-76.

62. Carlsson, A.: *Catecholamines: Basic and Clinical Frontiers*, Vol. I. S. Usdin, I. Kopin, and J. Barchas (eds.). England, Pergamon, Oxford; 1973.

63. Lake, C. R., Sternberg, D. E., van Kammen, D. P., Ballenger, J. C., Ziegler, M. G., Post, R. M., Kopin, I. J., and Bunney, W. E.: Schizophrenia: Elevated cerebrospinal fluid norepinephrine. Science 207:331-333, 1980.

64. Gomes, U. C. R., Shanley, B. C., Potgieter, L., and Roux, J. T.: Noradrenergic overactivity in chronic schizophrenia: Evidence based on cerebrospinal fluid noradrenaline and cyclic nucleotide concentrations. Br. J. Psychiatry 137:346-351, 1980.

65. Sternberg, D. E., van Kammen, D. P., Lake, C. R., Ballenger, J. C., Marder, S. R., and Bunney, W. E.: Effect of pimozide on CSF norepinephrine in schizophrenia. Am. J. Psychiatry 138:1045-1051, 1981.

66. Freedman, R., Kirch, D., Bell, J., Adler, L. E., Pecevich, M., Pachtman, E., and Denver, P.: Clonidine treatment of schizophrenia. Double blind comparison to placebo and neuroleptic drugs. Acta Psychiatr. Scand. 65:35-45, 1982.

67. Adler, L. E., Bell, J., Kirch, D., Friedrich, E., and Freedman, R.: Psychosis associated with clonidine withdrawal. Am. J. Psychiatry 139:110-112, 1982.

68. Antelman, S. M., and Caggiula, A. R.: Norepinephrine-dopamine interactions and behavior. Science 195:646-653, 1977.

69. Tassin, J. P., Simon, H., Herve, D., Blanc, G., Le Moal, M., Glowinski, J., and Bockaert, J.: Nondopaminergic fibres may regulate dopamine-sensitive adenylate cyclase in the prefrontal cortex and nucleus accumbens. Nature 295:696-698, 1982.

70. Bennett, J. P., Enna, S. J., Bylund, D. B., Gillin, J. C., Ryatt, R. J., and Snyder, S. H.: Neurotransmitter receptors in frontal cortex of schizophrenics. Arch. Gen. Psychiatry 36: 927-934, 1979.

71. Reisine, T. D., Rossor, M., Spokes, E., Iversen, L. L., and Yamamura, H. I.: Opiates and neuroleptic receptor alterations in human schizophrenic brain tissue. In: Receptors for Neurotransmitters and Peptide Hormones. G. Pepeu, M. J. Kuhar, and S. J. Enna (eds.). New York, Raven, 1980, pp. 443-449.

72. Kafka, M., and van Kammen, D. P.: α Adrenergic receptor function in schizophrenia. Arch. Gen. Psychiatry 40:264-270, 1983.

73. Sternberg, D. E., Charney, D. S., Heninger, G. R., Leckman, J. F., Hafstad, K. M., and Landis, D. H.: Impaired presynaptic regulation of norepinephrine in schizophrenia. Arch. Gen. Psychiatry 39:285-289, 1982.

74. Ko, G. N., Bigelow, L. B., Karoum, F., and Wyatt, R. J.: Acute Clonidine in Schizophrenia. New Research Abstract 117, 136th American Psychiatric Association Meeting, New York, May 5, 1983.

75. Redmond, D. E., and Huang, Y. H.: New evidence for a locus coeruleus-norepinephrine connection with anxiety. Life Sci. 25: 2149-2160, 1979.

76. Charney, D. S., Heninger, G. R., and Redmond, D. E.: Yohimbine induced anxiety and increased noradrenergic function in humans: Effects of diazepam and clonidine. Life Sci. 33:19-29, 1983.

77. Hoehn-Saric, R., Merchant, A. F., Keyser, M. L., and Smith, V. K.: Effects of clonidine on anxiety disorders. Arch. Gen. Psychiatry 38:1278-1282, 1981.

78. Liebowitz, M. R., Fyer, A. T, McGrath, P., and Klein, D. F.: Clonidine treatment of panic disorder. Psychopharmacol. Bull. 17:122-123, 1981.

79. Uhde, T. W., Boulenger, J. P., Siever, L. J., Dupont, R. L., and Post, R. M.: Animal models of anxiety: Implications for research in humans. Psychopharmacol. Bull. 18:47-49, 1982.

80. Granville-Grossman, K. L., and Turner, P.: The effect of propranolol on anxiety. Lancet 1:788, 1966.

81. Tyrer, P. J., and Lader, M. H.: Response to propranolol and diazepam in somatic anxiety. Br. Med. J. 2:14-16, 1974.

82. Tanna, V. T., Penningroth, R. P., and Woolson, R. F.: Propranolol in the treatment of anxiety neurosis. Compr. Psychiatry 18:319-326, 1977.

83. Kathol, R. G., Noyes, R., Jr., Slymen, D. J., Crowe, R. R., Clancy, J., and Kerber, R. E.: Propranolol in chronic anxiety disorders. Arch. Gen. Psychiatry 37:1361-1365, 1980.

84. Cameron, O. G., Hollingsworth, P. J., Nesse, R. M., Curtis, G. C., and Smith, C. B.: Platelet α_2 Adrenoreceptors in Panic Disorder. New Research Abstract No. 55, 136th American Psychiatric Association Meeting, New York, May 4, 1983.

85. Charney, D. S., Heninger, G. R., and Redmond, D. E.: Noradrenergic Function in Human Anxiety States. New Research Abstract No. 96, 136th American Psychiatric Association Meeting, New York, May 5, 1983.

86. Siever, L. J., Insel, T. R., Jimerson, D., Lake, C. R., Uhde, T. W., Aloi, J., and Murphy, D. L.: Blunted growth hormone response to clonidine in obsessive compulsive patients. Br. J. Psychiatry 142:184-186, 1983.

87. Nadi, N. S., Nurnberger, J. I., and Gershon, E. S.: Muscarinic cholinergic receptors on skin fibroblasts in familial affective disorder. N. Engl. J. Med. 11:225-230, 1984.

88. Wagner, H. N., Burns, H. D., Dannals, R. F., Wong, D. F., Langstrom, B., Duelfer, T., Frost, J. J., Ravert, H. T., Links, J. M., Rosenbloom, J. B., Luckas, S. E., Kramer, A. V., and Kuhar, M. J.: Imaging dopamine receptors in the human brain by position tomography. Science 221:1264-1266, 1983.

89. Garnett, E. S., Firnau, G., and Nahmias, C.: Dopamine visualized in the basal ganglia of living man. Nature 305:137-138, 1983.

Index

A

Acromegaly, adrenergic response in, 293

Adenylate cyclase
activation of, 161-163, 171-175, 205, 211, 215
in aging, 205
of circulating leukocytes, 169-170, 202, 205, 211, 215, 269-270, 286, 289, 293
in depression, 363
of human platelets, 180-188, 202, 293, 359
inhibition of, 178-185, 363
molecular interactions of, 172-175, 180-185, 211
in pseudohypoparathyroidism, 174-175, 295
sex steroid regulation of, 292, 293
in skin, 260, 261
in thyroid disease, 289-290
"uncoupling" of, 172, 185-187, 215

Adipocytes
adrenergic response of, 56, 171, 205
in obesity, 206

Adrenergic agonists
classification of, 2-3, 39, 41-42
pharmacological assessment of, 12-25

[Adrenergic agonists]
receptor binding of, 150-154, 171-174, 181-187
receptor regulation by, 212-218, 220, 265, 324-325

Adrenergic antagonists
alpha-, 119-129
for angina pectoris, use of, 125
for benign prostatic obstruction, use of, 128
for cardiac arrhythmias, use of, 125-126
classification of, 25, 42, 76, 120-122
for congestive heart failure, use of, 123-125
for hypertension, use of, 122-123
for pheochromocytoma, use of, 126
for pulmonary disorders, use of, 127-128
for shock, use of, 126-127
tolerance to, 124, 125
beta-, 69-107
active metabolites of, 78-81
adverse effects of, 98-102
alpha-adrenergic receptor blockade by, 76
for angina pectoris, use of, 85-87

377

[Adrenergic antagonists, beta-]
anti-arrhythmic effects of,
87-92
antihypertensive effects of,
82-85
approved uses of, 69
on bronchial airways, effects
of, 10,100,101,264
central nervous system effects
of, 102
choice of, 105-107
for dissecting aneurysm, use
of, 94
drug interactions of, 102-105
elimination half-lives of, 78-80
first-pass hepatic metabolism
of, 76-81
for glaucoma, use of, 97-98
glycemic effects of, 11,12,101
for hypertrophic cardiomyopathy, use of, 93,94
intrinsic sympathomimetic
(partial agonist) activity
of, 72,74-76
ischemic myocardium, protection by, 92,93
lipid solubility of, 77,80
membrane stabilizing activity, 71
for migraine headache prophylaxis, use of, 96,97
for mitral valve prolapse, use
of, 94
myocardial failure, promotion
of, 99,240
oculomucocutaneous syndrome,
induction by, 102
peripheral vascular effects
of, 16-18,74,83,84,101
pharmacokinetic properties
of, 76-81
potency of, 9-10,70-72
for QT-interval prolongation
syndrome, use of, 95
on renin activity, effects of,
22,83

[Adrenergic antagonists, beta-]
selectivity of, 5,41,72,74
structure-activity relationships of, 71
for tetralology of Fallot, use
of, 95
for thyrotoxicosis, use of,
95,96
withdrawal syndrome of,
100,219,252-253
classification of, 38,41,42
norepinephrine clearance,
regulation by, 7
pharmacological assessment
of, 9,10,23,70
receptor regulation by, 218-
220
Adrenergic receptor subtypes
in brain, 354-357
classification of, 5,6,39-43,
120,146
desensitization of, 217,218,
250-252
quantitation of, 153-154,165
second messengers of, 39,
151,161,176
Adrenocorticotrophic hormone
(ACTH) release, adrenergic
regulation of, 57
Affective disorders, adrenergic
receptors in, 358-363
Aging, adrenergic receptors
and response in, 175,
202-205
Allergic rhinitis, adrenergic
receptors and response
in, 25,261,268,276
Alpha$_1$-adrenergic receptors
in adipocytes, 56,205
in autonomic dysfunction,
343,345
in central nervous system,
354-355
classification of, 5,6,41-43,
120

[Alpha₁-adrenergic receptors]
in gastrointestinal tract, 54, 55
in genitourinary tract, 52, 53,
208, 209
in heart, 48, 49
human tissue, 188-189
molecular mechanisms of, 161,
162, 176, 188
regulation of, 207-210, 215, 217,
287
in respiratory tract, 264, 273
in skeletal muscle, 42, 48
in smooth muscle, 45-47, 120,
239, 345
Alpha₂-adrenergic receptors
in adipocytes, 56, 171, 205, 287
in aging, 205
circadian changes in, 206
classification of, 5, 6, 41-43, 120
in endocrine secretion, 58
in the eye, 62
in gastrointestinal tract, 54,
55
in genitourinary tract, 52, 53,
208, 209
human tissue, 176-178, 187-188
molecular mechanisms of, 161,
162, 176, 178-179
norepinephrine release, regulation by, 41, 43, 84, 239, 356
nutritional alterations in, 206-
207
platelet, 58, 177, 180-188, 202-
203, 205, 206, 209, 210, 216-
217, 220, 293, 308-310, 322,
324-325, 359-361, 365, 366
regulation of, 207-210, 214-217,
219-221, 251, 252, 287, 293,
324, 325
in respiratory tract, 264
in smooth muscle, 45-47, 120,
239, 345
Amiloride, receptor blockade by,
221

Angina pectoris
alpha-adrenergic antagonist
treatment of, 125
beta-adrenergic antagonist
treatment of, 85-87
Anti-receptor antibodies, 157-
158, 276-277
Anxiety syndromes, 365-366
Aqueous humor
adrenergic regulation of, 60,
61
effects of beta-adrenergic
antagonist on, 97, 98
Arachidonic acid, 315
Asthma, adrenergic receptors
and response in, 10, 25,
50, 216, 264-276
Atopic dermatitis, 261, 269
Atopy
adrenergic receptors and
response in, 261-262,
264-274
"alpha-beta imbalance" theory
of, 266-269
Autonomic dysfunctional states
adrenergic receptors in,
343-347
incidence of, 339
symptoms of, 340-341
taxonomy of, 341
treatment of, 345-347

B

Bartter's syndrome, 323
Benign prostatic obstruction,
treatment with alpha-
adrenergic blockers, 128
Bernard Solier syndrome, 304
Beta₁-adrenergic receptors
in adipocytes, 56, 205
in autonomic dysfunction, 343
in brain, 354-355

[Beta₁-adrenergic receptors]
in the cardiovascular system,
10, 14-17, 45-49, 74
classification of, 5, 8, 39-41, 153
distribution of, 8, 39, 165-167,
263, 354-355
in the gastrointestinal system,
54
hypokalemia promoted by, 23
metabolic effects promoted by,
22, 56, 206
molecular mechanisms of, 161,
164, 167, 174-176
quantitation of, 153
regulation of, 217, 287, 292
renin release promoted by, 22,
52
selective blockers of, 23, 72, 74
in vascular smooth muscle, 45,
46
Beta₂-adrenergic receptors
autoantibodies to, 276, 277
in autonomic dysfunction, 343
in beta-adrenergic antagonist
therapy, 74
in blood cells, 58, 160, 202-203,
218
in brain, 354-355
in the cardiovascular system,
13-18, 45-47, 74, 84, 343
classification of, 5, 8, 39-41, 153
distribution of, 8, 40, 165-167,
260, 262, 354-355
in endocrine secretion, 57-58
in the eye, 61
in the gastrointestinal tract,
54, 55
in the genitourinary tract, 52-
54
hypokalemia promoted by, 23
metabolic effects promoted by,
21, 22
molecular mechanisms of, 161-
164, 167, 174-176
quantitation of, 153
regulation of, 214, 217, 265

[Beta₂-adrenergic receptors]
renin release promoted by,
22, 52
in skeletal muscle, 20, 47, 84
in smooth muscle, 16, 45-47,
51-54, 74, 84, 262
Blood pressure
adrenergic regulation of,
49, 50, 241-247, 250-251,
294
alpha-adrenergic receptor-
mediated changes in, 23,
24, 248, 250-251, 344-345
in autonomic dysfunction,
340-343, 346, 347
beta-adrenergic antagonist-
mediated effects on, 17,
82-85
beta-adrenergic receptor-
mediated changes in, 15-
17, 21
isoproterenol-mediated changes
in, 12, 15-17
Bradbury-Eggleston syndrome,
341-344, 346, 347
Bronchial airways
adrenergic receptor-mediated
regulation of, 50-52,
262-264
alpha-adrenergic antagonist-
mediated effects on, 127
in asthma, 100, 264-268, 273-274
beta-adrenergic antagonist-
mediated effects on, 10,
100-101, 264
desensitization of, 214, 216,
267-268
exercise-induced dilatation of,
10

C

Calcium
in mediating responses of
platelets, 306, 311-319

[Calcium]
mobilization by alpha$_1$-adrenergic receptors, 162, 176,188,354
Calcium entry blockers
adrenergic receptor blockade by, 220
treatment with beta-adrenergic antagonists and, 87,93
Cardiac automaticity
adrenergic agonist-mediated effects on, 48
beta-adrenergic antagonist-mediated effects on, 88,89
Cardiac contractility
adrenergic agonist-mediated effects on, 12,48,248
beta-adrenergic antagonist-mediated effects on, 82, 85, 240
in heart failure, 240-244
Catecholamines (*see also* Epinephrine, Norepinephrine)
cardiac, 240,241,244-245
endogenous release of, 6-8,204
in plasma, 202,220,320
in aging, 8,205
in asthma, 265,269
in autonomic dysfunction, 342-344
clearance of, 6-8
in congestive heart failure, 239-240,244,245
effects of dietary sodium on, 8,206
with exercise, 207
in hypertension, 247,249,250
methylxanthine-mediated increases in, 8,202,218,269
receptor regulation by, 219, 220,240-242,244-245,324
in platelets, 217,306,357-358
receptor regulation by, 212-219,220,240-242,244-245, 285-288,324-325
responses to, 8-12,32-62,260, 261-264,353 (*see also*

[Catecholamines]
Adrenergic agonists, Alpha$_1$-adrenergic receptors, Alpha$_2$-adrenergic receptors, Beta$_1$-adrenergic receptors, and Beta$_2$-adrenergic receptors)
tissue exposure to, 202-203, 220,240,245
uptake inhibitors of, 218
Cerebellum, adrenergic receptors in, 354
Cholera toxin, 174-175,179,184, 185,189
Circadian rhythm
in adrenergic receptors, 205, 206
of catecholamines, 265
Clonidine
in anxiety syndromes, 366
autonomic dysfunction, treatment with, 345-347
growth hormone levels, enhancement by, 358,361-366
properties of, 12,25,177,216, 219,308,309
withdrawal syndrome, 102,219
Congenital norepinephrine dysfunction, 342,346,347
Congestive heart failure, 239-242
alpha-adrenergic antagonist treatment of, 123-125,129
alpha-adrenergic receptors in, 218,245,246
beta-adrenergic antagonists, effects on, 99
beta-adrenergic receptors in, 218,240-245
pathophysiology of, 239-242, 245
Coronary artery spasm
alpha-adrenergic antagonist treatment of, 125,252
alpha-adrenergic receptors in, 252

Corticosteroids
  catecholamine-mediated in-
    creases in, 22
  regulation of beta-adrenergic
    receptors by, 169, 204, 210,
    211, 275, 291
Cyclic adenosine 3', 5'-
    monophosphate (cyclic AMP)
  in affective disorders, 359, 361
  in allergy, 267, 269-271, 274
  in autonomic dysfunction, 344
  in brain, 354
  in congestive heart failure,
    240-241
  in desensitization, 213-216, 274-
    275, 287
  discovery of, 39
  glucocorticoid-mediated regula-
    tion of, 211, 275
  isoproterenol-stimulated plasma
    levels of, 20-21, 267
  in platelet function, 319, 320
  in pseudohypoparathyroidism,
    294, 295
  second messenger role of, 2,
    161, 260, 263, 354
Cyclic AMP-dependent protein
    kinase, 2, 320
Cyclic nucleotide phospho-
    diesterase, 211, 213
  in atopic dermatitis, 261
Cystic fibrosis, adrenergic recep-
    tor autoantibodies in, 276

D

Deoxycorticosterone, regulation
    of adrenergic receptors
    by, 291
Depression, 358-363
  adrenergic receptors in, 359-
    361
  catecholamine hypothesis of,
    358-359

Dermis, adrenergic receptors
    in, 260
Desensitization (see also Down-
    regulation of adrenergic
    receptors)
  of adrenergic receptor sub-
    types, 217, 218, 250-252
  in asthma, 265, 269, 274, 275
  in congestive heart failure,
    240-245
  effects of glucocorticoids on,
    211, 275
  heterologous, 212, 213, 270
  homologous, 212, 213, 221
  in hypertension, 250, 251
  molecular mechanisms of, 173,
    174, 212-213
  in pheochromocytoma, 214, 217,
    287
  in physiological studies, 4, 5,
    206, 324
  of platelets, 215, 324
Diabetes mellitus, adrenergic
    receptors in, 293, 294
Diacylglyceride, in platelets,
    315-317
Dietary sodium, effect on
    adrenergic receptors of,
    206, 207, 214, 286
[^{3}H]dihydroergocryptine, 177,
    189, 209, 216, 217, 271, 272,
    308, 345, 359-361, 365
Dissociation constant
  in physiological studies, 2, 3,
    19, 152
  in radioligand binding studies,
    142-144, 147, 152
Dopamine receptors in schizo-
    phrenia, 364
Dose-response patterns in physio-
    logical studies, 2-5, 12, 13,
    19, 26-28, 239
Down-regulation of adrenergic
    receptors, 150, 204, 213,
    214-217, 220, 269, 275, 366

E

Epidermis, adrenergic receptors in, 260
Epinephrine (*see also* Catecholamines and Norepinephrine)
at adrenergic receptors, potency of, 5, 38, 39, 42, 308, 320
plasma concentrations of, 6-8, 11, 206, 207, 265, 320
asthmatic symptoms, relation to, 265
clearance of, 6-7
platelet aggregation and secretion stimulated by, 303, 308, 310-315, 320-326
receptor regulation by, 220, 285-287, 324-325
Erythrocytes
adrenergic receptors of, 60, 171
guanine nucleotide binding (G) protein of, 175
Essential thrombocythemia, 322
Estrogen, adrenergic receptor regulation by, 208-210, 292-293
Exercise
adrenergic receptor changes with, 207, 216, 324
beta-adrenergic antagonists, effects on, 76, 86
bronchodilation induced by, 10
in heart failure, 244, 246
nasal patency increased by, 261, 262
renin levels increased by, 22
sympathetic nervous system activation by, 9, 10, 16, 17, 207

F

Familial dysautonomia (Riley-Day syndrome), 342

Fat cells, adrenergic receptors and response of, 56, 205, 206, 222
Fibrinogen and platelet function, 310-311
Fibrinogen receptors of platelets, 311-313, 321
Fibroblast, beta$_2$-adrenergic receptors of, 260
Forearm blood flow, isoproterenol-mediated increase in, 17

G

Gastrin, adrenergic receptor-mediated-regulation of, 58
Glucocorticoids
beta-adrenergic receptor regulation by, 169, 204, 210, 211, 275, 291
catecholamine-mediated increase in level of, 22
Gluconeogenesis, adrenergic receptor-mediated stimulation of, 23, 52, 55
Glycogenolysis, adrenergic receptor-mediated stimulation of, 23, 47, 55
Granulocytes (*see also* Peripheral blood cells, Lymphocytes)
beta-adrenergic receptors of, 60, 166, 168-170, 202-203
regulation of, 173, 202-203, 205
Growth hormone
adrenergic receptor regulation of, 11, 288, 358
increase in depression by clonidine, 361, 367
Guanine nucleotide binding (G) proteins
adrenergic receptor interaction with, 151-152, 211
alpha$_2$-, 178-187, 309
beta, 162-164, 171-175

[Guanine nucleotide binding (G)
   proteins]
   regulation of adenylate cyclase
      by, 163,174,175,178-180,
      184-185,213,309
   identification of, 174-175,178-
      180
Guanine nucleotides, regulation
   of adrenergic receptor
   binding by, 151-154,162-
   164,182-187,207,212-214,
   309
Guanosine triphosphatase
   (GTPase) activity, 163,
   174,184,185

                  H

Heart rate
   adrenergic agonist-mediated
      effects on, 12-15, 19-20,
      24,48,237,240,248,343
   alpha-adrenergic antagonist-
      mediated effects on, 122-124
   beta-adrenergic antagonist-
      mediated effects on, 10,
      14-15,19,82,85-86,99
   in heart failure, 239,242
   in hypertension, 248-249
   response in autonomic dysfunc-
      tion and orthostatic hypo-
      tension, 287,343
High affinity state of adrenergic
   receptors, 152,172-174,
   178,182
6-Hydroxydopamine, 354
Hyperglycemia, adrenergic
   receptor-mediated, 21
   in insulin-dependent diabetes
      mellitus, 294
Hyperlactatemia, adrenergic
   receptor-mediated, 21
Hypertension (see also Blood
   pressure

[Hypertension]
   adrenergic receptors and
      response in, 247-252
   alpha-adrenergic antagonist
      treatment of, 82-85
   in autonomic dysfunction, 340
   beta-adrenergic antagonist
      treatment of, 82-85
   sympathetic nervous system
      in, 246-247
Hypocalcemia, adrenergic
   receptor-mediated, 21
Hypoglycemia
   beta-adrenergic antagonist
      treatment and, 101
   catecholamine release stimulated
      by, 11-12,22
Hypokalemia, adrenergic receptor-
      mediated, 21,23,47
Hypomagnesemia, adrenergic
   receptor-mediated, 21
Hypophosphatemia, adrenergic
   receptor-mediated, 21
Hypotension, orthostatic (see
   Orthostatic hypotension)
Hypothyroidism, adrenergic
   receptors in, 212,290

                  I, K

Indomethacin
   increase in leukocyte beta-
      adrenergic receptors by, 221
   inhibition of platelet fibrinogen
      receptor expression by, 311
Inositol triphosphate, 317
Insulin
   deficiency and altered adrener-
      gic response, 293,294
   release by adrenergic receptors,
      22,23,58
Intraocular pressure, adrenergic
   regulation of, 60-61,97-98
Intrinsic activity, definition of,
   2,3

Ischemic heart disease
alpha-adrenergic receptors in,
252
beta-adrenergic antagonist
treatment of, 92, 93
Islet-activating protein (IAP),
179, 180, 185, 189
Isoproterenol (*see also* Adrenergic agonists)
responses to, 12-20, 38, 40, 285-
286
in allergic disorders, 265, 268-
271, 275, 276
in autonomic dysfunction and
orthostatic hypotension,
287, 343
in congestive heart failure,
240
Ketonemia, adrenergic receptor-
mediated, 21

L

Law of mass action, 141-143, 148
Leukocytes (*see* Peripheral blood
cells, Granulocytes, and
Lymphocytes)
Limbic system, 353, 354
Locus ceruleus, 353, 356, 365, 366
Low-sodium diet, effect on
adrenergic receptors of,
214, 286
Lung, adrenergic receptors and
responses in, 50-52, 263,
272, 273
Lymphocytes (*see also* Peripheral
blood cells)
beta-adrenergic receptors of,
60, 167-170, 202-205
in affective disorders, 361
in asthma, 269-272, 274-276
in autonomic dysfunction, 344
in cardiovascular disorders,
238, 240, 242, 249-250

[Lymphocytes]
in diabetes mellitus, 294
physiological and pharmacological regulation of, 173,
202-205, 214-216, 219, 238,
249, 250, 285-287, 291, 293
in T and B lymphocytes, 170
in thyroid disorders, 289-290
guanine nucleotide (GTP)
binding proteins of, 175

M

Mast cells, adrenergic receptor-
mediated regulation of,
262-264
Megakaryocytes, 209, 304
Melatonin, adrenergic regulation
of, 58
Menstrual cycle, adrenergic
receptors in the, 202, 208,
209, 293
Methoxyhydroxyphenylglycol
(MHPG) in psychiatric
disease, 359, 362-365
Methylxanthines, increase in
plasma catecholamines
by, 8, 202, 218, 269
Migraine headaches, beta-
adrenergic antagonists
for prophylaxis of, 96, 97
Mineralocorticoids, regulation
of adrenergic receptors
by, 291
Monoamine oxidase inhibitors,
changes in adrenergic
receptors by, 218, 356
Mononuclear leukocytes (*see*
Lymphocytes and Peripheral
blood cells)
Mucus, adrenergic regulation
of, 261, 263, 264
Multiple system atrophy (Shy-
Drager syndrome), 26,
27, 341-342, 344, 345

Mydriasis, adrenergic receptor
    regulation of, 23, 25, 60, 61

N

Nasal mucosa, adrenergic recep-
    tor regulation of, 261, 262
Neonatal respiratory distress
    syndrome, 204
Nonlinear regression techniques
    for analysis of radioligand
    binding, 148-149, 153
Norepinephrine (see also Cate-
    cholamines, Epinephrine)
    at adrenergic receptors,
        potency of, 5, 24-28, 38-39,
        42, 264, 308, 320
    in diabetes mellitus, 294
    in orthostatic hypotension,
        26-28, 342, 343
    plasma concentration of, 6, 8,
        11, 26, 206, 207, 320
    clearance of, 6-7
    in psychiatric disorders, 359,
        363, 364
    receptor regulation by, 217,
        220, 286-287, 324, 325
    release regulated by adrenergic
        receptors, 41, 43, 84, 239,
        346, 356
    tissue concentration of, 240,
        245, 261, 353, 354, 364
    tyramine-induced release of,
        27, 28
Nucleus tractus solitarius, 353

O

Obesity, adrenergic receptors
    in, 206
Oculomucocutaneous syndrome,
    102
Olfactory bulb, alpha$_1$-adrenergic
    receptors in, 354

Ontogeny, adrenergic receptors
    in, 202, 203
Oral contraceptives, alterations
    in adrenergic receptors
    produced by, 202, 209
Orthostatic hypotension
    adrenergic receptors in, 26-
        28, 287, 343-347
    in autonomic dysfunction,
        340-343
    in Parkinson's disease, 339
    plasma norepinephrine levels
        in, 26-28
    symptoms of, 340

P

Parathyroid hormone, adrenergic
    regulation of, 57, 293, 294
Partial agonists, 2, 74-76, 182,
    216, 308
Penicillins, changes in platelets
    produced by, 220, 324
Peripheral blood cells (see also
    Lymphocytes, Granulo-
    cytes, and Platelets)
    adrenergic receptor identifica-
        tion in, 149, 167-171, 177,
        202
    as models for studying tissue-
        bound adrenergic recep-
        tors, 167-170, 187-189,
        202-203, 205, 214, 216, 222,
        285-287, 357-358
    in allergic diseases, 268-271,
        275
    in cardiovascular diseases,
        238
    with corticosteroid treatment,
        291
    in thyroid disease, 289-290
Pertussis toxin (see Islet-
    activating protein)
Phentolamine (see also Adrenergic
    antagonists, alpha-), 42, 187

Phenylephrine (*see also* Adrenergic agonists)
in adrenergic receptor classification, 5, 41-42
responses to, 5, 23-25, 260, 264
in allergic disorders, 265, 268, 274-275
in autonomic dysfunction and orthostatic hypotension, 287, 343, 345
Pheochromocytoma
adrenergic receptor downregulation and desensitization in, 217-218, 287-288
alpha-adrenergic antagonist treatment of, 126
blood pressure in, 214
catecholamine levels in, 320
Phosphatidic acid, 315
Phosphatidylinositol turnover
in alpha$_1$-adrenergic receptor action, 56, 162, 176, 188
in platelets, 316-317
Photoaffinity labelling of adrenergic receptors, 156, 157
Platelet activating factor, 303, 308, 310
Platelet glycoprotein IIb/IIIa, 311, 313
Platelets (*see also* Peripheral blood cells), 303-326
adhesion of, 304-305
aggregation of, 305-308, 310-313
adenosine diphosphate (ADP)-mediated, 303-308, 310, 312, 315, 321
pathological changes in epinephrine-mediated, 321-324
physiological role of catecholamines in, 320-321, 324-325
thrombin-mediated, 303-308, 310
alpha$_2$-adrenergic receptors of, adenylate cyclase inhibition by, 180-187, 309

[Platelets]
agonist-mediated desensitization of, 215-217, 324-325
guanine nucleotide binding protein, interaction with, 178-185, 309
identification of, 177-178, 308-310
to monitor alpha$_2$-adrenergic receptors in other tissues, 187-188, 202-203, 205, 357-358
in psychiatric disorders, 357-361, 365-366
responses mediated by, 58, 305-308, 310-315, 320-321
sex steroid regulation of, 208-210, 293
sodium and agonist interaction with, 185-187
beta-adrenergic receptors of, 168
calcium in function of, 312-315
catecholamines in, 217, 306, 357-358
cyclic AMP-mediated regulation of, 319-320
phospholipid metabolism in, 315-318
secretion of, 305-308, 313-319
shape change of, 304, 305, 307
synergy (potentiation) of agonists in the activation of, 320-321
Plethysmography, assessment of adrenergic response with, 17, 58
Polymorphonuclear leukocytes (*see* Granulocytes)
Positron emission tomography, 367
Postural hypotension (*see* Orthostatic hypotension)
Potassium disposal, adrenergic receptor-mediated, 40, 47, 48

Prazosin (*see also* Adrenergic
    antagonists, alpha-)
  classification of adrenergic
    receptors with, 25, 42, 188,
    189, 208, 264
  clinical use of, 129
  first-dose phenomenon of,
    129
Pregnancy, adrenergic receptors
    in, 207-210
Progesterone, changes in adrener-
    gic receptors mediated by,
    208-210, 292, 293
Prolactin
  beta-adrenergic response in
    animals with tumors secret-
    ing, 288, 289
  release regulated by dopamine,
    288
Propranolol (*see also* Adrenergic
    antagonists, beta)
  anxiety syndromes, treatment
    with, 366
  beta-adrenergic receptor-
    mediated responses, as
    defined by use of, 14-20,
    22, 38, 48
  pulmonary function, alteration
    by administration of,
    264
  up-regulation of beta-adrenergic
    receptors by, 219, 286
  withdrawal syndrome (*see*
    Adrenergic antagonist,
    beta, withdrawal
    syndrome)
Prostaglandin
  adenylate cyclase stimulation
    by, 182, 319
  in atopic dermatitis, 269-270
  in psychiatric disorders, 359,
    361, 365
  production in Bartter's syn-
    drome, 323
  synthesis and platelet function,
    307, 311, 315-318, 323

Protein kinase C in platelet
    activation, 318
Protein phosphorylation in
    platelet activation, 318-319
"Pseudo down-regulation" of
    adrenergic receptors, 150,
    217
Pseudohypoparathyroidism,
    174, 175, 294, 295

                    Q

Quinidine
  adrenergic receptor blockade
    by, 221
  beta-adrenergic antagonist
    effects similar to those of,
    71, 84

                    R

Radioligand binding, 139-155
  characterization of human
    tissues using, 149, 150,
    164-168, 176-178, 189, 208,
    238-239, 272, 308, 309, 354
  competition studies of, 144-
    147, 150-153, 163-164, 172,
    181-185
  data analysis of, 147-149, 164
  kinetic studies of, 141-142
  nonspecific, 140-141
  steady state studies of, 142-144
  techniques of, 139-140, 164-165,
    167
  uses of, 149-155, 172
Radioreceptor assay, 155
Rauwolscine (*see also* adrenergic
    antagonists, alpha-),
    classification of adrenergic
    receptors with, 42, 177,
    208, 216

Raynaud's syndrome, effects of
  beta-adrenergic antagonists
  on, 18,101
Receptor redistribution, 155,213
Receptor refractoriness (*see*
  Desensitization)
Receptor sequestration, 213
Receptor solubilization, 156, 183-
  184,187
Receptor turnover, 155,213
Receptor "uncoupling," 172-174,
  212
Renin release
  adrenergic receptors that
    mediate, 22,52,248
  effects of beta-adrenergic an-
    tagonists on, 22,83

S

Scatchard plot, 145,148
Schild plot, 19,20
Schizophrenia, 364-365
Sex steroids, changes in adrener-
  gic receptors produced by,
  202,207-210,292,293
Shy-Drager syndrome (*see*
  Multiple system atrophy)
Skeletal muscle tremor, beta-
  adrenergic receptor-
  mediated, 20,47,214
Skin, adrenergic regulation of,
  18,23,25,260-261
Sodium ion
  alpha$_2$-adrenergic receptor
    regulation by, 185-187
  dietary manipulation of, 206-208,
    214,286
Somatostatin, adrenergic regulation of, 58
Spare receptors, 3
Supraventricular arrhythmias,
  use of beta-adrenergic
  antagonists in treatment of,
  89-91

T

Tachyphylaxis (*see also* Desensitization), 173,212,267
Testosterone, changes in adrenergic receptors produced
  by, 292,293
Thermogenesis, adrenergic
  regulation of, 56,216
Thyroid hormone
  adrenergic regulation of re-
    lease of, 57
  regulation of adrenergic re-
    ceptors and response by,
    211-212,289-290
Thyrotoxicosis (hyperthyroidism)
  adrenergic receptors in, 212,
    289,290
  treatment with beta-adrenergic
    antagonists of, 95,96
Tricyclic antidepressants, decreases in beta-adrenergic
  receptors produced by,
  356
Tyramine, assessment of norepinephrine release by, 25,27-
  28,245

U

Up-regulation of adrenergic
  receptors
  antagonist-mediated, 150,219,
    220,286
  in autonomic dysfunction, 344-
    345
Uterine myometrium, adrenergic
  receptors in, 189,208-210

V

Vasoconstriction
  alpha-adrenergic agonist-
    mediated, 24-25,45,46,260

[Vasoconstriction]
   in atopic dermatitis, 261
   in autonomic dysfunction,
      343, 347
   alpha-adrenergic antagonist-
      mediated effects on, 121-
      125, 127, 128
   beta-adrenergic antagonist-
      mediated effects of, 101,
      252
   in coronary arteries, 252
   in hypertension, 247, 250
Vasodilation
   alpha-adrenergic agonist-
      mediated, 24
   alpha-adrenergic antagonist-
      mediated, 121-126, 128, 245
   beta-adrenergic receptor
      mediated, 14-18, 45, 46, 84,
      252, 260

[Vasodilation]
   in autonomic dysfunction, 343
   in hypertension, 247, 250
Ventricular arrhythmias, beta-
   adrenergic antagonists in
   treatment of, 91, 92

Y

Yohimbine (see also Adrenergic
   antagonists, alpha-)
   anxiety syndromes, assessment
      with, 366
   autonomic dysfunction, treat-
      ment with, 346, 347
   in classification of adrenergic
      receptors and response,
      25, 42, 176, 177, 187-188,
      216, 264, 308